Bioengineering of the Skin: Cutaneous Blood Flow and Erythema

CRC Series in
DERMATOLOGY: CLINICAL AND BASIC SCIENCE
Edited by Dr. Howard I. Maibach

The CRC Dermatology Series combines scholarship, basic science, and clinical relevance. These comprehensive references focus on dermal absorption, dermabiology, dermatopharmacology, dermatotoxicology, and occupational and clinical dermatology.

The intellectual theme emphasizes in-depth, easy to comprehend surveys that blend advances in basic science and clinical research with practical aspects of clinical medicine.

Published Titles:

Hand Eczema
Torkil Menne and Howard I. Maibach

Health Risk Assessment: Dermal and Inhalation Exposure and Absorption of Toxicants
Rhoda G. M. Wang, James B. Knaak, and Howard I. Maibach

Handbook of Mouse Mutations with Skin and Hair Abnormalities:
Animal Models and Biomedical Tools
John P. Sundburg

Pigmentation and Pigmentary Disorders
Norman Levine

Protective Gloves for Occupational Use
Gunh Mellström, J.E. Walhberg, and Howard I. Maibach

Forthcoming Titles:

Bioengineering of the Skin: Water and the Stratum Corneum
Peter Elsner, Enzo Berardesca, and Howard I. Maibach

Bioengineering of the Skin: Methods and Instrumentation
Enzo Berardesca, Peter Elsner, Klaus P. Wilhelm, and Howard I. Maibach

The Contact Urticaria Syndrome
Arto Lahti and Howard I. Maibach

Dermatologic Research Techniques
Howard I. Maibach

Handbook of Contact Dermatitis
Christopher J. Dannaker, Daniel J. Hogan, and Howard I. Maibach

Human Papillomavirus Infections in Dermatovenereology
Gerd Gross and Geo von Krogh

The Irritant Contact Dermatitis Syndrome
Pieter Van der Valk, Pieter Coenrads, and Howard I. Maibach

Skin Cancer: Mechanisms and Relevance
Hasan Mukhtar

Bioengineering of the Skin: Cutaneous Blood Flow and Erythema

Edited by
Enzo Berardesca, M.D.
Peter Elsner, M.D.
Howard I. Maibach, M.D.

CRC Press
Boca Raton   Ann Arbor   London   Tokyo

Library of Congress Cataloging-In-Publication Data

Bioengineering of the skin: cutaneous blood flow and erythema/edited by Enzo Berardesca, Peter Elsner, and
    Howard I. Maibach. p. cm.—(CRC series in dermatology)
    Includes bibliographical references and index.
    ISBN 0–8493–8371–4
    1. Erythema. 2. Skin—Blood-vessels. 3. Blood flow—Measurement.
I. Berardesca, Enzo. II. Elsner, Peter, 1955–. III. Maibach, Howard I. IV. Series.
    [DNLM: 1. Erythema. 2. Skin—blood supply. 3. Blood Flow
Velocity. WR 150 B615 1994]
RL271.B55 1994
616.5—dc20
DNLM/DLC
for Library of Congress   94–4460
                                                                                                               CIP

THE EDITORS

Enzo Berardesca, M.D., is Senior Dermatologist and Professor at the School of Dermatology of the University of Pavia, Pavia, Italy. He obtained his training at the University of Pavia and received his M.D. degree in 1979. He served as resident and dermatologist at the Department of Dermatology, IRCCS Policlinico S. Matteo, Pavia, from 1982 to 1987, and as assistant researcher in the Department of Dermatology, University of California School of Medicine in San Francisco, California, in 1987. In 1988 he assumed his present position.

Dr. Berardesca is Chairman of the International Society for Bioengineering and the Skin and is also a member of the Society for Investigative Dermatology, the European Society for Dermatological Research, the International Society for Androgenic Disorders (ISAD), and the Italian Group for Research on Contact Dermatitis (GIRDCA). He is also a member of the Committee of Bioengineering Standardization of the European Society for Contact Dermatitis.

Dr. Berardesca's current major research interests are irritant dermatitis, barrier function, and noninvasive techniques to investigate skin physiology with particular regard to racial differences in skin function. He is the author of more than 130 papers and book chapters.

Peter Elsner, M.D., is Senior Dermatologist and Chief, Laboratories of Skin Physiology, Dermatology and Patch Testing, Department of Dermatology, University of Zurich, Switzerland, and Lecturer in Dermatology, University of Zurich, Switzerland and University of Wuerzburg, Germany.

Dr. Elsner obtained his M.D. degree in 1981 from Bavarian Julius-Maximilians-University, Wuerzburg, Germany. In 1987 he was awarded the degree of lecturer by the same institution. From 1988 to 1992 he was Visiting Research Dermatologist at the Department of Dermatology, University of California School of Medicine in San Francisco. In 1993 he became lecturer at Zurich University.

Dr. Elsner is a member of 25 national and international scientific societies and acts as Secretary and Treasurer of the International Society for Bioengineering and the Skin. He has published 3 books and more than 100 scientific papers, and he serves on the editorial board of several dermatological journals. His research interests include skin physiology and pharmacology—especially using bioengineering techniques—occupational and environmental dermatology, allergology, and sexually transmitted diseases.

Howard I. Maibach, M.D., is Professor of Dermatology, School of Medicine, University of California, San Francisco. Dr. Maibach graduated from Tulane University, New Orleans, Louisiana (A.B. and M.D.) and received his research and clinical training at the University of Pennsylvania, Philadelphia. He received an honorary doctorate from the University of Paris Sud in 1988.

Dr. Maibach is a member of the International Contact Dermatitis Research Group, the North American Contact Dermatitis Group, and the European Environmental Contact Dermatitis Group. He has published more than 1100 papers and 40 volumes.

CONTRIBUTORS

Pierre G. Agache, M.D.
Department of Dermatology
Hospital Saint-Jacques
Besançon, France

Tove Agner, M.D.
Department of Dermatology
Blspebjerg Hospital
Copenhagen, Denmark

Hans-Peter Albrecht, M.D.
Department of Dermatology
University of Erlangen
Erlangen, Germany

Peter H. Andersen, M.D.
Department of Dermatology
Marselisborg Hospital
Aarhus, Denmark

Enzo Berardesca, M.D.
Department of Dermatology
University of Pavia
Pavia, Italy

Luciano Bernardi, M.D.
Department of Internal Medicine
University of Pavia
Pavia, Italy

Andreas J. Bircher, M.D.
Department of Dermatology
Kantonsspital
Basel, Switzerland

Peter Bjerring, M.D.
Department of Dermatology
Marselisborg Hospital
Aarhus, Denmark

Irwin M. Braverman, M.D.
Department of Dermatology
Yale University School of Medicine
New Haven, Connecticut

Sarah Brenner
Department of Dermatology
Ichilov Hospital
Tel Aviv, Israel

Derk P. Bruynzeel, M.D.
Department of Occupational Dermatology
Free University Academic Hospital
Amsterdam, The Netherlands

John A. Cotterill, M.D.
Department of Dermatology
General Infirmary
Leeds, England

Edith M. De Boer, M.D.
Department of Occupational Dermatology
Free University Academic Hospital
Amsterdam, The Netherlands

Anne-Sophie Dupond
Department of Dermatology
Hospital Saint-Jacques
Besançon, France

Peter Elsner, M.D.
Department of Dermatology
University of Zurich
Zurich, Switzerland

Véronique Epstein-Drouard
Scientific Regulatory Affairs Department
Yves Rocher Research Center
Arcueil, France

Dorothea Hiller, M.D.
Department of Dermatology
University of Erlangen
Erlangen, Germany

Sean W. Lanigan, M.D.
Department of Dermatology
Ogwr Health Unit
Bridgend General Hospital
Bridgend, Wales

Stefano Leuzzi
Department of Internal Medicine
University of Pavia
Pavia, Italy

Magnus Lindberg, M.D.
Section of Occupational Dermatology
Department of Dermatology
University Hospital
Uppsala, Sweden

Alain Lucas
Department of Dermatology
Hospital Saint-Jacques
Besançon, France

Howard I. Maibach, M.D.
Department of Dermatology
University of California at San Francisco
San Francisco, California

Jørgen Serup, M.D., Ph.D.
Department of Dermatological Research
Leo Pharmaceutical Products
Ballerüp, Copenhagen, Denmark

Eric W. Smith, Ph.D.
School of Pharmaceutical Sciences
Rhodes University
Grahamstown, South Africa

Anat Tamir, M.D.
Department of Dermatology
Ichilov Hospital
Tel Aviv, Israel

Gavriel Tamir, M.D.
Department of Plastic Surgery
Beilinson Hospital
Tel Aviv, Israel

Ethel Tur, M.D.
Department of Dermatology
Ichilov Hospital
Tel Aviv University
Tel Aviv, Israel

Jan E. Wahlberg, M.D.
Department of Occupational Dermatology
Karolinska Hospital
Stockholm, Sweden

Volker Wienert, M.D.
Department of Dermatology
RWTH
Aachen, Germany

Klaus-P. Wilhelm, M.D.
Department of Dermatology
Medical University
Lübeck, Germany, and
proDERM Applied
Dermatological Research
Hamburg, Germany

PREFACE

The second volume of our series on Bioengineering of the Skin is devoted to the methods for blood flow and erythema quantification.

Skin microcirculation is an important system for the maintenance of skin metabolism, body temperature, and defense mechanisms. The objective assessment of functional properties of blood vessels is of fundamental importance not only in dermatology, toxicology, and related sciences, but also in general medicine and clinical practice. Objective techniques allow functional and dynamic assessment of blood flow and erythema often undetectable by the naked eye.

This second volume maintains the general outline of the series by focusing on the instrumentation and the techniques available to measure blood flow and erythema, with special regard to *what* these instruments do measure and *why* and *when* to use them in skin research and product testing.

Since all of us are currently active in clinical and investigative dermatology, particular attention has been paid to the assessment of common skin diseases such as psoriasis, scleroderma, irritant dermatitis, and skin vascular disease as well as corticosteroids and sunscreens testing.

We thank all the contributors of this book. Most of them are not only friends but also pioneers in the noninvasive approach to skin bioengineering, and due to their enthusiasm this work has been possible.

We would also like to acknowledge the important help in discussing and reviewing this project to Dr. Klaus Wilhelm, M.D., and to Paul Petralia and Tia Atchison at CRC Press for keeping tight the schedule of the editorial process.

We hope that this project can help in spreading the familiarity with such instrumentation for a different approach to skin investigation.

Enzo Berardesca
Peter Elsner
Howard I. Maibach

CONTENTS

Part I   Introduction

Chapter 1

Anatomy and Physiology of the Cutaneous Microcirculation

Irwin M. Braverman

CONTENTS

I. ORGANIZATION OF THE DERMAL MICROCIRCULATION

The cutaneous blood supply is a microcirculatory bed composed of three segments—arterioles, arterial and venous capillaries, and venules. The arterioles and venules form two important plexuses in the dermis: an upper horizontal network in the papillary dermis from which the capillary loops of the dermal papillae arise and a lower horizontal plexus at the dermal-subcutaneous interface (Figure 1). The lower plexus, formed by perforating vessels from the underlying muscles and subcutaneous fat, gives rise to arterioles and venules that directly connect with the upper horizontal plexus and also provide lateral tributaries that supply the hair bulbs and sweat glands. There are some interconnections among the ascending arterioles and descending venules within the dermis, but these two horizontal plexuses represent the physiologically important areas in the skin. Most of the microvasculature is contained in the papillary dermis 1 to 2 μm below the epidermal surface. In areas of skin where the dermal papillae are not well developed, arterioles connect with capillaries that course close to the dermal-epidermal junction before the latter move deeper into the dermis to join the postcapillary venules in the upper horizontal plexus.

Using 1-μm plastic embedded sections, arterioles can be identified with certainty by the presence of an internal elastic lamina, capillaries on the basis of a thin vascular wall containing pericytes, and venules on the basis of thicker walls without elastic fibers. Unfortunately, these distinctions cannot be made reliably in 5-μm paraffin sections. Electron microscopy is able

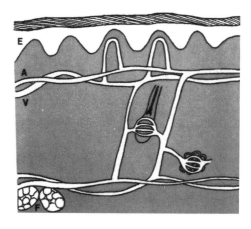

Figure 1   Schematic drawing of micro-vascular organization in human skin. E = epidermis, A = arteriole, V = venule, F = subcutaneous fat. (From Braverman, I. M., *J. Invest. Dermatol.*, 93, 2S, 1989. With permission.)

to identify definitively the different segments of the microvasculature and must be used in correlative studies of structure with normal and abnormal function for confirmation of the observed light microscopic findings. Ultrastructural criteria for identifying the three segments of the microvasculature were proposed by Braverman and Yen[1,2] after reconstructing portions of the horizontal plexus and capillary loops in normal human skin using a combination of serial 1-μm plastic sections and ultrathin sections as the vessels coursed from the arterial to the venous side. This approach eliminated the inconsistencies and errors of interpretation in earlier studies of the dermal microvasculature that had been based on ultrastructural descriptions of individual vessels as they were randomly encountered in human and animal tissues without any consideration being given to their location within the microvascular bed. The criteria proposed by Braverman and Yen have been confirmed by Higgins and Eady[3] and parallel the seminal work of Rhodin[4,5] on the subdermal microcirculatory bed in rabbits.

A.   GENERAL OBSERVATIONS

The microvessels in the papillary dermis vary in diameter from 10 to 35 μm, but most are in the 17 to 22 μm range. In the mid- and deep dermis the microvascular diameters range from 40 to 50 μm, although rarely a vessel as large as 100 μm can be found.

B.   MICROVASCULAR SEGMENTS

The arterioles in the papillary dermis vary from 17 to 26 μm in diameter and represent terminal arterioles. They most likely function as part of the resistance vessels in the skin. In these arterioles the endothelial cells are surrounded by two layers of smooth muscle cells. The cells in the inner layer are oriented longitudinally and those in the outer layer form a spiral. The vascular wall is composed of basement membrane material that has a relatively homogeneous appearance and completely surrounds and encompasses the elastin and smooth muscle cells. The inner smooth muscle cells and endothelial cells send cytoplasmic processes toward each other to make frequent tight junctional contacts through breaks in the basement membrane. The subendothelial elastic lamina appears as an interrupted layer between the endothelial and smooth muscle cells in cross section. In longitudinal sections and in three-dimensional (3-D) reconstructions, the elastic lamina can be seen to be made up of interwoven longitudinal fibers resembling a wire mesh fence and presumably performing a function analogous to steel rods embedded in concrete. Each bundle of longitudinal elastic fibers lies in a space formed by the cytoplasmic processes of the inner layer of smooth muscle cells and endothelial cells that have made contact with one another.

As the diameter of the arteriole decreases from 26 to 15 μm, the elastic fibers assume more and more of a peripheral position in the vascular wall. At the 15 μm level, the elastic fibers have disappeared from the vascular wall and are present only as an incomplete sheath between the wall and the surrounding adventitial (veil cell) layer. This external elastin sheath disappears at the 10 to 12 μm level, which corresponds to the beginning of the capillary bed. The basement membrane retains its homogeneous appearance during this transition. Smooth muscle cells—identified by numerous dense bodies and myofilaments—are not found below the 15 μm level. Their place is taken by a single cell with less well-developed dense bodies and many fewer filaments. This cell has many circumferential arms, some of which almost completely encircle the endothelial cell tube. The arms overlap each other slightly. The configuration of the cell and the localization of vasomotor activity to this terminal arteriolar segment by laser Doppler velocimetry suggest that this cell may function both as a precapillary sphincter and as the pacemaker for vasomotor activity.

The next contiguous microvascular segment, the arterial capillary, has an outside diameter of 10 to 12 μm and an internal endothelial tube diameter of 4 to 6 μm. The basement membrane material retains its homogeneous appearance in the cutaneous microvascular bed. Pericytes form tight junctions with endothelial cells through breaks in the basement membrane. Pericytes differ from smooth muscle cells by being thinner, lacking dense bodies, and having fewer myofilaments. However, pericytes do have the contractile proteins necessary for cellular contraction. The walls of these capillaries vary from 2 to 3 μm in thickness, but in a few instances the endothelial tube is surrounded by a wall only 0.5 to 1 μm wide. This is in contrast to capillaries elsewhere in the body where the vascular wall is usually only 0.1 μm wide.

As the arterial capillary is traced, the basement membrane material begins to develop lamellae within its previously homogeneous framework until a segment is reached in which the entire vascular wall is multilaminated. Dense layers of basement membrane material, 25 to 100 nm thick, alternate with less dense zones. As many as ten lamellae may be present. The outside diameters of the vessels remain at 10 to 12 μm, and their endothelial tubes at 4 to 6 μm. Pericytes and veil cells remain the important cellular elements in the vascular wall and in the immediately surrounding dermis, respectively.

The venous capillary connects with the postcapillary venule, a vessel whose external diameter increases from 12 to 35 μm and whose endothelial tube diameter enlarges from 8 to 26 μm. Most of the postcapillary venules seen in the papillary dermis measure 18 to 23 μm in external diameter and 10 to 15 μm in endothelial tube diameter. Pericytes form two to three layers around the vascular wall, in contrast to only one layer in the venous capillary, and cover approximately 80% of the endothelial cell tube surface. The basement membrane of the vascular wall is multilaminated. The wall is usually 3.5 to 5.0 μm wide. Collagen fibrils may be present between the lamellae or may form a thin sheath in the outer layer of the vascular wall.

The papillary dermal vessels are composed entirely of terminal arterioles, arterial and venous capillaries, and postcapillary venules. The majority of vessels, however, are postcapillary venules. The same spectrum of vessels is present as far as the mid-dermis. Some of the vessels in the lower third of the dermis are twice as large as the upper dermal vessels and represent arterioles and collecting venules (40 to 50 μm). The postcapillary venules are physiologically the most reactive segment of the microcirculation. This segment is the site where inflammatory cells migrate from the vascular space into the tissues and where endothelial cells develop gaps that result in increased vascular permeability during acute inflammation and in response to histamine release.

The homogeneous and multilaminated appearance of the vascular basement membrane is much more distinct in tissues fixed in Karnovsky's fixative than in those fixed in buffered osmium tetroxide. Osmium fixation may produce a pseudolaminated appearance.

The ultrastructural differences in basement membrane material between the arterial and venous components of the microcirculation are present in all areas of the skin (Figure 2).

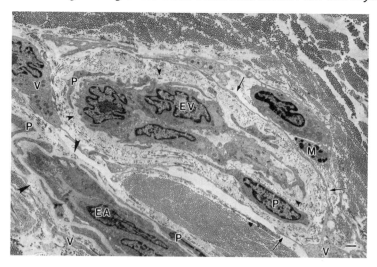

Figure 2   Electron micrograph of postcapillary venule and adjacent terminal arteriole. EV = venular endothelial cell, EA = arteriolar endothelial cell, P = pericyte, V = veil cell with thin cytoplasmic processes (arrows), M = mast cell. Small arrowhead indicates multilayered basement membrane material in venule. Large arrowhead points to homogeneous-appearing basement membrane material in arterioles. Bar = 1 μm. (From Braverman, I. M., *Dermal Immune System,* CRC Press, Boca Raton, FL, 1993. With permission.)

C. THE CAPILLARY LOOP

The capillary loop arises from a terminal arteriole in the horizontal papillary plexus and is composed of an ascending limb, an intrapapillary loop having a hairpin turn, and a descending limb that connects with a postcapillary venule in the horizontal plexus.[2] Each dermal papilla is supplied by a single capillary loop which has an intrapapillary and an extrapapillary portion defined by an imaginary line drawn between the deepest points of adjacent rete ridges. The ascending limb of the loop arises from the horizontal plexus and either passes directly to a papilla or subdivides once, twice, or three times before giving rise to the final ascending limbs that enter individual papillae. The ascending limb of the capillary loop in its extra- and intrapapillary portions has the characteristic of an arterial capillary—homogeneous-appearing basement membrane material in the vascular wall. The endothelial tube at the crest and in the intrapapillary descending portion is 1 to 1.5 μm wider than that of the ascending limb. The mural basement membrane material of the capillary loop retains its homogeneous appearance throughout its intrapapillary course. At the apex of the loop the endothelial cells become attenuated; in some areas cell wall thickness measures 11 to 30 nm, but true bridged fenestrations are rarely seen in healthy skin except in the zones adjacent to the eccrine sweat gland and dermal papilla of the hair.

The character of the descending limb changes abruptly either at the border or 10 to 30 μm distal to the border, where the loop becomes extrapapillary. The endothelial tube becomes wider, and the basement membrane material in the wall loses it homogeneous appearance and develops multilayers characteristic of a venous vessel before it connects with the horizontal plexus. There are no ultrastructural differences between the capillary loops at different sites in the skin. Bridged fenestrations are normally found in capillaries where there is a need for rapid exchange of molecules between the vascular system and the tissues (renal glomeruli, endocrine glands, intestinal lamina propria, choroid plexus of the brain, and the ciliary body of the eye). The capillary loops in psoriatic lesions have bridged fenestrations and a multilayered basement membrane. Fenestrations are believed to be the morphological equivalent of the large pore system originally proposed by physiologists.

D. DEEP DERMAL AND SUBCUTANEOUS VESSELS

The ultrastructure of the arterioles and venules in the mid- and lower dermis and in the lower horizontal plexus differs from the ultrastructure of comparable vessels in the superficial horizontal plexus in several ways.[6] The diameters are wider (50 μm vs. 25 μm); the walls are thicker (10 to 16 μm vs. 4 to 5 μm); smooth muscle cells or pericytes are present as four to five layers, in contrast to one to two layers in similar vessels of the superficial plexus; and bundles of collagen fibrils in arterioles are present in subendothelial positions rather than in the periphery of the vascular wall.

The arterioles and venules in the fat lobules are identical in structure and size to those of the lower horizontal plexus. Rarely, an arteriole 100 μm in diameter is seen. The walls of the capillaries in the fat are only 0.1 to 0.3 μm thick, in contrast to upper dermal capillaries, in which the walls are characteristically 2 to 3 μm thick. In addition, the veil cells that encircle all arterioles, capillaries, and venules in the dermis are not always present around the corresponding vessels in the subcutaneous layer. Venus capillaries with bridged fenestrations are found in close proximity to eccrine sweat glands and hair bulbs.

The endothelial cells of the arterioles in the mid- and lower dermis and fat contain bundles of actin filaments 4 to 7 nm wide that tend to course along the abluminal border of the cells. These bundles exhibit transverse linear wavy bands of increased density resembling the Z bands of skeletal muscle. In longitudinal sections, these intracellular bundles appear to be associated with extracellular filaments measuring 10 to 20 nm in diameter. A dense linear zone can be found in the region of the cell membrane where the intracellular filaments and extracellular filaments meet. These extracellular filaments form four kinds of linkages: (1) a connection is formed between the intracellular bundles of two adjacent endothelial cells; (2) in endothelial cells with an irregular abluminal border, they link one intracellular bundle with another in the same cell after coursing through the underlying basement membrane material; (3) they also run from the intracellular bundle to the underlying mural basement membrane material or (4) to an adjacent elastic fiber to which they connect through the microfibrillar component of these fibers. This complex of intracellular and extracellular filaments is not present in the elastin-free arterioles or postcapillary venules of the superficial plexus and is only rarely seen in the larger venules (50 μm) of the deep dermis and subcutaneous layer. Because these cytoskeletal filaments occur only in the high-pressure arterial vessels, they probably serve to maintain cellular shape and to increase the adhesion between adjacent endothelial cells and between the endothelial cells and the basement membrane material of the vascular wall.

E. COLLECTING VENULES WITH VALVES

Valve-containing vessels pass from the deep dermis into the superficial layer of the fat. These vessels range from 70 to 120 μm in diameter and can be identified easily with a dissecting microscope.[7] Postcapillary venules (25 to 50 μm) join these large vessels from all directions. Valves are found at most places where the small vessels join the larger ones, but valves are also present within the large vessels not associated with branch points. All of the valves have two cusps with associated sinuses (pockets formed between each valve leaflet and the adjacent vessel wall). The free edges of the valves are always directed away from the smaller vessel and toward the large one at the branch points.

Valves imply a mechanism involved in the forward propulsion of blood. Although the work of Landis[8] and Eichna and Bordley[9] in 1930 and 1939, respectively, strongly suggested that the blood flow in the capillary bed was pulsatile, the recent work by Mahler et al.[10] in human skin has established this point. Direct cannulation of human finger nailfold capillaries demonstrated that the blood pressure was pulsatile in both the arteriolar and venular limbs and that systolic pressures fluctuated from 11 to 75 mmHg. With pulsatile blood pressure fluctuations of such magnitude it is easy to reconcile the presence of valves in veins at the dermal-subcutaneous junction. The pulsatile flow detected in the nailfold capillary loops would be transmitted to the lower horizontal plexus by venules that run directly from the upper horizontal plexus, where the capillary loops connect, to the lower horizontal plexus, which feeds into these valve-containing veins. Valves in collecting veins at the dermal-fat interface would seem to be appropriately placed to insure the forward motion of the blood.

F. VEIL CELLS

The veil cells are flat adventitial cells that surround all the dermal microvessels.[1,11] The exact nature and function of these cells are still undetermined. They do not have cells markers for T, B, or Langerhans cells, nor do they stain for HLA-DR. However, they are strongly positive when stained for antibody against Factor XIIIa, which indicates that they are one component of the dermal dendrocyte system.[12] Their ultrastructure differs from fibroblasts by having extensive caveolae and a less prominent endoplasmic reticulum, but their function is not understood. Unlike pericytes, which are an integral component of the vascular wall and are enmeshed in the mural basement membrane material, the veil cells are totally external to the wall, demarcating the vessel from the surrounding dermis. Perivascular mast cells are usually situated in a space between the vascular wall and the surrounding veil cells. The veil cells are seen infrequently around the microvessels in the subcutaneous fat, and it is not known whether they are present in significant numbers around microcirculatory vessels in other organs.

G. CONTRACTILE CELLS OF THE MICROVASCULAR WALL

Smooth muscle cells and pericytes comprise the contractile elements of the microvascular wall.[13] In the upper horizontal plexus they form unique configurations in each functional segment of the microvasculature. The elastin-containing arteriole has two layers of smooth muscle cells: an inner layer of elongate cells beneath the endothelial cells, parallel to the long axis of the vessel, that serve as the scaffold for the elastic fibers that they have synthesized, and an outer layer of spirally arranged smooth muscle cells that produce contraction and relaxation of the endothelial cell tube.

In the arteriolar segment just before the capillary bed, the contractile cells resemble pericytes more than smooth muscle cells. They are arranged as single cells with multiple circumferential arms that completely or almost completely encircle the endothelial tube. The multiple arms overlap each other slightly. There are junctional contacts between this cell and the endothelial cell, but not to the extent found in the postcapillary venule described later. The length of an individual contractile cell in the terminal arteriole has not been established, but from our reconstructions they are at least 6.5 µm long. Such a cell could act as a sphincter, making this segment a functional precapillary sphincter.

In the postcapillary venule, the pericytes form one to two layers depending on the degree of overlap. They have two shapes: (1) elongate cells with short circumferential arms that lie parallel to the long axis of the vessel; and (2) short cells with long circumferential arms that almost, but never completely, encircle the endothelial cell tube. Pericytes and endothelial cells make innumerable contacts over their entire surfaces. The pericyte contacts range from cytoplasmic "fingers" and "ridges" that insert into deep endothelial cell invaginations to "fingers" that merely touch the endothelial cell surface. The pericytes grip the endothelial cell tube much like a vise-grip tool. Three-dimensional reconstructions show that a single pericyte makes contact with the two to four endothelial cells it covers. Endothelial cell invaginations into pericytes are much less frequent. This intimate interdigitation has been shown to be rich in fibronectin at the contact points. The presence of the necessary contractile proteins in pericytes and this complex interdigitation support the concept that pericytes are the major contractile cells responsible for producing the endothelial cell gaps in the postcapillary venular segment, where the physiological and biochemical events of inflammation occur.

The ultrastructural and organizational features of the normal dermal microvessels have begun to prove useful in understanding the role of the microvasculature in normal physiology and in pathologic states such as actinically damaged skin,[14] chronological aging of skin,[14] psoriasis,[15,16] and the formation of telangiectases.[17] The upper plexus is recognized to be the major site of migration of blood cells from the microvasculature into the tissues during inflammation. We will now be able to understand better the pathogenetic events in inflammation because of recent advances in the culture techniques and biochemistry of microvascular endothelial cells and the discovery of adhesion molecules and their regulatory cytokines.

Knowledge about the ultrastructure and organization of the cutaneous microvasculature has been obtained only in the past 20 years. Prior to that time the cutaneous vessels, with the exception of those in the sole, were believed to exist as a randomly anastomosing network without stratification. Physiologists had recognized for some time that the cutaneous blood flow was far greater than that needed for epidermal nutrition and had correctly surmised that the excess must be related to regulation of heat loss and temperature control. The superficial horizontal plexus is the radiator, but the exact sites of control for heat regulation remain to be determined. The putative control factors—arteriovenous communications—have not been identified in the skin except in the digits, nose, and ears.

Blood flow cannot be studied in individual cutaneous vessels as it is performed in animal mesentery, hamster cheek pouch, and the bat wing. The blood flow in nailfold capillaries which has been directly visualized and measured is not representative of the flow elsewhere in the skin. In virtually all physiological experiments the cutaneous vessels can be studied only as a ''black box'' phenomenon. However, with the advent of laser Doppler flowmetry (LDF), it has become possible to study the collective activity and physiological responses of the arterioles and venules which are present in the 1 mm^3 volume of tissue measured with the LDF probe. Based on studies correlating LDF patterns with the morphology of the underlying vasculature, described below, it is now possible to identify, from the LDF patterns, 1 mm^2 areas of skin that have a predominantly arteriolar composition, a predominantly capillary/venular composition, or are relatively avascular. These correlations will allow experiments to be designed to test mechanisms underlying dermal perfusion in normal skin and in microvascular disorders encountered in clinical medicine.

II. LASER DOPPLER FLOWMETRY STUDIES

A. BACKGROUND

LDF is a noninvasive technique that provides continuous measurement of cutaneous blood flow. Most investigators consider the results obtained by LDF to be comparable to the techniques of radioactive xenon washout and photopulse plethysmography,[18-22] although Kemp and Staberg believe that this correlation is only true at high rates of cutaneous blood flow.[23] The volume of tissue measured by LDF has a surface area of 1 mm^2 and a penetrating depth of 1 to 1.5 mm. The 1 mm depth effectively includes the entire upper horizontal plexus composed of arterioles, capillaries, and postcapillary venules. The deep horizontal plexus at the subcutaneous dermal junction is not measured.

The major drawback to LDF, which has been pointed out by Tenland et al.[22] and other early workers,[18,20,21] is the enormous variation in the RBC flux from comparable sites between individuals, from different sites in the same individual, and even from the same site in the same individual at intervals of hours, days, and weeks. Tenland and co-workers hypothesized that the great variations in blood flow at adjacent sites might be the result of inhomogeneities in the microvasculature, but they did not clearly define whether these inhomogeneities were structural or functional. Current investigators simply take multiple readings over an area and calculate the mean.

B. CORRELATION OF LDF PATTERNS WITH UNDERLYING MICROVASCULATURE

In preliminary experiments on blood flow in psoriatic and normal skin, we also noted that there was a variation of 100% in the RBC flux as the probe was moved over distances of 2 to 6 mm, but this variation was accompanied by distinctive wave patterns. The instrument used in these studies[24] was a Medpacific LD5000 laser Doppler flowmeter (Medpacific Corporation, Seattle, WA) with a standard 1.5-cm flat opaque tissue probe which had a 1 mm² circular detection window on its under surface. The flat tissue probe was lightly attached to the subject's skin with circular, open-centered, double-sided adhesive tape of the same diameter.

Three types of wave patterns were recognized. For each wave pattern of interest the probe was gently lifted from the adhesive tape, which remained on the skin as a circle with an open center measuring 4 mm. The sampling window of the probe (1 mm²) was at the center of the circle. A 2-mm punch biopsy of the center was performed after a ring of anesthesia consisting of 1% lidocaine without epinephrine was placed around the biopsy site. Ten biopsies from four volunteers were performed and analyzed. Each biopsy was cut into three vertical pieces of approximately equal thickness under a dissecting microscope, fixed in half-strength Karnovsky's fixative, and processed for embedding in Spurr's epoxy resin.[1] The orientation of each piece in relation to the original biopsy was maintained. Each piece was cut serially in 1-µm sections and stained with 1% azure blue. Three-dimensional spatial reconstructions of the ascending arterioles with their immediate branches were made in the microscope by tracing the vascular paths in the semithin sections and recording their positions in each section. This was possible because ascending arterioles, which give rise to the non-elastin-containing terminal arterioles, contain an elastic lamina easily seen in 1-µm plastic sections. Each segment of an observed arteriole was labeled 1, 2, or 3 depending on whether its position was found in the outer two thirds of the biopsy (1, 3) or in the central third (2). Three-dimensional computer reconstructions of the entire horizontal plexus were also made from these serial sections.[25]

Three types of wave patterns were detected: (1) high flux (40 to 80 mV), with prominent pulsations corresponding to the cardiac pulse superimposed on a slower undulating wave that had a periodicity of 6 to 10 cpm and represented vasomotor activity; (2) a low-flux (20 to 40 mV) wave pattern with no pulsatile or vasomotor activity; and (3) a low-flux (20 to 40 mV) wave with pulsatile activity but no apparent vasomotion. Figure 3 shows that high-flux pulsations with vasomotor activity were recorded when the 1-mm-diameter window of the probe was situated directly over the ascending arteriole with its immediate four to five elastin-containing branches (labeled 2). When the probe window was not centered over the arteriole only low-flux pulsations were recorded, and when the arterioles were present in the outer two segments of the trisected biopsy a low-flux, relatively flat wave was recorded by the centrally placed probe (Figure 4). The arrowheads indicate the 1 mm depth below the surface of the skin.

Figures 5 and 6 are computer reconstructions of the upper plexus. In Figure 5 the zone between layers 1 and 2 contains the dermal capillary loops that arise from the vessels of the horizontal plexus present between layers 2 and 3. The ascending arteriole (dark, thin line A) and the accompanying descending venule (lighter, broad line V) are shown. Figure 6 shows the same reconstruction tilted 70° on the x-axis so that the microvasculature is seen en face. The thinnest dark and light lines represent the capillary loops (arrows) in the dermal papillae. There is a void (asterisk) of arteriolar vessels that measures approximately 0.6 mm in diameter and a void of postcapillary venules measuring 0.3 mm. There are eight to ten postcapillary venular channels that eventually converge to form the descending venule (V) and four to

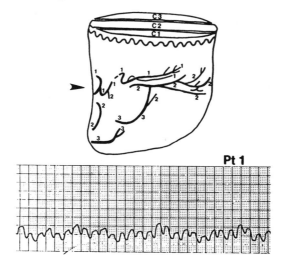

Figure 3 Correlation of laser Doppler flowmetry (LDF) wave patterns with location of ascending arterioles in trisected biopsies. Numbers on arterioles indicate in which trisected portion they were found. Arrow indicates 1 mm depth below skin surface. Trisected slices are numbered on surface of biopsy which has been drawn to scale. Description of LDF data. Paper speed = 5 mm/s (one major division on tracing). Pulsatile pattern is superimposed on a slower undulating wave that varies from 30 mV to approximately 45 mV and has a frequency of 6 cpm. The faster pulsatile pattern represents the cardiac pulse and the slower wave the vasomotor activity. (From Braverman, I. M., *J. Invest. Dermatol.*, 95, 283, 1990. With permission.)

Figure 4 Correlation of laser Doppler flowmetry wave patterns with location of ascending arterioles in trisected biopsies using the same parameters as in Figure 3. Relatively flat waves are seen without the undulating pattern of vasomotor activity. (From Braverman, I. M., *J. Invest. Dermatol.*, 95, 283, 1990. With permission.)

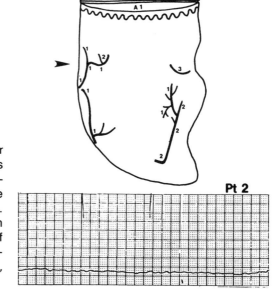

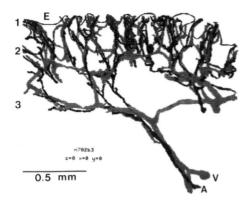

Figure 5 Computer reconstruction of vessels present in papillary dermis. E = epidermis. Zone 1 to 2 contains dermal capillary loops and zone 2 to 3 contains vessels of horizontal plexus. A = ascending arterial with five immediate branches and V = accompanying descending venule formed by nine converging postcapillary venules. (From Braverman, I. M., *J. Invest. Dermatol.*, 95, 283, 1990. With permission.)

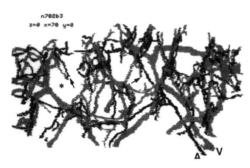

Figure 6 Computer reconstruction in Figure 5 rotated 70° on x-axis. Asterisk indicates space empty of arterioles and postcapillary vessels. Arrows indicate capillary loops in dermal papillae. A = ascending arteriole, V = descending venule. Width of reconstruction = 2 mm. (From Braverman, I. M., *J. Invest. Dermatol.*, 95, 283, 1990. With permission.)

five main arteriolar branches that come off the ascending arteriole (A) to form the terminal arterioles and capillaries and, eventually, the postcapillary venules of the horizontal plexus.

The variation in RBC flux detected by LDF is a function of the geometry between the measuring window of the probe and the underlying ascending arteriole and its branches. When it is directly over the arteriole the flux is high, pulsations are prominent, and vasomotor activity is detectable. When the probe catches the edge of an arteriolar field the flux is low and the vasomotor activity may or may not be detected, but cardiac pulsations are recorded. When the probe window is in a zone devoid of ascending arterioles and their immediate branches, low flux with no pulsations or vasomotor activity is found.

The correspondence of vasomotor activity to the ascending arterioles and their immediate elastic branches strongly suggests that these vessels are the morphological sites for the generation of vasomotion. The arteriolar continuations of these immediate branches lose both their elastic lamina and two layers of smooth muscle cells, which are replaced by a single smooth muscle cell that completely encircles the endothelial cell tube.[13] This configuration strongly suggests that the encircling smooth muscle cell functions both as a sphincter and as site of origin for the vasomotor activity detected by LDF, as has been demonstrated for the precapillary sphincters in cat mesentery.[26]

C. TOPOGRAPHIC MAPPING OF MICROCIRCULATION BASED ON FLUX

It is possible now to correlate the wave patterns from LDF with the approximate mixture of microvascular elements beneath the probe. It provides a way to determine the density of the ascending arterioles in a defined area of skin, as well as defining the areas of capillary and postcapillary venular beds in between, thus permitting noninvasive analysis of the microvasculature under experimental conditions and in disease states. However, the standard opaque flat probe, which is 1.5 cm in diameter and which has on its undersurface the 1-mm-diameter sampling window (not visible from above), can neither be moved as precisely as one would like nor be relocated on the same spot for experimental manipulations, after it has been removed, without the use of a stereotaxic device. To overcome this difficulty and to quantitate LDF data better, we devised a computer-linked system that incorporated a narrow (1.24 mm) probe in a simple probe holder that allowed for precise sequential movements and relocation of the probe on the skin after intervals of hours to days.[27] With this system, RBC flux was acquired, graphed, and stored for further analysis by computer; the exact same site could be sampled repeatedly and accurately before and after manipulations to the skin; and any site or sites within a defined area could be located accurately to within 1 mm for sampling at intervals of hours to days. The probe holder allowed the 1.24-mm probe to be moved in increments of 1.25 mm, approximately the width of the probe window itself, so that four contiguous 1 mm^2 spots in a 2.29 mm × 2.29 mm area could be sampled. An area as large as 2.48 cm × 2.48 cm could be measured with this device. The flux was fed into a computer by an analog/digital board and visualized as a wave form on the monitor. Each spot sampled was given a unique x,y coordinate, and the mean amplitude of the flux was designated as the z coordinate. With the aid of commercially available software, the values of these three coordinates were mathematically processed to produce contour graphs with shading that represented a map of the arteriolar circulation in the skin (Figures 7 and 8). Video images of the superficial plexus at the sites where the maps were made confirmed the topographic appearance of these maps.

The volunteers sat comfortably in a chair with their forearms supported at heart level. The room temperature was 20° C, and the subjects had rested for 20 min before the recordings began. To determine the reproducibility of the method, RBC flux was mapped on days 1, 4, and 7 in two healthy women, 23 and 46 years old, at comparable areas on the flexor forearm, 5 cm distal to the antecubital fossa crease. In two other female volunteers, 44 and 46 years old, mapping was performed on days 1, 2, 5, and 8. In a fifth volunteer, a 25-year-old male, mapping was performed on days 1, 4, 10, and 12. Fiducial marks were placed on the skin with indelible ink so that the probe holder could be accurately relocated on successive days.

The sequential maps in all five subjects showed that there were sites of high flux and low flux that were constantly present over the 7 to 12 days of mappings, as well as areas of high flux that were intermittently present. Figures 7 and 8, from one subject, are representative topographic maps. With this computerized system one can construct reliable topographic maps of the cutaneous microcirculation that indicate areas of either arteriolar or capillary/venular predominance.

Figures 9 and 10 demonstrate the 1-min variations in mean flux of two high-flux sites, separated by 2 to 7 mm in different experiments, that were simultaneously measured. In Figure 9 the flux changed dramatically every 12 to 20 min, whereas in Figure 10 changes occurred at intervals of 70 to 90 min.

The overall vascular patterns mapped by the laser Doppler machines were identical to those seen on the video monitor. The gradient of high to low flux within a vascular zone cannot be distinguished by the video camera—only the overall vascular outline of the superficial plexus in a given area can be visualized. The lowest-flux areas on the maps appear as relatively avascular areas in the video images. The relatively avascular areas correspond to the 0.3 to 0.6-mm-diameter spaces devoid of vessels which had been found previously in 3-D reconstructions of the horizontal plexus[24] (Figure 6). The accumulated evidence indicates that the upper horizontal plexus has a configuration resembling a "micro-livedo" pattern. These various experiments have established the making of contour maps as a reliable and interpretable procedure for identifying arteriolar-rich and relatively avascular zones within the horizontal plexus.

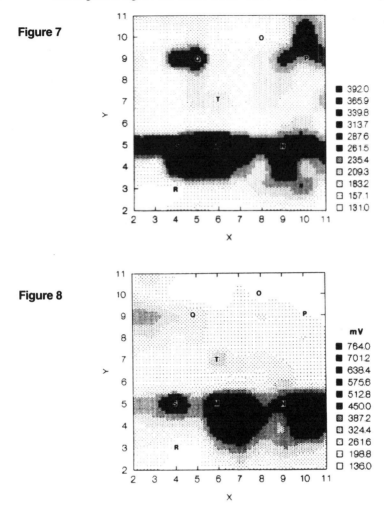

Figure 7

Figure 8

Figures 7 and 8   Results of serial topographic mapping of a 1 cm × 1 cm area on flexor forearm on days 1 (Figure 7) and 2 (Figure 8). High-flux areas (dark shading with gradations) are sites of ascending arterioles and their immediate branches in upper horizontal plexus. Sites M, N, and S consistently showed high flux. Site O had a consistently low flux, and sites P, Q, R, and T showed varying fluxes. These maps indicate that there may be arterioles with relatively constant flow patterns while there are others exhibiting intermittent flow. (From Braverman, I. M., *J. Invest. Dermatol.*, 97, 1013, 1991. With permission.)

The major drawback to LDF, which has been pointed out by Tenland et al.[22] and other early workers,[18,20,21] is the enormous variation in the RBC flux from comparable sites between individuals, from different sites in the same individual (spatial heterogeneity, and even from the same site in the same individual at intervals of hours, days, and weeks (temporal heterogeneity). It is now possible to explain the basis for spatial heterogeneity, which serves as the basis for topographic mapping. The data from the long-term simultaneous recordings from two adjacent sites explain, in part, the basis for the temporal heterogeneity.

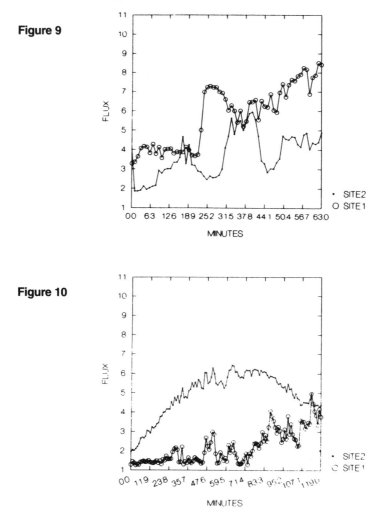

Figure 9

Figure 10

Figures 9 and 10 Continuous and simultaneous flux measurements from two sites. Each point on the graph is the mean of the flux measured during that minute. Y-axis values are arbitrary units of flux in millivolts. Figure 9: sites are 8.9 mm apart; flux changes every 12 to 20 min. Figure 10: sites are 6 mm apart; flux changes at intervals of 70 to 90 min. (From Braverman, I. M., *J. Invest. Dermatol.*, 97, 1013, 1991. With permission.)

D. TOPOGRAPHIC MAPPING OF MICROCIRCULATION BASED ON FLUX AND CONCENTRATION OF MOVING RED BLOOD CELLS

Next, we sought to determine whether it was possible to identify areas of capillary/postcapillary venular predominance in the upper plexus. The instrument used was the Perimed PF2B (Perimed Co., Stockholm, Sweden), which allowed two factors in the flux equation to be recorded simultaneously from individual sites: the number of moving red cells in a measured volume and their average velocities.[28] The Perimed machine produces a recordable signal that is proportional to the number of red cells moving through the tissues.[29] This number contributes to the flux value and is called the concentration of moving blood cells (CMBC). Using our system, two topographic maps were created at each site, in areas measuring

Table 1   **Laser Doppler Flow Patterns and Their Predicted Anatomical Correlations**

Topographic Description	Anatomical Prediction
High-medium flux/high-medium CMBC	Arterioles and venules
High-medium flux/low CMBC	Arteriolar predominance
Low flux/high-medium CMBC	Venular predominance
Low flux/low CMBC	Relative avascularity

Note: CMBC = concentration of moving blood cells.
From Braverman, I. M., Schechner, J. S., Siverman, D. G., and Keh-Yen, A., *Microvasc. Res.*, 44, 33, 1992. With permission.

CMBC

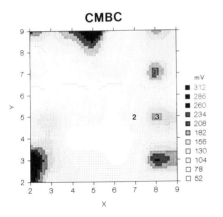

FLUX

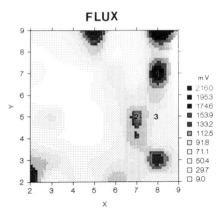

Figure 11   **Figure 12**

Figures 11 and 12   Topographic contour maps of flexor forearm skin. Coordinates of needle probe placement shown on x,y axes. Each point is 1.24 mm away from an adjacent coordinate. Scale of LDF values is in arbitrary units of millivolts. Figure 11 is a concentration of moving blood cells (CMBC) map; Figure 12 is a flux map. Biopsy sites are labeled 1, 2, and 3. Areas that appear clear or have the least dense stipples are relatively avascular. (From Braverman, I. M., Schechner, J. S., Siverman, D. G., and Keh-Yen, A., *Microvasc. Res.*, 44, 33, 1992. With permission.)

1.24 cm × 1.24 cm, one based on flux and one on CMBC. Sites with four different combinations of values could be distinguished. Reasoning that mean velocity was highest in arterioles and CMBC highest in capillary/postcapillary venular beds, it was hypothesized that these four sites should be indicators of the anatomical composition of the microvascular bed beneath them (Table 1). This method allows one to identify these different microvascular areas so that they can be selectively studied physiologically.

Three men ages 25, 40, and 61 served as subjects.[28] In each subject a 1 cm × 1 cm area was mapped on the flexor surface of the forearm 5 cm below the antecubital fossa. At each template hole the probe was moved to each of the four possible positions, the flux and CMBC were recorded simultaneously for 20 s (gain 5, frequency 20 Hz, time constant = 0.2 s for flux, 1.5 s for CMBC), and the data were saved to the disk along with the x,y coordinates. Contour maps for CMBC and flux were then created by computer (Figures 11 and 12, respectively).

After the maps were processed, sites of varying combinations of flux and CMBC values were identified. The flux and CMBC values were categorized as high, medium, or low depending on whether the value at a specific recorded site fell into the ranges of 76% to 100%, 26% to 75%, or 25% or less, respectively, of the maximally recorded value for that parameter in the topographic map. Correlative biopsies were then performed on three sites showing different combinations of values for flux and CMBC. A 2-cm-diameter ring of 1% lidocaine anesthesia without epinephrine was placed around each site. Each 1 mm^2 site was removed by a 2-mm-diameter trephine punch. In this way there was no needle injury to the microvessels in the probed areas. Of the four topographic patterns described, the only sites not biopsied were those of low flux/low CMBC. The tissue was fixed in half-strength Karnovsky's fixative, processed as for electron microscopy, embedded in Spurr's resin, and sectioned serially at 2 μm horizontally beginning at the stratum corneum down through the mid-dermis. The sections were stained with methylene blue/azure II as previously described.[1] The upper horizontal plexus was identified as the zone bounded by the origin of the dermal capillary loops to the boundary where the vessels were no longer oriented horizontally and the deeper ascending and descending vessels first appeared. From either camera lucida drawings or from microphotographs of the serial sections, the upper plexus was then reconstructed to show the courses of the capillaries, postcapillary venules, and arterioles. These vessels can be distinguished from one another in 2-μm thin sections by morphological criteria. For purposes of illustration in the reconstruction of the horizontal plexus, the arterioles are presented by solid lines and the postcapillary venules by open lines. The entire 2-mm-diameter biopsy is outlined, the horizontal plexus is drawn as if it were perfectly flat without its true undulations, and a central dotted circle is placed to represent the 1 mm diameter of dermis under the probe window which is situated in the center of the 2-mm biopsy (Figures 13 and 14).

The plexuses were reconstructed without the microscopist knowing the flux and CMBC values obtained at that site. The analysis of the final drawings and their predicted relative flux and CMBC values were made by the investigators in a blinded fashion. Quantitation of the vascular reconstructions was made in the following way. Most of the vessels could be categorized into main trunks with primary and secondary branches that supplied or drained a specific region. In a minority of cases a network of vessels was present in such an area. In such instances the longest linear portion was considered as the trunk and the rest of the vessels as branches. The dermal depth below the stratum corneum of any linear segment of a vessel was known by the serial section number in which it was found. For each major trunk and its individual branches, the span of highest to lowest points in the dermis was determined. These numbers provided information about the thickness of the upper plexus as well as the relative location of vessels within this plexus. Bar graphs were generated to indicate both the number and the level of the arteriolar and venular trunks and branches found at each biopsy site under the 1 mm^2 area of the probe window. For each vascular segment the highest and lowest levels below the stratum corneum were plotted. These points were connected to produce a bar which indicated whether the vessel was ascending from lower to higher levels in the plexus (longer bars) or whether the vessels were mostly in a horizontal orientation (shorter bars; Figures 15 and 16).

The findings in subject 1 are illustrated in Figures 11 and 12. The contour maps show two main features: sites of active flow (darker areas) and a large oval-shaped zone of very low flow (clear and least densely stippled space). The very low-flow area measures approximately 3.75 mm × 7.5 mm. The highest CMBC value (coordinate x8,y7 on the contour map) was associated with the highest density of venules present under the probe window, as seen in Figure 13. The highest flux value (also at x8,y7) was associated with the greatest number of arterioles under the probe window (Figure 13). This biopsy site (x8,y7) also contained a greater number of vessels coursing in the upper third of the plexus (400 to 650 μm below the stratum corneum; Figures 15 and 16) than did the other two biopsy sites in this subject. These vessels tended to be oriented horizontally (shorter bars) rather than vertically. The empty spaces in the drawings represent a true absence of vessels.

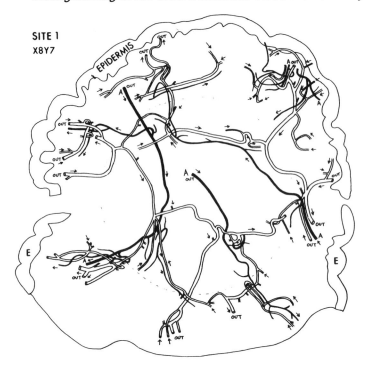

Figure 13

Figures 13 and 14   Reconstruction of microvessels in upper horizontal plexus at sites 1 and 2 shown in maps in Figures 11 and 12. Open lines are postcapillary venules. Closed lines are arterioles. Dotted circle represents the 1 mm² area under the probe window within the 2-mm trephine biopsy. Because biopsies tend to curl over at the edges like an umbrella, horizontal serial sections will show epidermis on the periphery of serial sections deep in the block. Arrows indicate presumed direction of flow based on vessel size. "Out" indicates where vessel ended in the serial sections. Figure 13 = biopsy site 1; Figure 14 = biopsy site 2. Scale: diameter of circle = 1 mm. (From Braverman, I. M., Schechner, J. S., Siverman, D. G., and Keh-Yen, A., *Microvasc. Res.*, 44, 33, 1992. With permission.)

The correlation of laser Doppler topographic maps based on relative flux and CMBC with histologic sections has provided additional evidence for the relationship between laser Doppler readings and the underlying microvascular anatomy. It has also emphasized the importance of precise probe placement. It was easy to predict the areas of highest flux or highest CMBC in a set of three reconstructions because they contained the largest number of vessels—some of which were located centrally under the probe window. Quantitative methods (bar graphs) showed additional correlative features not evident from the drawings alone. Sites of highest flux and CMBC were characterized by many horizontally oriented vessels in the upper third of the plexus. These correlative anatomical studies also confirm the inferences by Svensson and Jonsson,[30] Druce et al.,[31] and Bonner et al.[32] from their studies that CMBC could be used as a sign of flow through the upper capillary/venular bed of skin.

Our studies strongly suggest that the vessels in the upper third of the plexus make the largest contribution both to the flux and CMBC signals. Nilsson et al. investigated the depth

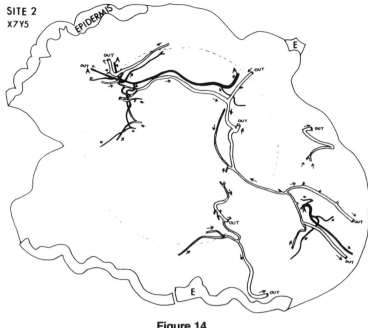

Figure 14

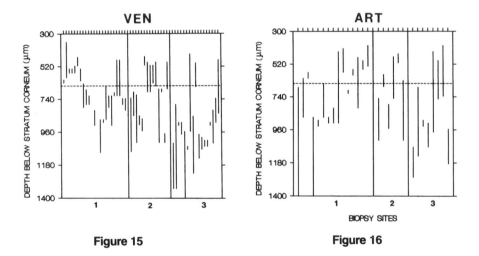

Figure 15   **Figure 16**

Figures 15 and 16   Bar graphs indicating depths (in micrometers) of vessels below the stratum corneum for the reconstructions in Figures 13 and 14. The upper plexus is approximately 800 μm thick and is situated approximately 400 to 1200 μm below the stratum corneum. A dotted line is placed at the level of 650 μm and delimits the upper third of the plexus. Figure 15 shows the depth of the venules (VEN) at biopsy sites 1, 2, and 3. Figure 16 shows the depths of the arterioles (ART) at biopsy sites 1, 2, and 3. Length of bars indicates orientation of individual vessels within the plexus, not the actual length of a vessel. The bar represents the distance between the highest and lowest points of an individual vessel within the plexus: horizontal orientation (short bars) and vertical orientation (longer bars). (From Braverman, I. M., Schechner, J. S., Siverman, D. G., and Keh-Yen, A., *Microvasc. Res.*, 44, 33, 1992. With permission.)

of measurement by using an *in vitro* model that simulated skin tissue. Polyacetyl discs were used to separate an LDF probe from cells moving within a hydraulic model.[33] They found the sensitivity to become maximal at 0.6 mm and to fall by 50% at distances of 0.2 mm and 1.2 mm. Their experimentally determined depth of maximum sensitivity corresponds to our correlation of maximal flux and CMBC values with horizontally oriented vessels in the upper third of the plexus (0.4 to 0.65 mm), below the stratum corneum. The horizontal orientation of vessels in the upper third, as opposed to the vertical, means that a greater volume of vessel with red cells lies within the area of highest sensitivity.

Flux depends not only on the number of arterioles under the probe window, but also on whether there is active perfusion at that time. The microscope cannot distinguish between vasoconstricted, relatively nonperfusing arterioles and dilated, actively perfusing vessels. Nevertheless, topographic mapping can be used reliably to identify three microvascular zones in the upper horizontal plexus: the point at which the ascending arteriole enters the plexus, the areas of relative avascularity, and the peripheral portion of the vascular territory supplied by the ascending arteriole.

III. CONCLUSIONS

The topographic maps of flux and CMBC can be directly related to our studies of the 3-D organization of the upper plexus (Figures 5 and 6). It was shown that at the point where the ascending arteriole from the deep dermis began to divide into the four or five branches that spread out to form the arterial component of the plexus there was also a high density of venules that were merging to form the accompanying descending venule. These main trunks and their branches were present at every level of the upper plexus. At a distance from the arteriolar entry point, however, the plexus had a different organization. Previous 3-D computer reconstructions of the upper plexus showed that there were 0.3-mm to 0.6-mm-diameter spaces within the plexus that were devoid of arterioles and venules.[24] The empty spaces in some of the reconstructions in the current experiments most likely represent portions of those voids. In addition, there were focal collections of only venules occupying the space from the upper border to the middle third of the plexus. The arterioles became paired with them only at the mid-plexus.[24]

From our reconstructions and topographic maps we also know that the ascending arterioles that give rise to the upper plexus may be separated from one another by distances ranging from 1.5 to 7 mm. The video images of the upper plexus show that there are relatively avascular spaces between the vascular zones. Our evidence thus far suggests that each vascular zone is fed by a single ascending arteriole. It is now possible to construct a reasonably accurate morphological and functional picture of the upper plexus. The upper plexus consists of vascular territories surrounded in part by relatively avascular areas, but joined to one another by vascular bridges. Such a configuration would produce a livedo pattern following vasodilatation. Such a picture was produced by raising the temperature of the skin with radiant energy.

Because of such a 3-D organization, the 1-mm-diameter window of the laser Doppler probe may be situated by chance over at least three different microvascular sites: the arteriolar- and venular-rich zones where the ascending arteriole begins to form the upper plexus, the relatively avascular zones, and the less dense vascular areas found in the periphery of these vascular units where venules may focally predominate. It is even possible for the probe to fall on an area that is primarily devoid of all vessels except for an arteriole and a venule which are present on the edge of a void.

The correlative integration of LDF patterns, topographic maps based on flux and CMBC, and the ultrastructure and 3-D configurations of the underlying microvessels has opened up the "black box" of skin microcirculation to daylight.

ACKNOWLEDGMENT

This work was supported by National Institutes of Health Grant AM 15739.

REFERENCES

1. Yen, A. and Braverman, I. M., Ultrastructure of the human dermal microcirculation: the horizontal plexus of the papillary dermis, *J. Invest. Dermatol.*, 66, 131, 1976.
2. Braverman, I. M. and Yen, A., Ultrastructure of the human dermal microcirculation. II. The capillary loops of the dermal papillae, *J. Invest. Dermatol.*, 68, 44, 1977.
3. Higgins, J. C. and Eady, R. A. J., Human dermal microvasculature: a morphological and enzyme histochemical investigation at the light and electron microscope levels, *Br. J. Dermatol.*, 104, 117, 1981.
4. Rhodin, J. A. G., The ultrastructure of mammalian arterioles and precapillary sphincters, *J. Ultrastruct. Res.*, 18, 181, 1967.
5. Rhodin, J. A. G., Ultrastructure of mammalian venous capillaries, venules, and small collecting veins, *J. Ultrastruct. Res.*, 25, 452, 1968.
6. Braverman, I. M. and Keh-Yen, A., Ultrastructure of the human dermal microcirculation. III. The vessels in the mid- and lower dermis and subcutaneous fat, *J. Invest. Dermatol.*, 77, 297, 1981.
7. Braverman, I. M. and Keh-Yen, A., Ultrastructure of the human dermal microcirculation. IV. Valve-containing collecting veins at the dermal-subcutaneous junction, *J. Invest. Dermatol.*, 81, 438, 1983.
8. Landis, E. M., Micro-injection studies of capillary blood pressure in human skin, *Heart*, 15, 209, 1930.
9. Eichna, L. W. and Bordley, J., III, Capillary blood pressure in man. Comparison of direct and indirect methods of measurement, *J. Clin. Invest.*, 18, 695, 1939.
10. Mahler, F., Muheim, M. H., Intaglietta, M., Bollinger, A., and Anliker, M., Blood pressure fluctuations in human nailfold capillaries, *Am. J. Physiol.*, 236, H888, 1979.
11. Braverman, I. M., Sibley, J., and Keh-Yen, A., A study of the veil cells around normal, diabetic, and aged cutaneous microvessels, *J. Invest. Dermatol.*, 86, 57, 1986.
12. Braverman, I. M., The dermal microvascular unit: relationship to immunological processes and dermal dendrocytes, in *Dermal Immune System*, Nickoloff, B. J., Ed., CRC Press, Boca Raton, FL, 1993, 91.
13. Braverman, I. M. and Sibley, J., Ultrastructural and 3-dimensional analysis of the contractile cells of the cutaneous microvasculature, *J. Invest. Dermatol.*, 95, 90, 1990.
14. Braverman, I. M. and Fonferko, E., Studies in cutaneous aging. II. The microvasculature, *J. Invest. Dermatol.*, 78, 434, 1982.
15. Braverman, I. M. and Sibley, J., Role of the microcirculation in the treatment and pathogenesis of psoriasis, *J. Invest. Dermatol.*, 78, 12, 1982.
16. Braverman, I. M. and Yen, A., Ultrastructure of the capillary loop in the dermal papillae of psoriasis, *J. Invest. Dermatol.*, 68, 53, 1977.
17. Braverman, I. M. and Keh-Yen, A., Ultrastructure and three-dimensional reconstruction of several macular and papular telangiectases, *J. Invest. Dermatol.*, 81, 480, 1983.
18. Johnson, J. M., Taylor, W. F., Shephard, A. P., and Park, M. K., Laser-Doppler measurement of skin blood flow: comparison with plethysmography, *J. Appl. Physiol.*, 56, 798, 1984.
19. Huether, S. E. and Jacobs, M. K., Determination of normal variation in skin blood flow velocity in healthy adults, *Nurs. Res.*, 35, 162, 1986.
20. Tur, E., Tur, M., Maibach, H. I., and Guy, R. H., Basal perfusion of the cutaneous microcirculation: measurements as a function of anatomic position, *J. Invest. Dermatol.*, 81, 442, 1983.

21. Sundberg, S., Acute effects and long term variations in skin blood flow measured with laser Doppler flowmetry, *Scand. J., Clin. Lab. Invest.*, 44, 341, 1984.

22. Tenland, T., Salerud, E. G., Nilsson, E. G., and Oberg, P. A., Spatial and temporal variations in human skin blood flow, *Int. J. Microcirc. Clin. Exp.*, 2, 81, 1983.

23. Klemp, P. and Staberg, B., The effects of antipsoriatic treatment on cutaneous blood flow in psoriasis measured by ^{133}Xe washout method and laser Doppler velocimetry, *J. Invest. Dermatol.*, 85, 259, 1985.

24. Braverman, I. M., Keh, A., and Goldminz, D., Correlation of laser Doppler wave patterns with underlying microvascular anatomy, *J. Invest. Dermatol.*, 95, 283, 1990.

25. Braverman, M. S. and Braverman, I. M., Three dimensional reconstruction of serial sections using a microcomputer graphics system, *J. Invest. Dermatol.*, 86, 290, 1986.

26. Johnson, P. C. and Wayland, H., Regulation of blood flow in single capillaries, *Am. J. Physiol.*, 212, 1405, 1967.

27. Braverman, I. M. and Schechner, J., Contour mapping of the cutaneous microvasculature by computerized laser Doppler velocimetry, *J. Invest. Dermatol.*, 97, 1013, 1991.

28. Braverman, I. M., Schechner, J. S., Siverman, D. G., and Keh-Yen, A., Topographic mapping of the cutaneous microcirculation using two outputs of laser Doppler flowmetry: flux and the concentration of moving blood cells, *Microvasc. Res.*, 44, 33, 1992.

29. Nilsson, G. E., Perimed's LDV flowmeter, in *Laser-Doppler Blood Flowmetry*, Shepherd, A. P. and Oberg, P. A., Eds., Kluwer Academic Publishers, Boston, 1990, 57.

30. Svensson, H. and Johnsson, B. A., Laser Doppler flowmetry during hyperaemic reactions in the skin, *Int. J. Microcirc. Clin. Exp.*, 7, 87, 1987.

31. Druce, H. M., Kaliner, M. A., Ramos, D., and Bonner, R. F., Measurement of multiple microcirculatory parameters in human nasal mucosa using laser Doppler velocimetry, *Microvas. Res.*, 38, 175, 1989.

32. Bonner, R. F., Clem, T. R., Bowen, P. D., and Bowman, R. L., Laser Doppler continuous real-time monitor of pulsatile and mean blood flow in tissue microcirculation, in *Scattering Techniques Applied to Supra-Molecular and Nonequilibrium Systems*, Vol. 73, Chen, S. H., Shu, B., and Nossal, R., Eds., Plenum Press, New York, 1981, 685.

33. Nilsson, G. E., Tenland, T., and Oberg, P. A., Evaluation of a laser Doppler flowmeter for measurement of tissue blood flow, *IEEE Trans. Biomed. Eng.*, 27, 597, 1980.

Chapter 2

Assessment of Skin Blood Flow—An Overview

Jan E. Wahlberg and Magnus Lindberg

CONTENTS

I. INTRODUCTION

The microcirculation of the skin is a rather complex and dynamic system which is important for skin metabolism and temperature regulation and is an important part of the organism's defense system against invaders. Several pathological conditions of a more general nature, e.g., arteriosclerosis, diabetes mellitus, Raynaud's phenomenon, and the circulation of the perinatal infant (Table 1), are associated with changes in skin microcirculation. Some skin disorders, both inflammatory and noninflammatory, such as psoriasis, rosacea, and morphea, are by themselves characterized by alterations in the microcirculation. The development of, or the recovery from, skin injuries such as burns can be monitored by determining the skin blood flow, and this is also the case in evaluating the vitality of skin flaps following plastic reconstructive surgery. Since the skin surface is directly accessible, the skin has become a valuable model organ for experimental purposes such as studies of irritant and allergic contact dermatitis or the evaluation of patch test reactions. Evaluations of this type of reaction often include analysis of changes in skin blood flow taken as a measure of the inflammation. The visual assessment of the degree of erythema or pallor/blanching (visual scoring)—reflecting skin blood flow—has been used for centuries in medicine and is still used in routine clinical work.

However, the introduction of objective and noninvasive techniques for the evaluation/ assessment of erythema has taught us that the naked eye is rather unreliable. In studies on solvents it has been demonstrated[1,2] that skin blood flow, objectively assessed by laser Doppler flowmetry, must be raised 3 to 4 times before the naked eye can detect erythema. There is also too much variation between different observers.[3] The scoring of erythema as faint, weak, moderate, strong, etc. or $+$, $++$, $+++$, etc. is obsolete in experimental work and does not allow accurate statistical treatment of the data. Thus, for scientific reasons also there is a great demand for objective scoring systems.

0-8493-8371-4/95/$0.00+$.50
© 1995 by CRC Press, Inc.

Table 1   Areas of Interest for Evaluation of Skin Blood Flow

Skin biology and physiology
Dermatology
    Pathology/diagnostics—psoriasis, rosacea, morphea, scleroderma
    Predictive testing of irritancy
    Evaluation of patch test reactions
    Evaluation of treatment—e.g., psoralen-ultraviolet A (PUVA), corticosteroids
    Evaluation of protection—sunscreens against UV light
Pharmacology
    Effects of drugs
    Transdermal drug delivery
Plastic surgery
    Burns
    Skin flaps
    Skin expansions
General medicine
    Perinatal surveillance
    Diabetes mellitus
    Sympathicus-parasympathicus disturbances
    Wound healing
    Collagen disorders
    Raynaud's phenomenon
    Circulatory disorders, arteriosclerosis

Table 2   Methods for Evaluation of Skin Blood Flow

Visual scoring
Capillary microscopy (capillaryoscopy)
Fluorescence microscopy—videomicroscopy and videodensitometry
Photoplethysmography
Isotope washout methods (^{133}Xe, ^{99}Tc)
Transcutaneous measurement of oxygen pressure (tcpO$_2$ measurement)
Laser Doppler flowmetry/velocimetry
Remittance spectroscopy
Chromametry techniques

The ease with which the skin is exposed to a potential investigator has also led to its being used for the evaluation of vasoactive drugs, including corticosteroids. The vasoconstriction test, introduced by McKenzie and Stoughton,[4,5] was initially used to rank commercially available corticosteroids in order of potency. This bioassay can be employed both to screen new steroids for clinical efficacy and to determine the bioavailability of steroids from various topical vehicles. The blanching/pallor observed, which reflects skin blood flow, was initially assessed visually against untreated skin on a 0 to 4 scale. Today several more sophisticated, objective, noninvasive techniques are available. Another aspect of this is that during the past decade there has been increasing interest in the use of the skin as the route of entry for drugs in transdermal drug delivery systems.[6]

There are numerous situations where the evaluation of skin microcirculation is a valuable tool for the investigator. Since the skin constitutes the barrier between the environment and the internal milieu of the individual, the skin blood flow is well suited for noninvasive evaluation (Table 2).

During the past few years several techniques, mostly noninvasive, have been developed for assessing the microcirculation of the skin. Some techniques give direct information on microvascular function (e.g., capillary microscopy or fluorescence microscopy), whereas others give an indirect measurement of the blood flow. Thus, laser Doppler velocimetry measures the movements of the red blood cells, while chromametry evaluates the redness of the skin as an indicator of skin blood flow. Current methods for studies of skin blood flow are listed in Table 2.

The application of the different techniques depends on the question under investigation. This chapter provides a brief description of the more common techniques. Laser Doppler velocimetry, photoplethysmography, remittance spectroscopy, and chromametry are described in more detail in the following chapters. The publication of several minireviews during recent years[7–10] demonstrates the increasing interest in, and use of, these techniques. The most reliable data will be obtained if some of these techniques are combined and used in a complementary and fruitful way, since they reflect different aspects of the pathological condition or the stimuli (e.g., drugs, irritants) under observation. The recent rapid development of personal and minicomputers has increased the potential of bioengineering devices. It is now possible to collect large amounts of data automatically over a long period. The data can be stored and subsequently processed within the computer-based recording systems.

II. CURRENT METHODS

A. CAPILLARY MICROSCOPY

With capillary microscopy[11–13] the capillaries are visualized using a light microscope. The picture is enhanced with an immersion objective and oil. The movements of the red blood cells can be further evaluated using videomicroscopy. Capillary microscopy has been most frequently applied for the evaluation of pathological changes reflected in the circulation in the nailfold capillaries. It is a sensitive method for evaluating the microcirculation in such conditions as diabetes mellitus, collagen disorders, arterial occlusive disease, and Raynaud's phenomenon and as an aid in diagnosing psoriasis.

B. FLUORESCENCE MICROSCOPY

Fluorescence microscopy[11–13] has been used for studies of morphological and dynamic changes in the microvessels of the skin. It can also be applied for analysis of transcapillary exchange. Injected intravenously, sodium fluorescein dye appears in the capillaries of the skin. The fluorescent light can be recorded on ordinary photographs or stored using videomicroscopy. The latter technique allows an evaluation of the dynamic changes in the diffusion of the fluorescent dye. The evaluation can be further enhanced with videodensitometry. This technique has been used for determining skin microvascular function in systemic diseases such as diabetes mellitus, arteriosclerosis, and scleroderma and is a useful tool for evaluating the skin microvasculature in leg ulcers of various causes (e.g., vasculites and arteriosclerosis) and also in studies on skin irritancy.[14]

C. PHOTOPLETHYSMOGRAPHY

Photoplethysmography allows the registration of pulsative changes in the dermal vasculature and is synchronized with the heartbeat. The apparatus consists of a transducer which leads infrared light to the skin. The light is absorbed by hemoglobin, and the backscattered radiation is detected and recorded. The backscattered light will depend on the amount of hemoglobin in the skin, and the result obtained will therefore reflect the cutaneous blood flow. The technique has been used for studies of dermatological disorders and systemic disease and for skin experimental purposes.

D. ISOTOPE WASHOUT METHODS

Different isotopes used have included ^{133}Xe and ^{99}Tc. The skin is loaded by injection or by epicutaneous application of the isotope in question. This is followed by a washout period during which the disappearance of the isotope is measured from the skin surface. The information obtained, i.e., how fast the isotope disappears, reflects the cutaneous blood flow and has been used to obtain quantitative data on skin blood flow. The techniques have been employed mostly for experimental studies.[15]

E. TRANSCUTANEOUS MEASUREMENT OF OXYGEN PRESSURE

Transcutaneous measurement of oxygen pressure (tcpO$_2$) measures the oxygen perfusion of the skin.[12,13] Since the normal pO$_2$ is close to zero at the skin surface, the skin is heated to a defined temperature by the electrodes used for the measurement of the tcpO$_2$. The usual electrodes consist of a platinum cathode with a silver ring anode. An electrolyte solution is placed between cathode and anode. The method is used both experimentally and clinically for the evaluation of conditions such as arterial diseases, for perinatal surveillance, for studies of various dermatoses, and for experimental skin investigations.

F. LASER DOPPLER FLOWMETRY/VELOCIMETRY

Laser Doppler flowmetry/velocimetry is based on measurement of the Doppler frequency shift in monochromatic laser light backscattered from moving red blood cells. It detects the frequency-shifted signal and derives an output proportional to the number of erythrocytes times their velocity in the cutaneous microcirculation.[16]

Possible fields of application are skin physiology, especially regional variations; diagnostics, especially scleroderma, Raynaud's disease, and patch test reactions; predictive testing of irritancy (topical drugs, cosmetics, detergents, cleansing agents, products used in industry); and effects of drugs (vasodilators, minoxidil, sunscreens and ultraviolet light, topical corticosteroids [blanching]).

G. REMITTANCE SPECTROSCOPY

Remittance spectroscopy is used to measure skin color. One part of the light shone on the skin surface is absorbed and one part is reflected. By comparing the absorption at different wavelengths, data are obtained from which changes in skin color (e.g., erythema and pigmentation) can be calculated. In dermatology, the method has been applied for evaluating pharmacological effects on the skin, for evaluations of treatment modalities, and for experimental purposes.

H. CHROMAMETRY

Chromametry is another method for determining skin color and/or erythema. The technique is based on the tristimulus values of the Commission Internationale de l'Éclairage (CIE). In short, the skin surface is illuminated by a pulsed xenon arc lamp. The lamp provides two CIE illuminant standards. The color tones of the illuminated skin area are recorded simultaneously.

III. CONCLUSIONS

In conclusion, interest in skin blood flow under normal and pathological conditions as well as in experimental studies has increased during the past two decades. This interest has been paralleled by the development of techniques used for evaluating skin blood flow. Older techniques have been refined and further developed (e.g., capillary microscopy, transcutaneous pO$_2$ measurements, and photoplethysmography), and new noninvasive techniques have been developed (e.g., laser Doppler velocimetry, remittance spectroscopy, and chromametry).

Some of the techniques, such as capillary microscopy and transcutaneous pO_2 measurement, are well established in both clinical and experimental work, while most of the others are still mainly experimental. Together the different methods have provided us with powerful tools for the evaluation of cutaneous blood flow under various clinical and experimental conditions.

REFERENCES

1. Wahlberg, J. E. and Nilsson G., Skin irritancy from propylene glycol, *Acta Derm. Venereol.,* 64, 286, 1984.
2. Wahlberg, J. E., Assessment of erythema: a comparison between the naked eye and laser Doppler flowmetry, in Frosch, P. J., Dooms-Goossens, A., Lachapelle, J.-M., Rycroft, R. J. G., and Scheper, R. J., Eds., *Current Topics in Contact Dermatitis,* Springer-Verlag, Berlin, 1989, 549.
3. Weil, C. S. and Scala, R. A., Study of intra- and interlaboratory variability in the results of rabbit eye and skin irritation tests, *Toxicol. Appl. Pharmacol.,* 19, 276, 1971.
4. McKenzie, A. W., Percutaneous absorption of steroids, *Arch. Dermatol.,* 86, 611, 1962.
5. McKenzie, A. W. and Stoughton, R. B., Method for comparing percutaneous absorption of steroids, *Arch. Dermatol.,* 86, 608, 1962.
6. Berardesca, E. and Maibach, H. I., Skin occlusion: treatment or drug-like device?, *Skin Pharmacol.,* 1, 207, 1988.
7. Guy, R. H., Tur, E., and Maibach, H. I., Optical techniques for monitoring cutaneous microcirculation. Recent applications, *Int. J. Dermatol.,* 24, 88, 1985.
8. Berardesca, E. and Maibach, H. I., Bioengineering and the patch test. Review article, *Contact Dermatitis,* 18, 3, 1988.
9. Agner, T. and Serup, J., Skin irritants assessed by non-invasive bioengineering methods, *Contact Dermatitis,* 20, 352, 1989.
10. Bongard, O. and Bounameaux, H., Clinical investigation of skin microcirculation, *Dermatology,* 186, 6, 1993.
11. Bollinger, A. and Fagrell, B., *Clinical Capillaroscopy,* Hogrefe and Huber, Bern, 1990.
12. Stuttgen, G., Ott, A., and Flesch, U., Measurement of skin microcirculation, in *Cutaneous Investigation in Health and Disease,* Leveque, J.-L., Ed., Marcel Dekker, New York, 1989, 359.
13. De LaCharrière, O. and Kalis, B., Measurement of cutaneous microcirculation, in *Cutaneous Investigation in Health and Disease,* Leveque, J.-L., Ed., Marcel Dekker, New York, 1989, 385.
14. Brown, V. K. and Clarke, R. A., Sulphan blue as an aid to the laboratory assessment of primary skin irritants, *J. Invest. Dermatol.,* 45, 173, 1965.
15. Kristensen, J. K. and Petersen, L. J., Measurement of cutaneous blood flow by the [133]Xenon washout method. Clinical applications in dermatology, *Acta Physiol. Scand. Suppl.,* 603, 67, 1991.
16. Öberg, P. A., Laser-Doppler flowmetry, *Crit. Rev. Biomed. Eng.,* 18, 125, 1990.

Part II   Laser Doppler Velocimetry and Photoplethysmography

General Aspects

Chapter 3

Laser Doppler Flowmetry and Photoplethysmography: Basic Principles and Hardware

Luciano Bernardi and Stefano Leuzzi

CONTENTS

0-8493-8371-4/95/$0.00+$.50
© 1995 by CRC Press, Inc.

ABSTRACT: Laser Doppler flowmetry (LDF) and photoplethysmography (PPG) are two methods capable of providing continuous noninvasive measurements related to changes in microvascular perfusion, in terms of relative changes of blood volume and velocity. The methods are based on the effect of the light on moving (mainly erythrocytes) and nonmoving components of a limited volume of tissue. Although LDF can provide an absolute measure of Doppler shift distribution, there is no simple method to obtain a quantitative continuous signal related to absolute blood flow in different tissues/sites, and various practical solutions so far can evaluate only relative blood flow changes. The origin of the PPG signal is not fully understood, although it also seems to be sensitive to changes in both volume and velocity of blood. The basic principles of the two techniques, together with their hardware and relevant aspects for practical use, are examined in this chapter.

I. INTRODUCTION

The need for a continuous, versatile, noninvasive method of measuring skin blood flow is paramount, not only for gaining knowledge about the pathophysiology of the human tissue, but also for the assessment and clinical management of microvascular skin disorders and for evaluating the effectiveness that a treatment (such as the application of a vasoactive drug) may have on the microcirculation. LDF and PPG are practical solutions to this need. Comprehensive aspects of both techniques have been published previously.[1,2]

II. LASER DOPPLER FLOWMETRY

When tissue is illuminated by a coherent, monochromatic, low-powered laser, light is scattered in static structures as well as in moving red blood cells (RBCs) within the microcirculatory beds (Figure 1). Light particles (photons) enter the body tissues and collide with the various cells and structures. Collisions with nonmoving structures/cells backscatter light that has the same frequency as the incident light, whereas collisions with moving particles, mostly RBCs, backscatter light which has a frequency altered according to the Doppler effect. Therefore, while the incident light spectrum generated by a laser source can be represented by a line, the light that can be detected from the illuminated tissue has a spectrum which is broadened around the frequency of the incident light (Figure 2).

Laser Doppler flowmeters produce an output signal that is proportional to the blood cell perfusion (or flux). This represents the movement of RBCs through microvasculature and is defined as:

microvascular perfusion (flux) = number of RBCs in the tissue sampling
volume × mean velocity of RBCs

Assuming a proportionality between RBC number and blood volume, the LDF signal should be linearly related to the volume-velocity product of blood in the measured volume. This relationship and the tissue volume under measurement are modified/influenced by a number of factors, including the characteristics of the laser light, the optical coupling to the tissues, the design of the photodetector circuitry and filters, and the final processing of the detected signal, together with the characteristics of the tissues and the amount/distribution of local blood flow.

A. MEASURING PRINCIPLES
1. Doppler Effect
All wave movements are characterized by the relation

$$v = f \times \lambda$$

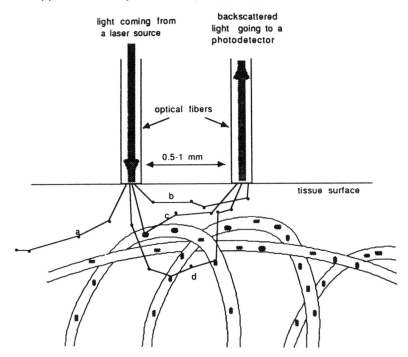

Figure 1 Schematic diagram of a typical two-fiber laser Doppler probe and light patterns in a vascularized tissue: **a,** photon is scattered way and/or absorbed; **b,** light is scattered by stationary tissue component and recaptured (no Doppler shift); **c,** light is scattered by a single moving RBC (single Doppler shift); **d,** light is scattered by two moving RBCs (multiple Doppler shift).

where v = propagation velocity (in meters per second), f = frequency (in Hertz), and λ = wavelength (in meters). Consider the situation of sound waves; if both a listener (l) and a sound source (s) are stationary, the listener receives v/λ waves per unit of time, or the frequency of the sensed wave is v/λ cps. Suppose the listener moves toward the source, encountering the various waves at earlier times. The number of waves sensed by him in the unit of time is higher, despite the fact that the propagation velocity and λ with respect to the medium remain the same (they are determined by the characteristic of the source and of the media between the listener and the source); the motion of the listener thus has the effect of shortening the propagation time, due to the shortening of the distance between him and the source. As a result, the listener velocity (V_1) is added to the original propagation velocity (V_p): the frequency of the wave λ at the source is then increased by the effect of V_1 to a new frequency, f_1:

$$f_1 = (V_p + V_1)/\lambda$$

By simple substitutions it is determined that the frequency perceived by the listener, f_1, is related to the frequency at the source, f_s, by the relation

$$f_1 = f_s (1 + V_1/V_p)$$

Similar considerations can be applied to the movement of the source.

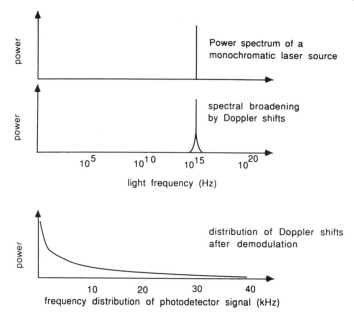

Figure 2   Schematic comparison of spectra of monochromatic laser source (showing a single spectral line), of the backscattered light after Doppler shifts (showing a spectral broadening), and of the Doppler shifts demodulated by the photodetector.

The situation is different for the Doppler effect with light waves. The speed of light is constant in all reference frames; light needs no material medium for its propagation, and its speed relative to the source or observer is always the same. Therefore, it is only the relative motion between the observer and the source that determines the Doppler shift frequency. The observed frequency (f_l) of the light modified by the Doppler effect is related to the source frequency of the light (f_s) by the relation

$$f_l = f_s \left[(1 + v/c)/\sqrt{1 - (v^2/c^2)} \right]$$

where c, the speed of light, here replaces the term V_p and v is the relative velocity of source vs. observer.

The previous equation can be applied to light entering a tissue. If light strikes a stationary object and is reflected directly to a receiving detector, the returning light will have the same frequency as the emitted light. However, if the returning light has been reflected from an object in motion, such as a RBC, the returning light will undergo a Doppler shift that is *twice* that predicted by the equation.[3] First, the light received by a RBC is Doppler shifted because of its motion with respect to the light emitter. Then the RBC also acts as a source in motion relative to the stationary receiver; this will cause a second Doppler shift of the same sign as the first. Only the component of the velocity vector directed toward or away from each receiver contributes to the Doppler shift frequency. Thus, one must actually scale the frequency shift by the cosine of the angle between the velocity vector of the RBC and the line connecting it to the emitter or the receiver.[3] Therefore, when the direction of propagation of light makes an angle θ with respect to the direction of the relative motion of two observers (here the RBC and the emitter/receiver system), the frequency shift will be modified by the factor cos θ as

$$f_l = f_s \left[(1 + (v/c) \cos \theta) / \sqrt{(1 - v^2/c^2)} \right]$$

2. Laser Light

Since the goal is to measure the velocity of the moving elements by the Doppler effect (i.e., by the frequency shift between input and output signals), this will be best achieved if the input signal is highly stable and monochromatic: laser light can fulfill these criteria. Laser light can be produced with different powers and wavelengths, according to a widespread range of applications. The power required for LDF instruments is fairly small, in the range of 1 to 2 mW, and it can be produced by gas lasers or, as demonstrated more recently, by laser diodes. Gas lasers were applied in the first experimental and commercial applications of the method and are still in use to generate wavelengths spanning from blue (441.6 nm helium/cadmium, 457.9 nm argon[4]) and green (543.5 nm helium/neon[5,6]) to the most commonly used red light (632.8 nm helium/neon). Due to their relatively large dimensions, the gas lasers are coupled to the measuring area (which is generally limited to the 1 mm range) through optical fibers. Laser diodes are, at present, manufactured to produce red or near-infrared light; the latter (780 to 805 nm) are now used in LDF (Section III.C). Diodes also can be coupled to the measuring site by optical fibers (Section III.C.2); however, due to their small dimensions they can be inserted into the probe, together with a miniature solid-state photodetector (Section III.C.4). Laser diodes in the near-infrared region appear far more responsive to blood flow changes in black skins than conventional red He-Ne systems, probably due to the reduced absorption of longer wavelengths by melanin.

3. Absorption and Scattering of Light in Biological Tissues

When a light beam hits a biological tissue, only a minor part (about 3 to 7%) is reflected back. The remaining 93 to 96% of the incident radiation, not returned by regular reflectance, in part is absorbed by various structures and in part undergoes single or multiple scattering.[7] A variable amount of the scattered light (more than 50% at 633 to 785 nm) is then re-emitted from the surface (and can be collected by a photodetector).

The amount of *absorption* depends on the wavelength of the incident light and on the ability of the tissue to absorb light. Particular molecules (chromophores such as melanin, hemoglobin, beta-carotene, and bilirubin) within the tissues have the specific capability of absorbing radiation and are major determinants of light absorption. For example, oxygenated and reduced hemoglobin have different, typical absorption curves as a function of light wavelength (Figure 3). Absorption of light by hemoglobin is much higher at shorter wavelengths (in the green range), thus reducing the proportion of scattering at these wavelengths.

Scattering results from collision of light photons with either static or moving components of the tissue: collision of one photon with a static structure determines a change in the direction of the photon without Doppler frequency shifting, whereas collision of one photon with a moving structure determines a change in the direction of the photon with Doppler frequency shifting (Figure 4). With short light path length and/or low RBC number/speed, the photon normally experiences a single collision with a moving RBC (single scattering). If the photon path length is long, or if the number of RBCs is high (as with high RBC number/speed), the photon may experience multiple collisions with more than one RBC (multiple scattering) and, hence, a further Doppler shift.

The stationary tissue matrix is assumed to be a strong scatterer of light, having more scattering power and causing much larger angular deviations, so that the tissue basically appears as a turbid medium. The diffusion of the laser light back to the photodetector thus

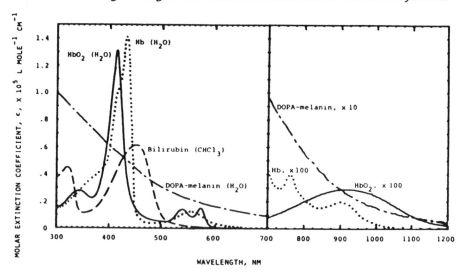

Figure 3 Absorption curves of major visible-absorbing pigments of human skin, oxygenated hemoglobin or HbO_2 (_), reduced hemoglobin or Hb (....), bilirubin (- - -), and DOPA-melanin (_._._), as a function of light wavelength. Notice the scale change above 700 nm. (From Anderson, R. R. and Parrish J. A., *J. Invest. Dermatol.*, 77, 13, 1981. With permission.)

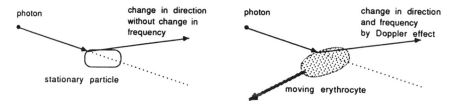

Figure 4 Collision (scattering) of a light photon with a stationary tissue component (left) and with a moving erythrocyte (right). Collision with the stationary particle changes the direction but not the frequency of the photon. Collision with the moving erythrocyte changes both direction and frequency of the photon, thus generating a Doppler shift proportional to the cosine of the angle generated by the directions of photon and erythrocyte.

depends to a much greater extent on these stationary tissue structures that do not cause significant Doppler shifts.

The characteristics of light scattering depend on the dimensions and shape of the structures hit by the photons. For molecules or small particles with dimensions less than roughly one tenth of the wavelength, scattering is generally weak, nearly isotropic (i.e., equally distributed spatially), and polarized and varies inversely with the fourth power of the wavelength (so-called Rayleigh scattering). For particles with dimensions similar to that of the light wavelength, scattering is much stronger, more forward-directed, and is still inversely related to wavelength, but to a lower extent; for particles with dimensions greatly exceeding the wavelength (such as human RBCs, with a diameter of ~7 to 8 μm), scattering becomes highly backward-directed.[7] Scattering in the bloodless dermis varies inversely with light wavelength.[7] This accounts for observations that longer wavelengths penetrate the dermis to a greater extent than shorter wavelengths.

4. Doppler Shifts Resulting from Light Scattering

As a photon diffuses through a vascularized tissue it experiences many collisions with essentially static tissue elements, but only occasional collisions with moving RBCs. Moreover, light principally is scattered by RBCs only through a small angle (on the order of 6°), with only approximately a 1/1000 chance of backscatter.[8] However, due to the complex tridimensional geometry of the tissues and superficial circulation, and due to the diffused nature of the light passing through such tissues, the angles between the RBC velocity vectors and the beam propagation vectors can be considered randomized: therefore, a continuous range of Doppler frequency shifts can be expected, regardless of external probe orientation.

The spectrum of the light diffused after scattering within a tissue with single or multiple Doppler shifts shows then a broadening of the frequency distribution compared to that of the incident light (which is basically a line, due to the monochromatic property of laser light; see Figure 2). Spectral broadening of the incident light laser can be attributed to either single or multiple photon scattering in moving RBCs, which in turn generate single or successive Doppler shifts. Therefore, in turbid tissues one cannot obtain a unique laser Doppler shift determined by particle velocity and external geometry. Rather, one detects a relatively small, symmetric Doppler broadening, the width and amplitude of which depend on mean speed and concentration of the RBCs relative to the wavelength of light (Figure 2).

Unlike ultrasound Doppler, where one might establish the velocity of the blood and its flow within a single vessel if a relatively limited number of factors related to the structure to be investigated are known (incidence angle, vessel diameter, velocity profile), in LDF methodology the extraction of RBC velocity and volume cannot be obtained directly. One must make a number of assumptions on the distribution of Doppler shifts and elaborate a mathematical model relating these measurements to meaningful physiological parameters.

Most of these models[8–12] have been elaborated by assuming that the transport of light can be accounted for by a diffusion model or an equivalent stochastic description of photon migration. According to this model, the propagating light is considered as a collection of discrete particle-like photons, and the direction of motion of these photons at any given point is supposed to be random.[11] It is also assumed that the network of microvessels in the tissue, while perhaps having local order, is random on a length scale defined by the mean distance between RBC scattering events. Consequently, when calculating the Doppler shifts of scattered photons, it has been assumed that the blood cell velocities also are randomly distributed in direction. Thus, the direction of scattered light resulting from collisions of photons with RBCs is randomized. Also, multiple Doppler shifts are uncorrelated when a particular photon collides with more than one moving cell. These assumptions seem justified by the common observation that the Doppler shifts obtained by the moving RBCs are distributed on a large bandwidth. A major portion of the light was shifted by much lower frequencies (30 to 500 Hz) than would be expected to arise from backscatter by RBCs moving at one 1 mm/s (i.e., 2.5 to 10 kHz).[8,11] Theoretical models[8] can explain this prevalence of low-frequency fluctuations from RBCs in the microcirculation by taking into account all scattering angles, weighted by their probability of occurring.

Bonner and Nossal[10] have demonstrated a dependence of the flux index (derived from the first moment of the power spectrum of the Doppler shifts, Section III.B.5) on both RBC speed (ranging from 0 to 2 mm/s) and hematocrit (ranging from 0.003 to 0.12). The relationship is also linear (and inverse) if one relates the Doppler shifts to the scattering cross section of the particle.[10] Due to their biconcave shape, RBCs cause more large-angle scattering than a perfect sphere of equal volume (as per the Mie theory of scattering).[13]

To a large extent, the feasibility of using a LDF for measurement of blood flow in the circulation rests on the linearity of the relationship between Doppler shift-derived information and RBC flux. Although for low perfusion volumes a linear relationship was found between most parameters when evaluating the Doppler shifts and RBC concentration or speed, this relationship deviated from linearity in the presence of higher flow rates, due to multiple scattering of light.[14-16] How the flux information can be actually obtained from the frequency distribution of Doppler shifts, and how corrections for higher blood flows can be achieved, is still a matter of research, and different algorithms have been proposed and implemented (Section III.B.5).

5. Relationship between Light Path Length, Absorption and Scattering, and Spacing of Optical Fibers

The relationship between mean number of collision between photons and moving cells and the number of RBCs within the tissue sample depends mainly on the photon path length in the tissue, which in turn is determined by the spacing of the emitting and receiving fibers. The average depth sampled by emergent photons increases as the fiber spacing is widened, but the use of light to probe tissues in effect limits the depth at which the microcirculatory flow can be measured (Section III.A.6). The detected intensity falls off very rapidly with increases in fiber separation. Conversely, the fraction that is Doppler shifted increases with increased fiber spacing.[8,17] With commercially available LDF probes the average depth of measurement is in the range of 1 to 2 mm for the skin, but it may be substantially different in other tissues; as an example, it may reach at least 4 to 6 mm in the intestine.[18]

6. Laser Light for Evaluating Depth of Probed Volume

The discrimination of different vascular beds may have practical relevance. For example, while the superficial dermal plexus and capillary loops (also called "nutritional blood flow") are often affected by diseases such as eczema, psoriasis, and acne,[6] the deeper circulation (deep dermal plexus, arterial-venous anastomoses) may be more responsive to reflex responses to more general cardiovascular adaptations.[19] Complex widespread diseases such as diabetes may act at both levels, possibly by different mechanisms (i.e., neural vs. metabolic). From a pharmacological point of view, these subpapillary plexuses are directly exposed to products diffusing from the skin surface after topical application. A weak vasoactive stimulus in the superficial dermis (capillary loops and superficial plexus) can thereby affect the microcirculation of the skin without influencing to a measurable extent the global dermal circulation.[4]

a. Penetration Depth of Light

In a dense medium, such as the human body, only surface layers can be easily penetrated by light: due to strong multiple scattering present in most biological structures, only a small portion of the light incident on a tissue surface will penetrate deeply and return to the surface. The penetration depth is generally measured in millimeters with wavelengths used in LDF applications (450 to 800 nm). In this region, absorption by chromophores (pigment molecules) is relatively weak and a substantial part is scattered away. Anderson and Parrish estimated the depth of penetration of various wavelengths in a vitiliginous skin (i.e., without pigmentation) *in vitro.*[7] Table 1 reports estimated depths for which the radiation is attenuated to 1/e (i.e., about 37%) of the incident radiation density.[7] The epidermis of pigmented individuals, however, can greatly reduce these values, especially at shorter wavelengths.

b. Analysis of Penetration Depth by Laser Doppler Techniques

One way to discriminate between total and superficial skin blood flow is selection of lasers with different suitable wavelengths which will be preferentially absorbed and/or scattered in different layers of dermis. Available wavelengths span from blue (458 nm) to infrared (around 800 nm) light. Obeid et al.[6] have compared the power spectra of Doppler shifts obtained with three light sources (green, $\lambda = 543$ nm; red, $\lambda = 632.8$ nm; near-infrared, $\lambda = 780$ nm) and an identical processing system on *in vitro* and *in vivo* models. They found (Figure 5) that the green light produced the narrowest and lowest spectrum, whereas the near-infrared

Table 1   **Approximate Depth for Penetration of Optical Radiation in Vitiliginous Caucasian Skin to a Value of 1/e (37%) of the Incident Energy Density**

Wavelength (nm)	Depth (μm)
250	2.0
280	1.5
300	6.0
350	60.0
400	90.0
450	150.0
500	230.0
600	550.0
700	750.0
800	1200.0
1000	1600.0
1200	2200.0

From Anderson, R. R. and Parrish J. A., *J. Invest. Dermatol.*, 77, 13, 1981. With permission.

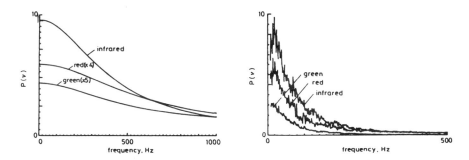

Figure 5   Photocurrent spectra obtained in *in vitro* (left side) and *in vivo* (right side) experiments with three different light sources (infrared, red, and green) and identical hardware. The green-light Doppler shifting has reduced high frequencies, suggesting an origin from the slower moving RBCs (which travel on the outside of the flow, near the tube wall). Notice the similarity of the two diagrams, suggesting that green light senses the slower RBC speed of more superficial vessels. (From Obeid, A. N., Boggett, D. M., Barnett, N. J., Dougherty, G., and Rolfe, P., *Med. Biol. Eng. Comput.*, 26, 415, 1988. With permission.)

light produced the broadest spectrum. The red light produced an intermediate spectrum, closer to that of near-infrared light. This indicates that the longer wavelengths generate relatively higher Doppler-frequency shifts that can be attributed to multiple scattering (produced by successive Doppler shifts occurring in a high concentration environment) and/or to a significant proportion of RBCs moving at high velocities. Conversely, the shorter green wavelength did not generate high-frequency shifts in the photocurrent spectrum, suggesting that the effect of multiple scattering is negligible and/or that the number of RBCs and/or their velocities probed with green light is smaller. The authors concluded that the characteristics of the photocurrent power spectra recorded with the green laser were consistent with a response limited to RBC activity in the capillary beds and superficial papillary plexus. This is supported by the observation that in these regions the concentration and mean velocity of moving RBCs are known to be small.[20] Similar conclusions were reached by Duteil et al.,[4] who compared the effects of classical red light (632.8 nm) to those of blue light (457.9 nm). Gush and King[5] also attempted to discriminate between capillary and arteriovenular blood flow in skin using a combination of green (543.5 nm) and red (632.8 nm) laser lights. Reconsidering the data from Reference 7 they observed, however, that while the green light penetration was just 0.6 times that of red light (Table 1), the absorption of green light by blood pigments was 20 times greater than that of red light (Figure 3). As a consequence, differences in light-scattered Doppler spectra using the two lasers were likely to be dependent mainly on differences in multiple scattering and absorption by blood, rather than tissue penetration. They also questioned a common assumption that the distribution of blood in tissue is dilute and homogeneous.[10,16] This assumption was necessary to invoke uncorrelated phase shifts and scattering vectors, but was not realistic, according to the authors, since in connective tissue blood is absent and in capillaries blood approximates a dilute solution (with the RBCs flowing in single file), whereas in venules/arterioles RBCs are distributed in a three-dimensional volume of blood, with varying velocity profiles (depending on vessel size and blood mean velocity) and different types of RBC motion within these vessels. They suggested that comparisons could not be made by simply considering depth of light penetration through homogeneous tissues, but that it was necessary to consider the combined effects of light absorption and multiple scattering at the individual microvessel level, together with the blood rheology and vessel morphology of the specific site of measurement. Separation of capillary from deeper flow changes could then be hypothesized by the combined contribution of light wavelength and frequency of Doppler shift (lower frequency shifts associated with capillary flow, higher frequency shifts associated with flow in deeper vessels).

B. STRUCTURE OF THE LASER DOPPLER FLOWMETER

The structure of the LDF apparatus can be considered to be made up of several distinct functional/physical modules in cascade: (1) light-emitting device, (2) coupling to the measuring area, (3) signal demodulation, (4) preamplification and filtering, and (5) signal processor.

1. Light-Emitting Device

This essentially is the laser light emitter with related circuitry. General characteristics of various laser types have been outlined previously. Since the Doppler shift generated by moving RBCs is in the range of 10^{-12} times the frequency of the emitting light, the noise characteristics and the frequency stability of the light source are of paramount importance. Also, the mode interference due to the internal modulation of different longitudinal modes within the laser cavity generates unstable amplitude conditions in earlier experiences, resulting in a slowly varying frequency component in the light-beating spectrum.[21,22] These problems were solved in part by adopting the single-mode laser as a strictly monochromatic light source and in part by common mode rejection methods in a differential input.[15,22] A drive circuit with automatic power control function is used in order to stabilize the power supply to the laser. This is particularly important with low-mass emitters (such as laser diodes) which would otherwise drift with fluctuations in ambient temperature.[23]

2. Coupling to the Measuring Area and Detection of Backscattered Light

Two methods are used in LDF (particularly when it is applied to the skin): optical fibers and direct coupling of laser source and photodetectors. Optical fibers are the only possible solution with the typical helium/neon laser due to the large dimensions of the emitter; appropriate optical coupling between laser source and afferent fibers is required. Flexible thin (around 100 μm) fibers, less prone to movement artifacts, are preferred. Laser diodes can also be coupled through optical fibers (Sections III.C.2 and III.C.4), but due to the small dimensions they can also be incorporated into the probe and positioned directly over the measuring area.[24,25] In this case a miniature photodetector system is employed, which is also incorporated into the probe. Compared to the optical fibers, the direct application of the diode to the skin has the advantage of reducing to a great extent the sensitivity to movement artifacts, which is one of the major problems encountered with optical fiber systems. In addition, the area of measurement is broadened. This seems to result in more stable and reproducible data (Section III.C.4), since spatial differences in skin blood flow may easily occur if a typical 1-mm-spaced, two-fiber probe is moved over a short distance.[25]

The number of optical fibers used in laser Doppler systems is variable. An early and still common disposition involves two fibers: one is used to emit the laser light, while the other collects the backscattered light and sends it to a photodetector. In some applications two receiving fibers (positioned at opposite sides of the light-emitting fiber) are connected to the same photodetector. Fibers are normally spaced 0.5 to 1 mm apart.

The use of three fibers (basically one emitter and two receivers, connected to two photodetectors; in some cases each photodetector is in fact connected to a couple of fibers, giving a total number of five) was found to provide substantial advantages and is now widely used. The random fashion in which the RBCs pass through the scattering areas generates randomly fluctuating phase shifts in the photodetector current. Since the two signals received are produced by adjacent but still separate scattering areas of the vascular bed, they represent two statistically independent realizations of the same stochastic process,[22] so that their difference reinforces the signal related to the distribution of Doppler shifts. Conversely, the noise generated by the laser (mode interference noise, amplitude instability, wide-band beam noise) is highly correlated in the two signals and affects them in the same way. As a result, the difference between the two signals virtually eliminates laser-generated noise (common mode rejection). In practice, however, the noise reduction is determined primarily by the crosstalk reduction and accurate matching of the two channels (in terms of both optics and electronics). Nilsson et al.[22] found a noise rejection of better than 30:1 with this method.

Single-fiber operation also has been proposed for special purposes (insertion in muscles for deep tissue perfusion measurement) due to the ability of light to travel through a linear medium without mixing its inherent frequencies.[26]

An integrating probe combining one emitter and seven receiving fibers has been proposed, with the aim of reducing the often relevant differences in laser Doppler signal recorded at adjacent parts of the skin when using a typical probe. This results in a reduced intrasubject variability.[27]

In the first experiments and applications of the laser Doppler technique a standard photomultiplier tube was the device of choice to detect the backscattered light and transform it into an electric signal. This has been replaced in more recent studies and commercial devices by solid-state miniature devices, photodiodes or phototransistors, with the advantage of low power consumption, reduced dimensions, and lower cost.

3. Signal Demodulation

The maximum frequency change ratio $\Delta f/f_0$ for near-infrared light (say $f_0 = 3.84 \times 10^{14}$ Hz, corresponding to $\lambda = 780$ nm) scattered off a particle moving with a velocity of 1 mm/s is approximately 6.7×10^{-12} and corresponds to a Δf of about 2600 Hz. With light sources of shorter wavelength applied to the same example, the Δf is proportionally higher.[4] In order to detect such very small frequency shifts, it is necessary to employ the technique of "optical beating." If the two optical frequencies (Doppler-shifted and unshifted light) are close (as in the example), then the optical "beating" produced from their difference is in the low frequency range (Figure 2, bottom). In practice, the output of the detector contains a fluctuating component related to the difference, or beat frequency, between the frequency-shifted and nonshifted beams, together with the unshifted light and a sum component.

The light frequency is so high (in the range of 10^{14} to 10^{15} Hz) that it is far above the frequency response of the photodiode and of the other electronic circuits. The unshifted light frequency and the sum component are therefore not detected, but leave a DC component proportional to the intensity of total backscattered light. Conversely, the frequency differences which are identical to the Doppler-evoked frequency changes (Figure 2, bottom), being in the range from 0 to about 15 to 20 kHz, are easily treated by electronic circuits and appear as an AC signal, the Doppler signal, at the photodetector output, superimposed on the DC component. Hence, the demodulation is inherently performed by the photodetector.

Contributions to this AC photocurrent signal arise from both *heterodyne detection,* which requires the presence of both shifted and unshifted laser light, and the so-called *homodyne detection,* which results from the self-mixing of the frequency-shifted light only. Theoretical studies indicate that heterodyne mixing predominates if the scattering is from tissues which have fairly low levels of blood perfusion. In highly perfused tissues, homodyne mixing processes become more important since the incidence of multiple photon scattering increases. This determines, however, a nonlinear relationship between RBC flux and mean Doppler shifts, and correction terms are required (Section III.B.5).

4. Preamplification and Filtering; Differential Mode
The output signal from the photodetector(s) is amplified and filtered into two components: the AC and the DC signals, the latter proportional to the total intensity of backscattered light. The AC signal is isolated by high-pass filtering with a cutoff frequency of around 18 to 75 Hz. If a differential method is being employed, two identical circuits comprised of photodetector, amplifier, and high-pass filters are required, and their outputs are fed to a differential amplifier.[15,22,23] Accurate matching of the two channels is also necessary in order to obtain an optimal signal/noise ratio. After this module the signal is normally low-pass filtered; the cutoff frequency of the commercial LDFs ranges from 12 to 20 kHz, but in some models can be reduced down to 1.28 kHz.

5. Processing of the AC Signal
The most appropriate representation of the signal at this point of processing is certainly obtained by a real-time spectrum analyzer, which generates a nearly continuous representation of the frequency distribution of Doppler shifts (Figure 2, bottom). Alternatively, the auto-correlation function of the photodetector signal (by means of an autocorrelator[4,8]) represents an equivalent estimation in the time domain. The power spectrum P(f) density of the fluctuating photocurrent has been evaluated by a spectrum analyzer or equivalent device in earlier studies or in basic experimental settings.[11,14] The effects of an acute physiological intervention are well evident as a modification of frequency distribution and/or of the spectral power (Figure 6). Nevertheless, practical considerations lead to a need to obtain a single analog

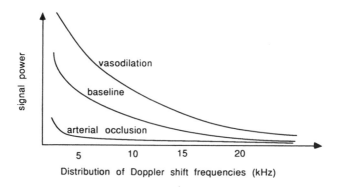

Figure 6 Schematic representation of Doppler shift power spectra. Notice the reduction (though incomplete) of all frequencies after arterial occlusion and the increase of all frequencies during skin vasodilation.

value related to RBC flux, rather than a function, and further signal processing of the power spectrum of Doppler shifts is necessary. Different methods for obtaining a blood flow index by mathematical elaboration of laser Doppler shifts have been proposed in the literature and have been compared recently in an *in vitro* test.[28]

The simplest value that can be obtained from the spectrum of laser Doppler shifts is the total Doppler-shifted power that is described by the following integral:

$$\left\langle \int_{f_l}^{f_u} P(f)d(f) \right\rangle \tag{A}$$

where P(f) = the total power spectral density of the amplifier output signal, f_u = upper cutoff frequency, and f_l = lower cutoff frequency. This value was related mainly to the RBC concentration,[8,16] although the relationship was linear only for low RBC concentrations.[28] The characteristic prevalence of low-frequency components in the laser Doppler shift spectra makes this representation minimally influenced by changes in RBC speed, which mainly influence the high-frequency components.

Therefore, "weighting" by the frequency was proposed as a solution to increase the sensitivity to RBC velocity. The blood flow indicator algorithm by Bonner and Nossal[10] produces an output proportional to the first moment of the power spectral density and is linear for low RBC concentrations. Its implementation requires the **f** weighting of the photodetector signal to modify formula A:

$$RBC\ flux = \left\langle \int_{f_l}^{f_u} f\ P(f)d(f) - N \right\rangle / i^2 \tag{B}$$

where the integral term is the first moment of the power spectral density, i^2 (a normalization factor) is the square of the mean photocurrent, and N is an offset voltage.

An empirical flow index, considered to vary linearly with changes in the root-mean-square (RMS) velocity of the RBCs, weights the Doppler shift spectrum by the square of the frequency;[14,23,28] in normalized form the algorithm is

$$Flow\ index: \quad \left\langle \int_{f_l}^{f_u} f^2\ P(f)d(f) - N \right\rangle^{1/2} / i \tag{C}$$

This algorithm can be implemented by differentiating the signal prior to RMS-to-DC conversion.

Finally, a simple empirical skin blood flow index algorithm, based simply on the normalized RMS of the detected spectrum, was used in earlier reports:[21]

$$\left\{ \left\langle \int_{f_l}^{f_u} P(f)d(f) \right\rangle^{1/2} - RMS\ noise\ /i \right\} \tag{D}$$

Algorithms A and D emphasize the effect of concentration (by relatively enhancing the effect of low-frequency shifts and reducing the effects of higher frequencies, which are more influenced by speed), while algorithms B and C emphasize the velocity effect because of weighting of the spectrum with the effect of factors **f** and **f²**, respectively. In all these algorithms the final information is obtained by passing the signal through a RMS-to-DC converter.

In vitro testing[28] against known RBC velocities indicated that algorithms B and C were both sensitive and proportional to RBC speed whereas algorithm D, like algorithm A, had an essentially flat response. Concentration measurements showed some degree of proportionality for all algorithms, however when comparing the measurements obtained in small and large tubes it was found that algorithm C had the best compromise and widest range of linearity (from 0 to 0.3%) with RBC concentration.

The presence of nonlinearities in the evaluation of RBC flux is well known, and empirical correction factors have been suggested.[16,23] According to Nilsson,[16] for low RBC volume fractions in a tissue the first moment of the power spectral density varies linearly with the RBC flux through the illuminated tissue matrix.[10,15] This value also varies linearly with the average absolute blood cell velocity provided the moving blood cell volume fraction is constant and the signal bandwidth is entirely covered by the bandwidth of the instrument. A high flux of cells through the illuminated tissue volume may, at its extremes, be formed either by a large number of blood cells moving with a low or moderate average velocity or by a low number of cells traversing the tissue at high speed. In the former case, however, the occurrence of multiple scattering and homodyne mixing violates the linear relationship between flux and the first moment of the unnormalized power spectral density. Since the volume fraction of moving RBCs can be determined from the total power of the fluctuating portion of the photocurrent (similarly to algorithm A), this factor may be used to obtain an output which, independent of RBC concentration, varies linearly with the flux of RBCs. An empirical correction factor $f(c)$ was then calculated from *in vitro* experimental data based on formula A and was multiplied by algorithm B to obtain

$$\text{RBC flux} = \left\langle k_0 f(c) \int_{f_l}^{f_u} f\, P(f)d(f) - V_{\text{offset}} \right\rangle \qquad (E)$$

where k_0 is an instrument constant; the term i^{-2} of equation B is contained in $f(c)$. This method resulted in an extended linear relationship between the output signal and the RBC flux,[16,30] but doubts of its validity were voiced[25,28] due to observations that algorithm A became rather insensitive to RBC concentrations above 0.2% by volume in *in vitro* models; the same author suggested that algorithm C should perform better in an extended RBC concentration range.

6. Relationship between Processing Algorithms and Signal Bandwidth

Although an increase in RBC speed increases the high-frequency component of the Doppler-shifted spectrum, the high-frequency components are small in amplitude and barely contribute to the overall RMS computation in algorithms A and D. The weighting of the power spectral density by a frequency or a frequency squared factor (f in B and f^2 in C, respectively) thus has the effect of increasing the relative importance of high-order components and, hence, results in a higher sensitivity to RBC speed. This points to the importance of the frequency bandwidth selected, since elimination of the higher frequencies results in a likely reduction of the most important components related to RBC speed. It has been shown[28] that an increase in the upper frequency cutoff from 5 kHz to 20 kHz augments the signal output in all algorithms, particularly in those weighted for the frequency (C more than B, and B more than D), when considering *in vitro* changes of either speed or concentration of latex microspheres. The linearity of the relationship between signal output and microsphere velocity was greatly extended for all algorithms by increasing the bandwidth up to 20 kHz.[28] A similar increase in linearity with increased bandwidth was found in *in vivo* tests.[30]

These results point to the utility of setting the upper LDF bandwidth as high as possible (in those instruments provided with variable bandwidth), despite the apparent and misleading "better" signal stability of low bandwidth settings.

C. LASER DOPPLER FLOWMETERS
1. Medpacific Models

Medpacific LD-5000 and LD-6000 models (Seattle, WA, now no longer manufactured) are based on the original work of Holloway and Watkins.[21,29,31] The light source in both these systems is a 2-mW, 632.8-nm (red) helium-neon gas laser. The probe is composed of two glass graded-index optical fibers, one transmitting and one receiving, each having a core diameter of 100 μm. The two fibers enter the probe head parallel to each other and are separated by 0.5 mm at the probe surface. Both models use single-channel operation (Section III.B.4). The configuration of the probe head is of several types. The *standard probe* can be manually held on the tissue or, more appropriately, glued to it. In the *needle probe* the fibers are embedded in 18-gauge needle stock, and it can be used in situations where space is limited or inserted into the tissue. The *flat probe* is only 2.5 mm thick and can be used in situations in which a low probe profile is desired. A heater can be added to permit warming and, hence, active temperature control of the skin beneath the probe.

In the articles describing this system[21,29] a simple RMS algorithm (algorithm D) was used; in subsequent systems[31] the Doppler-shift spectrum was weighted by the square of the frequency or by the frequency itself (algorithms C and B).

The instruments are factory calibrated with a standardized suspension of latex microspheres. The LD-5000 instrument is zeroed by placing the probe on a nonmoving surface that has optical properties similar to human tissues. This provides a noise component which will be subtracted from the signal and gives an absolute zero reference. The LD-6000 instrument provides an automatic zeroing circuit and an automatic gain control (vs. the manual gain control of the LD-5000). In addition, the LD-6000 provides continuous readout of algorithm A, which is considered an indicator of the relative number of RBCs in the area being evaluated. A standard RS-232 serial output port allows interfacing with external computers.

2. Perimed Models

Instruments marketed by Perimed (Stockholm, Sweden) are the newer PF3 (which was substituted for the PF1 and PF2 models) and, recently, the PF4001 Master/4002 Satellite.

The light source of the PF2 and PF3 is a 2-mW He-Ne laser (632.8 nm), connected to the measuring tissue via three optical fibers. All probes utilize a step-index silica fiber with a core diameter of 120 μm, selected because of their flexibility (for the purpose of avoiding movement artifacts). At the probe tip, the three fibers are positioned symmetrically with a core center spacing of 250 μm. This choice of spacing is considered a reasonable compromise that will give a measuring volume of about 1 mm³ and a sufficient signal-to-noise ratio to record even minute perfusion rates. One optical fiber is used to transport the emitted light, and the others are connected to two photodetectors, used in a differential configuration (Section III.B.4).

The bandwidth ranges from 20 Hz to 12 kHz, but it can be limited to 4 kHz by manual operation. To make the signal independent of the total light intensity that impinges on the photodetectors, the output is divided by a signal proportional to the total light intensity. The signal processor provides an overall output related to the tissue perfusion according to algorithm E; an output considered relative to the concentration of moving RBCs is also available as an output, and it is obtained by computing the natural logarithm of formula A, corrected by a number of factors related to optical coherence of the signal at the detector and to the total photocurrent. A temperature module includes a thermostatic probe holder that contains a heater and a temperature sensor so that temperature can be set and maintained from 26 to 44° C. Compared to the PF2, PF3 has extended monitoring capabilities, but there is little difference in the main electronics. The output signal of the instrument represents a relative value, and it is recommended to express the observed changes in blood flow as percentages of the resting flow value.[32]

The new PF4001 Master now uses a near-infrared 780-nm laser diode, generating 2 mW maximum power, connected to the tissue to be measured via optical fibers. Probes carry three optical fibers, but only two are utilized. The signal processing seems similar to previous models (information obtained from the company in July 1993). The output is expressed in "perfusion units." The instrument has an extended bandwidth, from 20 Hz to 20 kHz. A standard RS-232 serial output port allows interfacing with external computers. The 4001 is available with one or two channels. In addition, up to two "4002 Satellite" modules can be added, allowing up to four additional channels. An optional "Green Wavelength Accessory Kit PF543" allows simultaneous use of the near-infrared (780 nm) and green (543 nm) wavelengths at a close site to distinguish between the circulatory changes of the more superficial (capillary or "nutrient") vs. the deeper (arteriolar/venular/anastomotic) circulation (Section III.A.6). A 4005 PeriTemp module can be connected to provide local tissue heating. The PF1000 calibration device allows matching of different PF instruments to give the same values in perfusion units. The company also supplies comprehensive software ("PeriSoft") for handling LDF data, mainly intended for monitoring and analysis of microvascular reaction tests.

3. TSI Models

The "Laserflo BPM" incorporates a semiconductor laser operating in a single longitudinal mode, coupled through optical fibers. The laser generates a near-infrared (780 nm) light and delivers a power of approximately 2 mW at the end of the fibers. Small-diameter fibers are used: a 50-μm graded-index fiber as transmitter and two 100-μm fibers as receivers, spaced 0.5 mm from the transmitter fiber. These fibers have been encapsulated into a variety of probes ranging from a typical flat design for skin evaluation to biocompatible implantable disk-shaped and needle probes. The two efferent fibers return to a single photodetector, and single (rather than differential) channel operation is used (Section III.B.4). High-pass filtering is performed at 30 Hz, while low-pass filtering can be varied from 1.28 kHz to 18.50 kHz.

To compute the blood flow parameter, the signal processor calculates the noise content of the signal. The initial estimate of blood volume is based on the normalized mean square high-pass filtered signal (similar to algorithm A). The mean Doppler frequency is obtained from the first moment of the spectral power of Doppler shifts (basically algorithm B) divided by the estimated blood volume (similar to algorithm A). The result of this correction is the final mean frequency estimate, expressed in kilohertz. However, the instrument outputs a flow index calibrated in absolute terms (ml $\times$ min^{-1} $\times$ 100g^{-1}) by scaling the above values. The outputs also include RBC volume (in percentage of volume) and RBC velocity (in millimeters per second). Built-in software incorporates several tests for internal diagnostics and fault checks, in case of suspected malfunction. The calibration of the instrument is factory determined according to comparative investigations performed in various tissues, such as comparison with hydrogen clearance in gastric mucosa.[33] Calibration capabilities of BPM thus are restricted to routines that set the output signal to either zero or full scale for an external recorder. The new "Laserflo BPM^2" has similar technical aspects; however, it has a LCD display for graphic trends. The outputs are still given as absolute flow, velocity, and volume units.

4. Oxford Optronics Models

The "Microvascular Perfusion Monitor" MPM 3S uses a near-infrared (805 nm) semiconductor laser diode with output power ranging from 1 to 2.5 mW. Unlike all other instruments, this system does not use optical fibers, but rather a direct coupling of the diode to the surface under measurement. In order to increase the signal-to-noise ratio the diameter and the distance of optical fibers (and hence the measured region) have been reduced in other instruments. However, this had the inconvenience of increasing the site specificity by reducing the effective tissue sampling volume to the point where flow measurements might be confined to only a few capillaries. The problem of artifact noise has been solved in the MPM 3S by replacing the conventional optical fiber light delivery and collection system with an optical sensor probe which contains both the laser diode and the sensitive pin photodetection system in a small, lightweight (7 g) package, with a large probed area (16 mm^2, thereby significantly minimizing the contribution of localized spatial flow variations) and low local pressure (1.7 mmHg). This results in more stable measurements, even if the subject is moving,[25] and has an effect similar to that of an integrated probe,[27] but is obtained with a simpler method. The light wavelength (805 nm) is coincident with the isobestic point of hemoglobin and thus avoids the possibility of signal variation due to oxygenation[2,7] (Figure 3).

The signal processing features a double channel with differential operation, in order to further improve the signal-to-noise ratio, and a 15 Hz to 15 kHz bandwidth. The signal processor is based on algorithm C. The output signal is expressed in arbitrary perfusion units. These measurements, in the designers' view, should be considered as strictly relative. A wide range of additional probes are available by the manufacturer, including endoscopic, implantable, and a bent-tip probe for dentistry.

The "Oxford array" is the multichannel extension of the MPM 3S. To date this is the only LDF system originally conceived on a multichannel basis. It provides continuous and real-time mapping of blood flow from up to 12 different sites and also features up to eight temperature measurements. The general design is similar to the MPM 3S, but in order to cope with such a high number of signals the instrument is fully computerized, and the data are displayed on a color computer screen. The Oxford array enables comparison of microvascular blood flow readings between control sites and a large number of other active sites under investigation, and it can use the full range of clinical and experimental probes of the MPM 3S. Due to the high number of measurement sites, the instrument also appears to be ideal for studying reflex changes in the microvasculature and for distinguishing between local and central effects. To this purpose, the extensive use of computer technology allows storage of data in MS-DOS format for postprocessing analysis, such as spectral analysis of spontaneous LDF fluctuations together with other physiological variables.

D. PRACTICAL ASPECTS OF LASER DOPPLER FLOWMETRY METHODOLOGY

1. Zeroing and Calibration of the Laser Doppler Flowmeter

Zero flux is normally obtained by placing the measuring head against an object where no movements occur, so that the backscattered light will contain no Doppler-shifted components. This is a true zero, indicating the absence of moving structures within the measuring volume. It has also been suggested that a condition of "zero" flux may be obtained by evaluating the LDF signal during suprasystolic occlusion at a proximal level of a limb. Nevertheless, this "occlusive" zero generally amounts to about 20% of the resting flow value in forearm skin. This discrepancy is attributed to the fact that, even though blood flow is arrested by the pressure cuff, the RBCs in the peripheral vessels are still moving randomly and producing minor Doppler components, which are recorded by the instrument.[32] The assessment of a reliable zero *in vivo* is still under investigation.

A calibration curve for laser Doppler instruments has been attempted by comparison with other methods evaluating various (and different) aspects of cutaneous circulation, such as venous occlusion plethysmography,[34] radioactive microsphere injection,[35] photoplethysmography,[35,36] and ^{133}Xe clearance.[14,29] Comparisons with the [85]Kr washout technique,[37] venous collection[38,39] in the intestine, and measurements in isolated organ preparations[40] have been performed. Although a good correlation was found with all methods, it is difficult to obtain from these comparisons a unique "calibration factor" because of high inter/intrasubject and interstudy variability of the relationship, due to the site specificity of the LDF signal, and because of the varying significance of reference methods. Engelhart and Kristensen[41] questioned the appropriateness of a comparison between LDF and the ^{133}Xe clearance method in the skin, since the LDF seems to measure blood flow in capillaries as well as in arteriovenous anastomoses while the ^{133}Xe method probably measures only capillary flow. A calibration of the LDF against the ^{133}Xe method would appear to be impossible in skin areas where arteriovenous anastomoses are present, whereas the changes in skin blood flow were found to be parallel when measured in skin areas without shunt vessels. These correlations are probably valid only for the area examined and, due to the large difference in various tissues, cannot be extended to other observations.[37]

In conclusion, although the LDF can be calibrated in absolute units for a particular measurement site on a specific preparation, there is general agreement that calibration factors probably cannot be used for other sites on the same tissue or in other tissues. In areas such as dermatology, the regional complexity of the microvasculature, its global variability also in specific regions (such as fingers[42]), and the complex nature of light scattering in tissue make LDF measurements suitable for characterizing only relative changes in blood flow. LDF data are also strictly site specific. Attempts of some commercial producers to market laser Doppler instruments calibrated in absolute terms rest on the assumptions that microcirculatory flow is essentially homogeneous for all tissue structures of the human body. These assumptions have been questioned[25,43] since these absolute "calibration factors" (mostly based on theoretical physical-optical models) are not necessarily valid for different tissues/ sites.[25,44]

2. Artifacts

Movement artifact is a common problem for clinicians using LDF based on fiber-optic probes. Changes in the blood flow signal unrelated to actual physiological changes are usually attributed to the movement of the optical fibers connected to the probe and are particularly severe in uncollaborative subjects or during long-term observations. Thinner and more flexible optical fibers have now reduced, but not abolished, the problem while accentuating the site specificity of the tissue sampling volume.[25] Two kinds of solutions have been proposed. The "mechanical" solution implies the elimination of optical fibers, which can be accomplished by including the laser source (laser diode) and the photodetector system within the probe (Section III.C.4). The "signal processing solution" is based on the observation that motion artifacts have a lower frequency distribution compared to the Doppler shift. Elimination or reduction of artifacts can be obtained by high-pass filtering of the signal (cutoff at 15 to 75 Hz), which is normally performed in all instruments. In addition, a motion artifact rejection system has been implemented in the Periflow PF3 (Section III.C.2). This compares the rate of change in the blood flow index with a maximum value expected for physiological changes; for "excessive" rates of change the output signal is discontinued. Problems may arise, however, with small artifacts when the systems may be unable to distinguish between changes in blood flow and motion artifacts. In addition, Gush and King[45] have suggested that part of the heart-synchronous pulsatile LDF signal can be due not only to a pulsatile flow in the capillaries, but also in part to a pulsatile movement of the tissue caused by arterial pulsations and/or to the method of normalization. This source of artifact has not yet been eliminated.

Laser instability also can be a problem. Besides typical aspects of laser stabilization (Section III.B.1), which have been solved in commercially available instruments, spontaneous fluctuations of He-Ne laser output intensity have been observed occasionally long after switching on the instrument, even with the probe positioned on a stationary scattering surface.[46]

Pressure of the probe must be appropriate. It has been shown by both LDF[25] and photoplethysmography[47] that skin blood flow is reduced dramatically by low (<15 mmHg) pressure of the probe on the skin. Lightweight probes, flexible leads, and application of the probe by biadhesive tape (rather than by wrapping it to its sides) may help to reduce gross variations in pressure.

PHOTOPLETHYSMOGRAPHIC PROBES

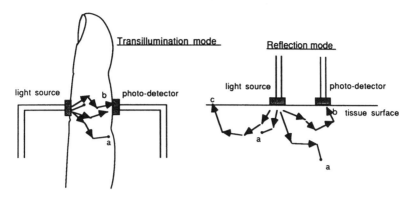

Figure 7 Positioning of light source and photodetector in the "transillumination" and "reflection" photoplethysmographic (PPG) probes: **a,** photon is absorbed; **b,** photon is scattered and recaptured by the photodetector; **c,** photon is scattered away from the photodetector. Changes in the hemoglobin content and/or erythrocyte orientation probably alter the relationship between absorbed and scattered light and cause the PPG signal fluctuation.

III. PHOTOPLETHYSMOGRAPHY

A. DEFINITION, BASIC PRINCIPLES, AND ORIGIN OF THE SIGNAL AND HARDWARE

Photoplethysmography (PPG) can be simply defined as the continuous recording of the light intensity scattered from a given source by the tissues and collected by a suitable photodetector. The terms "reflection PPG" and "transillumination PPG" are used to indicate that emitter and detector are positioned adjacent to each other (as for LDF, Figure 7, right) or at opposite sides of selected areas such as the ear lobe/pinna and the finger (Figure 7, left)), respectively. It was found,[2] however, that the significance of the signals obtained by the two methods is the same: in both cases it depends on the superficial circulation.[47] The PPG is an old technique (it preceded LDF by about 40 years[48]) which provides a simple and inexpensive method to continuously investigate changes in the peripheral circulation. A comprehensive review of methodology and applications was published by Challoner.[2]

The origin of the signal is not completely understood. One possible cause of the PPG signal lies in the reciprocal relationship between absorption and scattering of light by the blood. In other words, the more blood in the illuminated region, the more light is absorbed by the hemoglobin, the less light will be received by the detector, and vice versa.[2] This hypothesis supports the finding that PPG is sensitive to blood volume. Nevertheless, pulsatile records have been obtained from the blood supply in the dental pulp, in which volumetric changes of the vessels are not possible, and from glass tubes in which the flow was set to be pulsatile. These indicate that PPG is sensitive not only to the blood *volume,* but also to the *velocity* of RBCs, possibly through changes in the optical transmittance due to changes in orientation of the RBCs, by effect of changes in RBC speed.[2] What factor predominates it is not yet established.

Since PPG involves interaction between light and body tissues, most of the basic principles of physics underlying PPG are the same as for LDF. Similar considerations also apply to the light absorption and scattering and to the effect of various types of light and their interaction with body tissues and blood flow or volume. Similar considerations also apply in terms of choice of the emitting light and depth of measurement.

The light source was originally a bulb lamp, but in more recent applications light-emitting diodes (LEDs), providing a more selective light beam of red or (more often) infrared light, are now employed. The photodetector, which in the past was a photoresistance type, is now a photodiode or a phototransistor, and in some case it can be identical to that of LDF. This allows construction of very low-mass probes of small dimensions, resulting in signals of excellent stability and little susceptibility to movement artifacts. Probes can be built in any dimension and shape and are suitable for a wide range of applications and body surfaces, ranging from skin to nasal or intestinal mucosae. The use of low-power emitters has now eliminated the inconvenience, frequent with lamps, of heating the skin. Alternatively, the probes can be constructed with the possibility of controlling the temperature, similarly to LDF probes (Section III.C).

The electronics is very simple and inexpensive compared to the LDF, since the output signal is simply proportional to the intensity of the light collected by the photodetector. The signal is further divided into DC and AC components by appropriate filters or manual offset suppression. The faster AC components are related to the cardiac pulsations, while the slower AC component (below~0.4 Hz) reflects slow circulatory changes. Frequencies above 25 to 30 Hz are normally filtered.

B. APPLICATIONS

The PPG technique has found a wide range of applications,[2] particularly dealing with the possibility of following the reflex changes in peripheral circulation (such as during the Valsalva maneuver, in the cold, during exercise, and in reactive hyperemia), the effects of vascular surgery, and the effects of drugs acting on the skin.[2,35,49] The analysis of heart pulsations (particularly when recorded on the ear, so-called "ear densitography") reflects changes in a number of important cardiovascular variables (including stroke volume and ejection fraction[50]). When combined with the Valsalva maneuver, ear PPG showed a high correlation with left ventricular end-diastolic pressure.[51]

IV. FINAL REMARKS

Although the exact origin of PPG is not fully understood, the dependence of PPG on both RBC volume and speed makes comparison with LDF unavoidable. Similar considerations for the two techniques apply to the depth of measurement, since this is determined mostly by the geometry of the emitting/receiving device and by the wavelength of light, although PPG light does not have to be monochromatic (Doppler shifts are averaged or eliminated by filtering). It has been suggested that PPG may correspond to the low-frequency components of the Doppler shift spectrum,[45] while LDF has stimulated a general enthusiasm because of its capability of providing an absolute measure of Doppler shift distribution, related to the microcirculation. This is the most important technical difference between the two methods, but to what extent different technical aspects can provide different practical information is still not clearly established. If one does not consider differences in probe structure and light power and wavelength, clinical and experimental data suggest that both LDF and PPG provide a relative measure of a combination of RBC volume and speed.

How the Doppler shift distribution can be processed is still a matter of research; *in vitro* experimental tests have shown major differences in results obtained with different algorithms.[28] The comparison of different LDF instruments is further complicated by several factors (such as different methods of calibration and different light wavelengths and probe geometries), yet results obtained with different commercial instruments have been considered similar.[52] At present, it remains doubtful whether translation from absolute Doppler shift to absolute values of blood volume or flow (independent of tissue type and site) has been accomplished. Since neither technique can establish the absolute amount of blood (flow or volume, moving or stagnant) of a tissue in a clinical environment, their use is restricted to

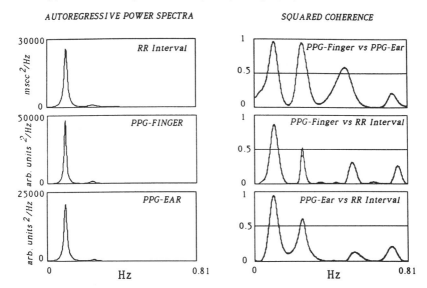

Figure 8 Power spectrum analysis (left side) and coherence functions (right side) of spontaneous R-R interval and photoplethysmographic (PPG) fluctuations recorded at the index finger and at the ear in a healthy subject while standing. Notice the similarity of spectra, all showing two main peaks, at 0.1 Hz (vasomotor) and at 0.25 Hz (respiratory). The high (>0.5) coherence values at the frequency of the spectral peaks indicate that these cardiac and microcirculatory (in two different areas) fluctuations are statistically related.

only relative circulatory changes. This apparent limitation still allows important clinical applications, also in view of the new methods of analysis of cardiovascular regulation, applied to the microcirculation. The importance of comparative measurements in different body areas is also well established, and in advanced research multisite recording has been attempted with a single movable LDF probe.[53] Simultaneous multichannel systems can be designed inexpensively with PPG, while one recent commercial LDF apparatus has been conceived for this purpose (Section III.C.4).

The study of the cardiovascular control mechanisms by spectral analysis of heart rate and blood pressure variabilities is now one of the most interesting and promising research fields. Applications span from early detection of autonomic dysfunction in diseases such as diabetes[54] to coronary artery disease, hypertension, heart failure, and others.[55] The method is based on the finding that the modulation of the cardiovascular system by the two branches of the autonomic nervous system can be studied by the analysis of spontaneous fluctuations induced by respiration (related to the vagus in the heart) and by vasomotor activity (around 0.1 Hz, related to the sympathetic system). A noninvasive technique capable of monitoring changes in circulatory parameters of peripheral vessels offers a unique possibility for studying the circulatory effects of the efferent pathway of the sympathetic nervous system. Since this approach does not require absolute calibration (one is interested in looking at the *distribution* of spontaneous fluctuations in the microvasculature), the LDF and the PPG appear adequate. In agreement with previous reports indicating the presence of non-thermoregulatory control of the skin vessels,[19] LDF and PPG undergo fluctuations that for some frequencies (particularly in the range of 0.1 Hz) are synchronous in different areas of the skin[56] and depend in part on changes in the sympathetic modulation, as elicited by tilting.[57] Figure 8 shows the spectra and the coherence functions of spontaneous PPG fluctuations recorded at the finger and at the ear in normal subjects, together with the heart period (R-R interval). The similarity of the spectra and the high coherence indicate a close relationship of these signals in different skin areas and demonstrate a common origin of such fluctuations. Combined analysis of cardiac and peripheral (in different areas) circulatory fluctuations thus provides an invaluable tool to monitor the effect of local vs. central changes in the microvascular regulation.[57] Studies with either LDF[58] or PPG[59] in diabetes have demonstrated an early and selective reduction of the 0.1-Hz component of the microcirculation, suggesting an early sympathetic impairment in these patients, even in the absence of clinical symptoms of autonomic dysfunction.

In practice, despite important theoretical and technical differences, the range of clinical applications for the two methods seems to overlap; studies comparing LDF to PPG[35,36,60] did not show major differences, suggesting that the two techniques are measuring rather similar (or two related) vascular phenomena. More specific comparisons of the two methods are necessary in order to establish their practical differences and attitudes and, hence, possible specific fields of application; comparable light sources and identical detectors should be used, or else similarities or differences obtained could be simply ascribed to the characteristics of probes and light wavelengths. Technical improvements seem possible, mainly in order to obtain signals sensitive to more specific and better identified sections of the microvasculature. It will also be important to further verify *in vivo* to what extent different algorithms applied to the Doppler shift distribution will enable one to study different aspects of microcirculation of clinical relevance.

ACKNOWLEDGMENT

We gratefully acknowledge Dr. Valerio Annovazzi-Lodi, from the Department of Electronics, University of Pavia, Pavia, Italy, for revising the manuscript and for providing informed technical advice.

REFERENCES

1. Shepherd, A. P. and Oberg, P. A., Eds. *Laser-Doppler Blood Flowmetry,* Kluwer Academic Publishers, Dordrecht, 1990.
2. Challoner, A. V. J., Photoelectric plethysmography for estimating cutaneous blood flow, in *Non-Invasive Physiological Measurements,* Rolfe, P., Ed., Academic Press, London, 1979, chap. 6.
3. Shepherd, A. P., History of laser-Doppler flowmetry, in *Laser-Doppler Blood Flowmetry,* Shepherd, A. P. and Oberg, P. A., Eds., Kluwer Academic Publishers, Kluwer Dordrecht, 1990, chap. 1.
4. Duteil, L., Bernengo, J. C., and Schalla, W., A double wavelength laser Doppler system to investigate skin microcirculation, *IEEE Trans. Biomed. Eng.,* BME-32, 439, 1985.
5. Gush, R. J., and King, T. A., Discrimination of capillary and arterio-venular blood flow in skin by laser Doppler flowmetry, *Med. Biol. Eng. Comput.,* 29, 387, 1991.
6. Obeid, A. N., Boggett, D. M., Barnett, N. J., Dougherty, G., and Rolfe, P., Depth discrimination in laser Doppler skin blood flow measurement using different lasers, *Med. Biol. Eng. Comput.,* 26, 415, 1988.
7. Anderson, R. R. and Parrish, J. A., The optics of human skin, *J. Invest. Dermatol.,* 77, 13, 1981.
8. Bonner, R. F. and Nossal, R., Principles of laser-Doppler flowmetry, in *Laser-Doppler Blood Flowmetry,* Shepherd, A. P. and Oberg, P. A., Eds., Kluwer Academic Publishers, Dordrecht, 1990, chap. 2.
9. Steinke, J. M. and Shepherd, A. P., Diffuse reflectance of whole blood: model for a diverging light beam, *IEEE Trans. Biomed. Eng.,* BME-34, 826, 1987.

10. Bonner, R. and Nossal, R., Model for laser Doppler measurements of blood flow in tissue, *Appl. Optics,* 20, 2097, 1981.

11. Stern, M. D., In vivo evaluation of microcirculation by coherent light scattering, *Nature,* 254, 56, 1975.

12. Stern, M. D., Laser Doppler velocimetry in blood and multiply scattering fluids, *Appl. Optics,* 24, 1968, 1985.

13. Steinke, J. M. and Shepherd, A. P., Comparison of Mie theory and the light scattering of red blood cells, *Appl. Optics,* 19, 4027, 1988.

14. Stern, M. D., Lappe, D. L., Bowen, P. D., Chimosky, J. E., Holloway, G. A., Jr., Keiser, H. R., and Bowman, R. L., Continuous measurement of tissue blood flow by laser-Doppler spectroscopy, *Am. J. Physiol.,* 232, H441, 1977.

15. Nilsson, G. E., Tenland, T., and Oberg, P. Å., Evaluation of a laser Doppler flowmetry for measurement of tissue blood flow, *IEEE Trans. Biomed. Eng.,* BME-27, 597, 1980.

16. Nilsson, G. E., Signal processor for laser Doppler tissue flowmeters, *Med. Biol. Eng. Comput.,* 22, 343, 1984.

17. Gush, R. J., King, T. A., and Jayson, M. I. V., Aspects of laser light scattering from skin tissue with application to laser-Doppler blood flow measurement, *Phys. Med. Biol.,* 29, 1463, 1984.

18. Johansson, K., Ahn, H., Lindhagen, J., and Lundgren, O., Tissue penetration and measuring depth of laser Doppler flowmetry in the gastrointestinal application, *Scand. J. Gastroenterol.,* 22, 1081, 1987.

19. Johnson, J. M., Nonthermoregulatory control of human skin blood flow, *J. Appl. Physiol.,* 61, 1613, 1986.

20. Ryan, T. J., The blood vessels of the skin, in *The Physiology and Pathophysiology of the Skin,* Vol. 2, Jarrett, A., Ed., Academic Press, London, 1973, chap. 16, 17.

21. Watkins, D. and Holloway, G. A., An instrument to measure cutaneous blood flow using the Doppler shift of laser light, *IEEE Trans. Biomed. Eng.,* BME-25, 28, 1978.

22. Nilsson, G. E., Tenland, T., and Oberg, P. Å., A new instrument for continuous measurement of tissue blood flow by light beating spectroscopy, *IEEE Trans. Biomed. Eng.,* BME-27, 12, 1980.

23. Obeid, A. N., Dougherty, G., and Pettinger, S., In vivo comparison of a twin wavelength laser Doppler flowmetry using He-Ne and laser diode sources, *J. Med. Eng. Technol.,* 14, 102, 1990.

24. de Mul, F. F. M., van Spijker, J., van der Plas, D., Greve, J., Aarnoudse, J. G., and Smits, T. M., Mini laser-Doppler (blood) flow monitor with diode laser source and detection integrated in the probe, *Appl. Optics,* 23, 2970, 1984.

25. Obeid, A. N., Barnett, N. J., Dougherty, G., and Ward, G., A critical review of laser Doppler flowmetry, *J. Med. Eng. Technol.,* 14, 178, 1990.

26. Salerud, E. G. and Oberg, P. Å., Single-fibre laser Doppler flowmetry. A method for deep tissue perfusion measurements, *Med. Biol. Eng. Comput.,* 25, 329, 1987.

27. Salerud, E. G. and Nilsson, G. E., Integrating probe for tissue laser Doppler flowmeters, *Med. Biol. Eng. Comput.,* 24, 415, 1986.

28. Obeid, A. N., In vitro comparison of different signal processing algorithms used in laser Doppler flowmetry, *Med. Biol. Eng. Comput.,* 31, 43, 1993.

29. Holloway, G. A., Jr. and Watkins, A. W., Laser Doppler measurement of cutaneous blood flow, *J. Invest. Dermatol.,* 69, 306, 1977.

30. Ahn, H., Lundgren, O., and Nilsson, G. E., In vivo evaluation of signal processors for laser Doppler tissue flowmeters, *Med. Biol. Eng. Comput.,* 25, 207, 1987.

31. Holloway, G. A., Jr., Medpacific's LDV blood flowmeter, in *Laser-Doppler Blood Flowmetry,* Shepherd, A. P. and Oberg P. Å. Eds., Kluwer Academic Publishers, Dordrecht, 1990, chap. 3.

32. Nilsson, G. E., Perimed's LDV flowmeter, in *Laser-Doppler Blood Flowmetry,* Shepherd, A. P. and Oberg P. Å., Eds., Kluwer Academic Publishers, Dordrecht, 1990, chap. 4.

33. Borgos, J. A., TSI's LDV blood flowmeter, in: *Laser-Doppler Blood Flowmetry,* Shepherd, A. P. and Oberg P. Å., Eds., Kluwer Academic Publishers, Dordrecht, 1990, chap. 5.

34. Johnson, J. M., Taylor, W. F., Shepherd, A. P., and Park, M. K., Laser-Doppler measurement of skin blood flow: comparison with plethysmography, *J. Appl. Physiol.,* 56(3), 798, 1984.

35. Hales, J. R. S., Stephens, F. R. N., Fawcett, A. A., Daniel, K., Sheahan, J., Westerman, R. A., and James, S. B., Observation on a new non-invasive monitor of skin blood flow, *Clin. Exp. Pharmacol. Physiol.,* 16, 403, 1989.

36. Tur, E., Tur, M., Maibach, H. I., and Guy, R. H., Basal perfusion of the cutaneous microcirculation: measurements as a function of anatomic position, *J. Invest. Dermatol.,* 81, 442, 1983.

37. Ahn, H., Lindhagen, J., Nilsson, G. E., Salerud, E. G., Jodal, M., and Lundgren, O., Evaluation of laser Doppler flowmetry in the assessment of intestinal blood flow in cat, *Gastroenterology,* 88, 951, 1985.

38. Ahn, H., Lindhagen, J., Nilsson, G. E., Oberg, P. Å., and Lundgren, O., Assessment of blood flow in the small intestine with laser Doppler flowmetry, *Scand. J. Gastroenterol.,* 21, 863, 1986.

39. Ahn, H., Lindhagen, J., and Lundgren, O., Measurement of colonic blood flow with laser Doppler flowmetry, *Scand. J. Gastroenterol.,* 21, 871, 1986.

40. Sheperd, A. P., Riedel, G. L., Kiel, J. W., Haumshield, D. J., and Maxwell, L., C., Evaluation of an infrared laser Doppler blood flowmeter, *Am. J. Physiol.,* 252, G832, 1987.

41. Engelhart, M. and Kristensen, J. K., Evaluation of cutaneous blood flow responses by [133]Xenon washout and a laser Doppler flowmeter, *J. Invest. Dermatol.,* 80, 12, 1983.

42. Hale, A. R., and Burch, G. E., The arteriovenous anastomoses and blood vessels of the human finger, *Medicine (Baltimore),* 39, 191, 1960.

43. Oberg, A. P., Innovations and precautions, in: *Laser-Doppler Blood Flowmetry,* Shepherd, A. P. and Oberg, P. Å., Eds., Kluwer Academic Publishers, Dordrecht, 1990, chap. 6.

44. Johnson, J. J., The cutaneous circulation, in *Laser-Doppler Blood Flowmetry,* Shepherd, A. P. and Oberg, P. Å., Eds., Kluwer Academic Publishers, Dordrecht, 1990, chap. 8.

45. Gush, R. J. and King, T. A., Investigation and improved performance of optical fibre probes in laser Doppler blood flow measurement, *Med. Biol. Eng. Comput.,* 25, 391, 1987.

46. Fairs, S. L. E., Observations of a laser Doppler flowmeter output made using a calibration standard, *Med. Biol. Eng. Comput.,* 26, 404, 1988.

47. Uretzky, G. and Palti Y., A method for comparing transmitted and reflected light photoelectric plethysmography, *J. Appl. Physiol.,* 31, 132, 1971.

48. Hertzmann, A., B. and Spealman, C., R., Observations on the finger volume pulse recorded photoelectrically, *Am. J. Physiol.,* 119, 334, 1937.

49. Tur, E., Guy, R. H., Tur, M., and Maibach, H. I., Noninvasive assessment of local nicotinate pharmacodynamics by photoplethysmography, *J. Invest. Dermatol.,* 80, 499, 1983.

50. Nakamura, Y., Haffty, B., Spodick, D. H., Paladino, D., Moreau, K., and Flessas, A., Responses of peak ear-pulse derivative to changing left ventricular function, *Am. J. Physiol.,* 238, H355, 1980.

51. Bernardi, L., Saviolo, R., and Spodick, D. H., Noninvasive assessment of central circulatory pressures by analysis of ear densitographic changes during Valsalva maneuver, *Am. J. Cardiol.,* 64, 787, 1989.

52. Fischer, J. C., Parker, P. M., and Shaw, W. W., Comparison of two laser Doppler flowmeters for the monitoring of dermal blood flow, *Microsurgery,* 4, 164, 1983.

53. Braverman, I. M. and Schechner, J. S., Contour mapping of the cutaneous microvasculature by computerized laser Doppler velocimetry, *J. Invest. Dermatol.,* 91, 1013, 1991.

54. Bernardi, L., Ricordi, L., Lazzari, P. L., Solda', P. L., Calciati, A., Ferrari, M. R., Vandea, I., Finardi, G., and Fratino, P., Impaired circadian modulation of sympatho-vagal activity in diabetes: a possible explanation for altered temporal onset of cardiovascular disease, *Circulation,* 86, 1443, 1992.

55. Malliani, A., Pagani, M., Lombardi, F., and Cerutti, S., Cardiovascular neural regulation explored in the frequency domain, *Circulation,* 84, 482, 1991.

56. Schechner, J. S. and Braverman, I. M., Synchronous vasomotion in the human cutaneous microvasculature provides evidence for central modulation, *Microvasc. Res.,* 44, 27, 1992.

57. Bernardi, L., Rossi, M., Fratino, P., Finardi, G., Mevio, E., and Orlandi C., Relationship between changes in human skin blood flow and autonomic tone, *Microvasc. Res.,* 37, 16, 1989.

58. Rossi, M., Ricordi, L., Mevio, E., Fornasari, G., Orlandi, C., Fratino, P., Finardi, G., and Bernardi, L., Autonomic nervous system and microcirculation in diabetes, *J. Autonom. Nerv. Syst.,* 30, S133, 1990.

59. Bernardi, L., Radaelli, A., Ricordi, L., Corbellini, D., Solda', P. L., Leuzzi, S., and Fratino, P., Spectral analysis of heart rate, blood pressure and peripheral circulation in diabetes: evidence for a global autonomic impairment?, *Diabetologia,* 34, A158, 1991.

60. Amantea, M., Tur, E., Maibach, H. I., and Guy, R. H., Preliminary skin blood flow measurements appear unsuccessful for assessing topical corticosteroid effect, *Arch. Dermatol. Res.,* 275, 419, 1983.

Chapter 4

Cutaneous Blood Flow and Erythema: Standardization of Measurements

Jørgen Serup

CONTENTS

I. INTRODUCTION

Methodologies used in dermatological research have improved substantially during the last few decades. Official guidelines for good clinical practices and rules for good laboratory practice have influenced the way studies are designed, conducted, and reported.[1,2] The emphasis on objectivity and accuracy is a special opportunity for the bioengineering techniques if they can meet modern standards for validation and standardization. Standard operating procedures will be a future vade mecum of any advanced laboratory or contract laboratory which conducts such measurements.

The purpose of the present chapter is to give the researcher a practical introduction to standardization of the laser Doppler method for measurement of cutaneous blood flow and to erythema measurement; it is not designed to give an extensive review of the literature.

II. VALIDATION—GENERAL ASPECTS

When a study is planned, it is recommended that the researcher ask himself a number of questions concerning the design, test individuals and their preconditioning, laboratory facilities, the technique being used, the qualification of the persons conducting the measurements, data handling, and reporting, as expressed in Table 1.

It is very important that the researcher is familiar with the theoretical background and the literature about the method being used, and the researcher must have some practical experience with the method beforehand. The literature will normally include at least some information about the validation of the method as defined in Table 2. However, it is exceptional if a method is fully validated in every respect, and the researcher often has to rely on his own sound evaluation of the usefulness of the method.

There is great variation in test conditions and measuring devices which are used. Around ten different laser Doppler devices are commercially available. Thus, in most situations the researcher himself must perform validation experiments with the device he is actually using, at his own laboratory facility, and express the result of such evaluation in a standard operating procedure.

Table 1   **Planning a Study Involving Bioengineering Technique: A Checklist**

- What information is expected?
- What is the most relevant variable to be measured, and which variables serve for description, comparison, support, or exclusion?
- What is the expected time course of variables, and when should the measurements be performed?
- Are variables expected to develop linearly or not?
- What are the ranges of variables in relation to the expected phenomenon or structure being studied, including inter- and intraindividual variation and dependence on anatomical site, sex, and age?
- What function or structure is actually being tested?
- What is the measuring area and, if small, do more recordings need to be taken and averaged to overcome local site variation?
- Are recordings with the equipment reproducible, and is the precision acceptable relative to variables being measured and their expected range?
- List measuring standards and calibration procedures.
- Determine environmental influences (including season) and the need for special laboratory room facilities.
- Is it necessary to precondition the individuals before testing?
- What keeps measurements from being performed?
- Has the researcher or technician both the educational background and enough practical experience to conduct the study?

Table 2   **Validation Requirements and Definitions**

	Definition
Accuracy	Closeness of agreement between the value which is accepted either as a conventional true value (in-house standard) or an accepted reference value (international standard) and the value found (mean) by performing the test procedure a number of times; provides indication of systematic error
Precision	Closeness of agreement (degree of scatter) between a series of measurements obtained from multiple sampling of the same homogeneous sample under prescribed conditions, expressed as repeatability and reproducibility
Repeatability	Expresses the situation under the *same* conditions, i.e., same operator, same apparatus, short time interval, identical sample
Reproducibility	Expresses the situation under *different* conditions, i.e., different laboratories or samples, different operators, different days, different instruments from different manufacturers
Range	The interval between the upper and lower levels for which the procedure has been demonstrated as applicable with precision, accuracy, and linearity
Linearity	Ability of the procedure (within a given range) to obtain test results directly proportional to true values
Sensitivity	Capacity of the procedure to record small variations or differences within the defined range
Limit of detection	Lowest change over zero which is just detectable
Limit of quantification	Lowest change over zero which can be quantitatively determined (not only detected) with defined precision and accuracy under the stated experimental conditions
Ruggedness	Evaluates the effects of small changes in the test procedure on measuring performance

III. STANDARDIZATION OF LASER DOPPLER FLOWMETRY

The standardization group of the European Contact Dermatitis Society recently published guidelines for measurement of cutaneous blood flow by laser Doppler flowmetry.[3] Readers are recommended to consult this paper for detailed information. Factors and variables with effects on cutaneous blood flow which should be taken into account when the laser Doppler method is used are listed in Table 3. Anatomical site, physical activity, mental activity, food and drugs, and temperature all have a considerable effect on the result.

The standardization group defined the following guidelines:

1. Use the flowmeter in accordance with manufacturer's and laser light safety directions and with a defined operating procedure including validation (determination of resting cutaneous blood flow and repeatability followed by a 3-min arterial compression of the upper arm with a cuff and determination of biological zero and peak cutaneous blood flow on flexor side forearm skin of at least three healthy adults); see Figure 1.
2. Allow the flowmeter to warm up. Apply the probe to the measuring site with no special pressure.
3. Make sure that the test subject has not taken any food or drugs which might influence cutaneous blood flow. Avoid any topical treatment of the test site prior to study unless it is a part of the experiment.
4. Make sure that the test subject has not deliberately exercised, been exposed to unusual temperatures, or been under mental stress immediately before cutaneous blood flow measurements.
5. Allow the test subject to rest for 15 min or more under quiet conditions, preferably in the laboratory room in the position in which recordings are going to be obtained, i.e., sitting or supine, and with the test site uncovered.
6. Take three or more recordings and average them if the flowmeter operates within a small measuring field, i.e., 1 to 3 mm^2. Avoid movement of the optical cable.
7. Use a pen recorder or a computer with appropriate software for data collection. Store the basic data safely.
8. Control the laboratory room and the measuring conditions, particularly with respect to temperature, convection of air, and noise. Avoid measurement under direct light, including sunshine, which might influence skin temperature. Perform measurements with the site studied at a standardized level relative to the level of the heart.

Table 3   **Factors and Variables with Effects on Cutaneous Blood Flow (CBF)**

Age	Widely age independent
Sex	Minor or no difference
Menstrual cycle	Minor or no difference
Race	Minor or no difference
Anatomical site	Considerable variation
Position	Orthostatic dependence
Temporal, diurnal	Minor or no effect
Temporal, day to day	May be significant
Physical activity	Considerable effect
Mental activity	Considerable effect
Food and drugs	Considerable effect
Temperature	Very significant effect

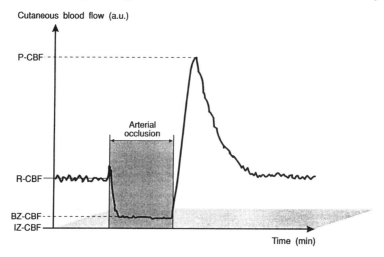

Figure 1 Cutaneous blood flow (CBF) in arbitrary units (a.u.) before and during arterial occlusion and postocclusive reactive hyperemia (schematic). R-CBF = resting flow; BZ-CBF = biological zero; IZ-CBF = instrumental zero determined on white porcelain; P-CBF = peak flow after occlusion and during reactive hyperemia. With the hyperemia experiment the range of operation of the instrument is described. In order to describe the repeatability of the instrument it is recommended that R-CBF be determined a number of times, and mean, SD, and coefficient of variation can be given. In the validation experiment a minimum of three healthy adults should be studied on flexor side forearm skin. Results of validation experiments shall ideally appear in every publication.

9. In reports and publications, clearly describe the instrument, including details about the probe and the effective measuring area laterally and in depth, and also describe the instrument setting and validation, handling of the test subject, and control of measuring conditions in a way which provides precise and detailed information about the experiment, so that it can be reproduced.

The standardization group introduced a 3-min arterial compression experiment as a standard procedure for the validation of the technique (Figure 1). This procedure can be performed easily in any laboratory working with any piece of equipment.

Detailed information on the methodology is found in the guideline paper,[3] and detailed information about the technical principles is found in the monograph by Shepherd and Öberg.[4]

IV. STANDARDIZATION OF ERYTHEMA MEASUREMENTS

Redness or erythema depends on the volume of blood under a given area of skin and not directly on blood flow or flux, which represents moving blood corpuscles. This can be demonstrated easily by conducting an orthostatic experiment with the arm. The cutaneous blood flow is less dependent on orthostatic position due to the venoarteriolar reflex (degree of dilation of arterioles as determined by venous pressure); in contrast, the skin becomes pale on elevation and red on lowering in accordance with measurements of cutaneous blood flow and erythema. Although flow and redness often correlate, they represent different features of cutaneous vascularity, and they are not expected to correlate automatically in any situation.

Redness or erythema is a dermatological term based on the light perception of the eye and the brain. The eye does not register colors of different wavelength in a linear fashion vs.

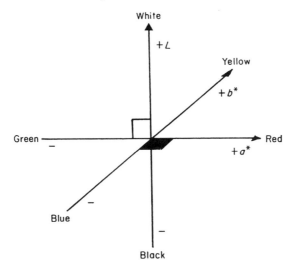

Figure 2 The Commission Internationale de l'Éclairage (CIE) color space. This system is a three-dimensional coordinate system with an L-axis, an a*-axis, and a b*-axis.

light energy. The eye is especially sensitive to red colors, and clinical evaluation is therefore a very precise method for assessment of minute differences or nuances in redness.

The color measuring system of the Commission Internationale de l'Éclairage (CIE) takes this nonlinearity of color perception of the human eye into account[5] (see Figure 2). Every color has a position in a three-dimensional color space. Two measuring devices based on the CIE system are available, i.e., the Minolta Chroma Meter® and the Lange Micro Color®.[6] Although the two devices correlate well, there are systematic differences in the actual readings.

Skin color also can be measured by a spectrophotometric principle where absorbance or remittance is determined either at selected wavelengths or within a given range—for example, from 400 to 800 nm, corresponding to visible light. Diffey and Farr have described a hemoglobin index where the overlap of melanin absorption on hemoglobin absorption of light is corrected.[7] The Derma-Spectrometer® is an example of a commercially available equipment based on this principle.

Despite the difference in principles, the Minolta Chroma Meter® and the Derma-Spectrometer® show a high degree of correlation (Figure 3). The two devices were compared and validated in three recent papers.[8–10]

The studies with the Derma-Spectrometer® and the Minolta Chroma Meter® have shown that skin color is not simply a question of pigments of the skin, i.e., hemoglobin, melanin, and bilirubin. It is useful in relation to color measurement to consider the skin a cross-sectional structure with the following three layers:

1. The surface itself, maybe with white scales
2. The brown melanocyte layer at the dermoepidermal junction
3. The red cutaneous vasculature in the outer dermis, including the dermis papillae, and in the lower part of the dermis at the interface to the fat

Variations in the skin surface, including changes in the optical properties of the interface due to ambient air and scaling in skin diseases, and variations in skin pigmentation for obvious reasons exert a major filtering and covering effect when redness, depending on the volume of red blood corpuscles in the vasculature in the outer dermis, is measured. Scales look white and are measured as white. They are white due to their content of air. If a drop of oil is put

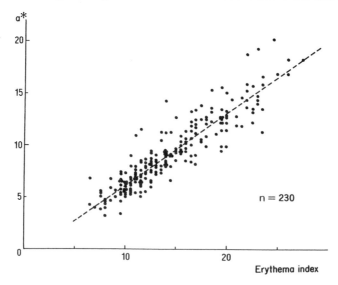

Figure 3 Correlation between a*, measured with the Minolta Chroma Meter®, and ery-thema index, measured with the Derma-spectrometer®. Normal skin of 23 different ana-tomical sites was studied. (From Takiwaki, H., Overgaard, L., and Serup, J., *Skin Pharmacol.*, S. Karger AG, BASEL, 7, 217, 1994. With permission.)

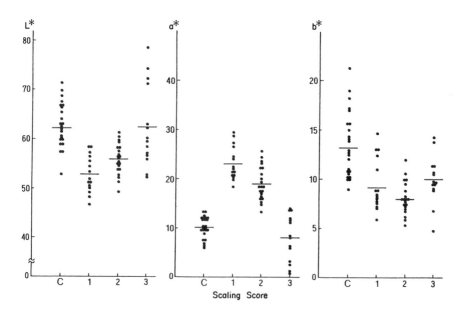

Figure 4 Psoriatic plaques studied with the Minolta® Chroma Meter; C, uninvolved skin of the same body region; 1, slight scaling; 2, moderate scaling; 3, heavy scaling. (From Takiwaki, H., Overgaard, L., and Serup, J., *Skin Pharmacol.*, S. Karger AG, BASEL, 7, 217, 1994. With permission.)

on the skin, the scales instantly become translucent. Using a Minolta Chroma Meter® it is easily demonstrated that redness (increase of a*, decrease of L) depends on the scaling score (Figure 4).

Thus, when skin lesions are studied and erythema is measured, the researcher must decide whether alterations of the skin surface and the epidermis are expected to interfere with the light emitted from the measuring device on its way into and out of the skin to the detection system. It may be useful as a standard to apply an oil to the skin surface.

Skin color measurements are like cutaneous blood flow vascular phenomena, and color measurements essentially depend on the same variables, as listed in Table 3.

There is no official guideline on the use of the Derma-Spectrometer,® the Minolta Chroma Meter®, or the Lange Colorimeter® except for the recommendations given in users' manuals. The equipment is delivered from the manufacturer with white and black color standards which can be used to verify that the equipment does not change over time. It is recommended that such verification is performed at regular intervals. It also may be useful to check the equipment on different standards of red.

The standardization group of the European Society of Contact Dermatitis previously developed guidelines for measurement of transepidermal water loss and laser Doppler flowmetry.[3,11] Standardization of skin color measurement is undertaken in a running project.

REFERENCES

1. CPMP working party, EEC note for guidance. Good clinical practice on medical products in the European Community, *Pharmacol. Toxicol.,* 67, 361, 1990.
2. Hirsch, A. F., *Good Laboratory Practice Regulations,* Marcel Dekker, New York, 1989.
3. Bircher, A., de Boer, E., Agner, T., Wahlberg, J. E., and Serup, J., Guidelines for measurement of cutaneous blood flow by laser Doppler flowmetry. A report from the standardization group of the European Society of Contact Dermatitis, *Contact Dermatitis,* 30, 65, 1994.
4. Shepherd, A. P. and Öberg, P. Å., Eds., *Laser-Doppler Blood Flowmetry,* 1st ed., Kluwer Academic Publishers, Boston, 1990.
5. Weatherall, I. L. and Coombs, B. D., Skin color measurements in terms of CIELAB color space values, *J. Invest. Dermatol.,* 99, 468, 1992.
6. Serup, J. and Agner, T., Colorimetric quantification of erythema. Comparison of two colorimeters (Lange Micro Color and Minolta Chroma Meter CR-200) with a clinical scoring scheme and laser Doppler flowmetry, *Clin. Exp. Dermatol.,* 15, 267, 1990.
7. Diffey, B. L., Oliver, R. J., and Farr, P. M., A portable instrument for quantifying erythema induced by ultraviolet radiation, *Br. J. Dermatol.,* 111, 663, 1984.
8. Takiwaki, H. and Serup, J., Orthostatic maneuver, skin color and cutaneous blood flow— comparison of tristimulus colorimetry, narrow-band reflectance spectrophotometry and laser-Doppler flowmetry, *Skin Pharmacol.,* 7, 226, 1994.
9. Takiwaki, H., Overgaard, L., and Serup, J., Comparison of narrow-band reflectance spectro-photometric and tristimulus colorimetric measurement of skin color—twenty-three anatomical sites evaluated by the DermaSpectrometer® and the Chroma Meter CR-200®, *Skin Pharmacol.,* 7, 217, 1994.
10. Takiwaki, H. and Serup, J., Measurement of color parameters of psoriatic plaques by narrow-band reflectance spectrophotometry and tristimulus colorimetry, *Skin Pharmacol.,* 7, 145, 1994.
11. Pinnagoda, J., Tupker, R., Agner, T., and Serup, J., Guidelines for transepidermal water loss (TEWL) measurement. A report from the standardization group of the European Society of Contact Dermatitis, *Contact Dermatitis,* 22, 164, 1990.

Contact Dermatitis

Tove Agner

CONTENTS

ABSTRACT: Within the area of contact dermatitis the laser Doppler flowmetry technique has been used for quantification of inflammatory skin reactions. Skin reactions can be followed for a period of time with no interference from the non-invasive laser Doppler flowmetry measurements. Acute and chronic irritant contact dermatitis, allergic contact dermatitis, and contact urticaria have been studied.

In the following chapter, utilization of laser Doppler flowmetry within the field of contact dermatitis is evaluated. Methodological aspects are discussed, and experimental studies attending different areas of contact dermatitis are reviewed.

I. INTRODUCTION

Until 10 years ago most research within the area of contact dermatitis was concentrated on allergic contact dermatitis, and only a few papers on irritant contact dermatitis appeared. A rational explanation for this was the lack of scientific methods and techniques accessible for irritant studies. Within the last 10 to 15 years a number of noninvasive bioengineering methods have become available and have brought a new dimension into the study of irritant contact dermatitis. One of the most commonly used methods is laser Doppler flowmetry (LDF).

The laser Doppler flowmeter is a useful instrument for evaluation of cutaneous microcirculation. The LDF technique combines the Doppler phenomenon with the spectral purity of laser light.[1] After partial absorption and diffuse scattering have taken place, the monochromatic laser light is reflected with Doppler-shifted frequencies from moving blood cells in the skin and with unshifted frequencies from stationary tissues. Cutaneous blood vessels are present in the dermis, while the epidermis is avascular, and blood flow in dermal vessels to a depth of approximately 1 mm is recorded by LDF. The dimensionless output signal is proportional to the cutaneous blood flow.

LDF is now being used extensively for quantification of inflammation in irritant as well as allergic contact dermatitis, in experimental studies and in clinical settings. In the following the methodological aspects of LDF within the area of contact dermatitis will be presented, and experimental studies attending different areas of contact dermatitis involving LDF will be reviewed.

II. METHODOLOGICAL ASPECTS

A. DOSE-RESPONSE STUDIES

The utility of LDF for assessment of inflammation in contact dermatitis has been noted in a substantial number of papers. Originally, Nilsson et al. studied skin irritancy caused by sodium lauryl sulfate (SLS) and found a clear relationship between applied doses of SLS and recorded blood flow values.[2] This finding was later confirmed by other groups.[3-5] Blanken et al., however, found no relationship between blood flow and SLS dose.[6] In this study only one measurement was obtained from each test site. Due to local variation in the microcirculation, this may give less valid results than repetitive measurements within the test area and may explain the reported lack of correlation between LDF and applied dose of SLS.[6]

LDF was recently used for evaluation of dose-response relationships for topically applied antigens.[7] Patch test responses to serially diluted nickel sulfate and potassium dichromate solutions in patients with nickel and chrome sensitivity were assessed by LDF, and a positive dose-response relationship was reported.[7]

B. COMPARISON TO CLINICAL EVALUATION

A clear relationship between LDF-recorded blood flow values and visual scores has been reported for SLS-induced skin irritancy.[2] Willis et al. examined irritant reactions from seven different irritants and found a close correlation between visual scoring and the recorded LDF values.[8] For patch test reactions with no visible erythema, blood flow values similar to those of normal skin were reported.[8] Increased blood flow in patch test reactions without changes observed by the naked eye was, however, reported by Wahlberg and Wahlberg,[9,10] implying that the eye may be less reliable than LDF for assessment of skin irritancy.

Staberg et al. found that differentiation between positive, doubtful, and negative allergic standard patch test reactions was possible by LDF.[11] This observation was recently supported in a patch test study on guinea pigs.[12] Due to its high sensitivity the LDF technique may be useful in detection of allergic contact dermatitis in cases of doubtful patch test reactions. However, one should be aware that the patch test procedure itself may give rise to increased blood flow for a period of 24 to 48 h after removal of patch tests.[9,10] With the introduction of the LDF technique the interpretation of positive, negative, or uncertain patch test reactions has moved to another level which may or may not be more reliable than observations obtained by visual assessment.

C. COMPARISON TO OTHER BIOENGINEERING METHODS

A number of different noninvasive bioengineering methods have been used for quantification of the inflammatory reaction in contact dermatitis.[13,14] In a study comparing LDF to visual assessment, evaporimetry, and measurement of skin color, Wilhelm et al. found evaporimetry to be the best-suited method overall for evaluation of SLS-induced skin damage.[3] This finding was supported by another study, also using SLS-induced skin damage as a model and

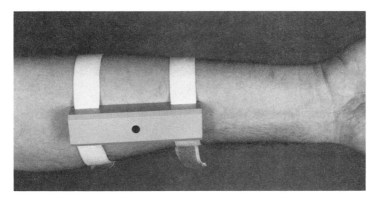

Figure 1 Plastic block used as a probe holder, 25 mm wide × 100 mm long × 15 mm high. The plastic block is fastened around the arm with Velcro® straps. (From Olivarius, F., Agner, T., and Menné, T., *Br. J. Dermatol.*, 129, 554, 1993. With permission.)

including comparison with ultrasound A-scan for quantification of the inflammatory response.[4] Evaporimetry was again found to be the most sensitive method, while the sensitivity of LDF or ultrasound A-scan varied according to time and dose of SLS. LDF was reported to be more sensitive for detection of skin damage than measurement of erythema by colorimetry.[4] However, this conclusion is limited to SLS-induced irritant contact dermatitis. Other irritants may affect the skin in different ways and may give rise to a different order of preference of the techniques. A chemical eliciting a high transepidermal water loss (TEWL) response is not necessarily erythematogenic, while another chemical may cause intense inflammation but hardly harm the barrier.[15]

LDF and colorimetry for assessment of skin irritation were compared in a SLS irritancy study on human volunteers.[16] For color measurements the a* value in the standard tristimulus system recommended by the Commission Internationale de l'Éclairage was used. Although the flowmeter and the colorimeter, as discussed below, measure somewhat different features—i.e., cutaneous microcirculation and erythema, respectively, which may often be but are not always identical—a significant correlation between LDF and color a* was found.[16] LDF and erythema index were recently compared in test reactions caused by SLS and benzalkonium chloride, and a significantly positive correlation was reported (r = 0.55, *p* < 0.001).[17]

D. VARIATION AND SOURCES OF ERROR

Reproducibility of LDF measurements on irritant contact dermatitis reactions was studied in different irritant patch test models, and coefficients of variation of 11.3% and 11.6% for normal and for SLS-exposed skin, respectively, were reported.[18] In a study by Freeman and Maibach, however, wide fluctuations in laser-Doppler-recorded blood flow values in response to SLS patches were found due to spotty erythema.[19] When using a probe covering only a small skin surface area, averaging of three or more recordings is necessary to overcome spatial variation in cutaneous blood supply. A probe with a skin-to-probe distance of 4 mm has been proposed to allow for a larger area of investigation.[20] However, this may cause scattering of light in the tissue surface directly into the fibers. Repeated applications of a probe with adhesive tape to the skin may lead to increased inflammation in the test area and should be avoided. In our laboratory we have used a small plastic block with a hole for the probe and Velcro® straps for fastening around the arm[21] (Figure 1). This probe holder ensures a stable position of the probe during measurements without the use of adhesive tape, but its use is clearly limited to measurements on the extremities.

Changes in the skin surface, such as vesicles, bullae, crusts, hyperkeratosis, and scaling, may influence the optics of the skin and the laser signal and thus impair the value and decrease the sensitivity of the measurements. For strong reactions edema may compress the vasculature and restrict the blood flow. In this case the recorded LDF value will underestimate the inflammatory response.

Cutaneous blood flow is easily influenced by physical activity and mental stress.[22] Preconditioning of subjects is necessary, preferably 30 min resting in a horizontal position. It is recommended that measurements are performed in a quiet environment in the absence of important audiovisual and other mental stimuli. In order to control the fluctuations due to temporal variations in skin blood flow it has been proposed to compare the reactivity of the tested skin each time with an untreated control site.

III. EXPERIMENTAL STUDIES

A. EVALUATION OF SKIN REACTIVITY

Skin reactivity to irritants is influenced by environment-related and individual-related physiological and pathophysiological variables. LDF has been used in a number of experimental studies for investigations in this field.

Regional variation in skin reactivity is well known, and a new aspect of this was recently looked into in a study comparing immediate contact reactions at different skin sites by use of LDF.[23] The special properties of vulvar skin were studied by Elsner et al., who found that blood flow in vulvar skin increased significantly less than in forearm skin after SLS exposure.[24] Spatial variation in skin reactivity within the same body region may be overcome in experimental studies by changing the position of the test sites and the control site according to a previously determined program for each individual. Contralateral observations, however, are equal.[18]

Influence of race on skin reactivity to different irritants was studied in black and white skin exposed to patch tests with SLS on normal and preoccluded skin. On the preoccluded test sites black skin was less reactive than white skin to low doses of SLS.[25] Examination of white and Hispanic skin revealed no differences in the microcirculatory response to SLS-induced irritancy between the two skin types.[26]

Differences in skin reactivity between males and females were investigated in several studies using LDF for quantification of inflammatory response to irritant stimuli.[27,28] No differences were found by LDF or by other bioengineering methods. In females a cyclic variation in skin reactivity to SLS was reported, with increased reactivity at day 1 in the menstrual cycle compared to days 9 to 11.[29] Although LDF values were increased at day 1 compared to days 9 to 11, the difference was not statistically significant, as was found for TEWL values and skin thickness.[29] This is in accordance with the finding that LDF is a less sensitive method for detection of SLS-induced skin irritation than evaporimetry and ultrasound scanning.[4]

Presence of an active eczema may, even if limited to a small skin area, cause a generalized hyperreactivity of the skin, including areas distant from the eczema. This was examined in a study including patients with acute and chronic hand eczema and a control group.[30] Patients with active eczema had an increased inflammatory response to SLS as compared to patients with chronic eczema or controls when evaluated by LDF and other bioengineering methods.[30]

B. EVALUATION OF SKIN RESPONSE TO MOISTURIZERS AND ANTI-INFLAMMATORY DRUGS

A recent study evaluated the effect of regular use of emollients on irritant contact dermatitis by measurement of TEWL and LDF.[31] A statistically significant difference in favor of the use of emollients was validated by TEWL; LDF recordings showed the same tendency, although statistical significance was not achieved.[31] The irritant dermatitis was caused by exposing the skin to a liquid detergent, and TEWL is a more sensitive method for measurement of skin response to this type of irritancy than is LDF.[3,4]

The irritant properties of urea in different vehicles were studied by several bioengineering methods, including LDF, and the irritant effect was found to be less pronounced with water as vehicle as compared to petrolatum.[32]

Evaluation of the anti-inflammatory effect of topical nonsteroidal anti-inflammatory drugs,[33] systemic nonsteroidal anti-inflammatory drugs,[34] and topical corticosteroids[35] by LDF was reported to be valuable.

C. ANIMAL STUDIES

The LDF technique also has been utilized for measurements on animal skin in experimental studies of contact dermatitis. In a guinea pig model, skin irritancy due to open application of low concentrations of SLS was evaluated by LDF and evaporimetry, as well as by macroscopic and microscopic examination, and results from the different assessment methods were compared.[36] Although an overall good correlation was found between the methods, examples of differences between macroscopic findings and LDF results were observed.[36] This emphasizes that LDF measures microcirculation in the skin and not erythema, although these parameters often run parallel. However, an increase in blood flow in the deeper vessels sometimes may be recorded in the absence of erythema, while erythema is probably caused more by capillary accumulation of blood. The difference may be more pronounced in animal studies, but has also been observed in a human model.[16] Recently, LDF was reported to be helpful in distinguishing positive from negative and uncertain patch test reactions when investigating the dose-response relationship in phases of induction and challenge in guinea pigs.[12]

D. DIFFERENTIATION BETWEEN IRRITANT AND ALLERGIC CONTACT DERMATITIS

Although many attempts to differentiate between allergic and irritant reactions have been accomplished, until now none has been successful.[37] Staberg and Serup studied skin reactions to SLS and nickel by LDF and found no consistent differences.[5] Gawkrodger et al.[17] compared irritant and allergic reactions by measurement of LDF and erythema index and concluded that the reactions could not be differentiated on the basis of these methods, although in mild irritant reactions a disproportionate increase in erythema index over laser Doppler flow was found.[17] Irritant skin responses depend on the chemical properties of the applied irritant and may differ in clinical and histological appearance as well as in functional and kinetic properties.[15]

E. INFLUENCE OF ULTRAVIOLET RADIATION

The influence of ultraviolet (UV) radiation on immediate contact urticaria reactions as well as irritant and allergic contact dermatitis was thoroughly studied by Larmi and associates using the LDF technique. Psoralen-ultraviolet A (PUVA) therapy was found to suppress non-immunologic immediate contact reactions induced by benzoic acid.[38] UVB was reported to diminish immediate and delayed response to dimethyl sulfoxide (DMSO) and delayed response to SLS, but not to phenol.[39] A diminished nonimmunologic contact reaction to benzoic acid was also reported after exposure to UVA alone, as well as after UVB.[40]

F. OCCUPATIONAL DERMATOLOGY

Within the area of occupational dermatology objective measurements often may be valuable to certify pathological changes in the skin following occupational exposure. For this purpose LDF may be useful. The irritancy potential of rock wool was recently studied.[41] Patch tests with rock wool fibers of different diameters were evaluated by LDF after 8, 24, and 48 h. After 48 h fibers with the largest diameter were found to cause significantly increased LDF values compared to the smaller fibers.[41]

de Boer et al.[42] studied irritant contact dermatitis from metalworking fluids by measurement of LDF and concluded that although a single exposure only causes marginal skin irritation, contact with metalworking fluids may play a crucial role in occupational contact dermatitis caused by cumulative subclinical insults.[42]

IV. COMMENTS

Today laser Doppler flowmetry is established as a precise and objective method for quantification of cutaneous microcirculation. Within the area of contact dermatitis the LDF technique has been used successfully for quantification of inflammatory responses in experimental studies on irritant and allergic contact dermatitis. Test reactions can be followed for a period of time without influencing the outcome of the test. The method has proved to be highly sensitive and specific.

With the increasing number of noninvasive measuring methods commercially available today, objective assessment rather than visual assessment of test results in experimental studies of contact dermatitis should be used as a general rule. For assessment of inflammatory response the LDF technique is an attractive alternative. In clinical settings, however, where preconditioning of subjects and adjustment of measuring circumstances generally are difficult to achieve, the role of LDF is more doubtful. If correct measuring circumstances cannot be secured the results will be biased by increased variation of measurements. This variation is related to physiological parameters, i.e., changes in the cutaneous microcirculation, and cannot be overcome by technical advantages. For experimental studies in the future the new scanning laser Doppler flowmeter may become an alternative. A recent report compares the scanning LDF with the traditional technique and finds the new LDF technique promising.[43]

REFERENCES

1. Shepherd, A. P., History on LDV flowmetry, in *Laser Doppler Blood Flowmetry,* Shepherd, A. P. and Öberg, P. Å., Eds., Kluwer Academic Publishers, Boston, 1990, 1.
2. Nilsson, G. E., Otto, U., and Wahlberg, J. E., Assessment of skin irritancy in man by laser Doppler flowmetry, *Contact Dermatitis,* 8, 401, 1982.
3. Wilhelm, K. P., Surber, C., and Maibach, H. I., Quantification of sodium lauryl sulphate dermatitis in man: comparison of four techniques: skin color reflectance, transepidermal water loss, laser Doppler flow measurement and visual scores, *Arch. Dermatol. Res.,* 281, 293, 1989.
4. Agner, T. and Serup, J., Sodium lauryl sulphate for irritant patch testing—a dose-response study using bioengineering methods for determination of skin irritation, *J. Invest. Dermatol.,* 95, 543, 1990.
5. Staberg, B. and Serup, J., Allergic and irritant skin reactions evaluated by laser Doppler flowmetry, *Contact Dermatitis,* 18, 40, 1988.
6. Blanken, R., van der Valk, P. G. M., and Nater, J. P., Laser-Doppler flowmetry in the investigation of irritant compounds on human skin, *Dermatosen,* 34, 5, 1986.
7. Eun, H. C. and Marks, R., Dose-response relationships for topically applied antigens, *Br. J. Dermatol.,* 122, 491, 1990.
8. Willis, C. M., Stephens, J. M., and Wilkinson, J. D., Assessment of erythema in irritant contact dermatitis. Comparison between visual scoring and laser Doppler flowmetry, *Contact Dermatitis,* 18, 138, 1988.
9. Wahlberg, J. E. and Wahlberg, E., Patch test irritancy quantified by laser Doppler flowmetry, *Contact Dermatitis,* 11, 257, 1984.

10. Wahlberg, J. E. and Wahlberg, E., Quantification of skin blood flow at patch test sites, *Contact Dermatitis,* 17, 229, 1987.

11. Staberg, B., Klemp, P., and Serup, J., Patch test responses evaluated by cutaneous blood flow measurements, *Arch. Dermatol.,* 120, 741, 1984.

12. Ll, Q., Aoyama, K., and Matsushita, T., Evaluation of contact allergy to chemicals using laser Doppler flowmetry (LDF) technique, *Contact Dermatitis,* 26, 27, 1992.

13. Serup, J., Characterization of contact dermatitis and atopy using bioengineering techniques. A survey. *Acta Derm. Venereol. Suppl.,* 177, 14, 1992.

14. Berardesca, E. and Maibach, H. I., Bioengineering and the patch test, *Contact Dermatitis,* 18, 3, 1988.

15. Agner, T. and Serup, J., Skin reactions to irritants assessed by non-invasive, bioengineering methods, *Contact Dermatitis,* 20, 352, 1989.

16. Serup, J. and Agner, T., Colorimetric quantification of erythema—a comparison of two colorimeters (Lange Micro Color and Minolta Chroma Meter CR-200) with a clinical scoring scheme and laser Doppler flowmetry, *Clin. Exp. Dermatol.,* 15, 267, 1990.

17. Gawkrodger, D. J., McDonagh, A. J. G., and Wright, A. L., Quantification of allergic and irritant patch test reactions using laser-Doppler flowmetry and erythema index, *Contact Dermatitis.,* 24, 172, 1991.

18. Agner, T. and Serup, J., Individual and instrumental variations in irritant patch-test reactions—clinical evaluation and quantification by bioengineering methods, *Clin. Exp. Dermatol.,* 15, 29, 1990.

19. Freeman, S. and Maibach, H. I., Study of irritant contact dermatitis produced by repeat patch testing with sodium lauryl sulfate and assessed by visual methods, transepidermal water loss and laser Doppler velocimetry, *J. Am. Acad. Dermatol.,* 19, 496, 1988.

20. de Boer, E. M., Bezemer, P. D., and Bruynzeel, D. P., A standard method for repeated recording of skin blood flow using laser Doppler flowmetry, *Dermatosen,* 37, 58, 1989.

21. Olivarius, F., Agner, T., and Menné, T., Skin barrier function and dermal inflammation, *Br. J. Dermatol.,* 129, 554, 1993.

22. Bircher, A., de Boer, E. M., Agner, T., Wahlberg, J. E., and Serup, J., Guidelines for measurement of cutaneous blood flow by laser Doppler flowmetry. A report from the standardization group of the European Contact Dermatitis Society, *Contact Dermatitis,* 30, 65, 1994.

23. Larmi, E., Lahti, A., and Hannuksels, M., Immediate contact reactions to benzoic acid and the sodium salt of pyrrolidone carboxylic acid. Comparison of various skin sites, *Contact Dermatitis,* 20, 38, 1989.

24. Elsner, P., Wilhelm, D., and Maibach, H., Multiple parameter assessment of vulvar irritant contact dermatitis, *Contact Dermatitis,* 23, 20, 1990.

25. Berardesca, E. and Maibach, H. I., Racial differences in sodium lauryl sulphate induced cutaneous irritation: black and white, *Contact Dermatitis,* 18, 65, 1988.

26. Berardesca, E. and Maibach, H. I., Sodium-lauryl-sulphate-induced cutaneous irritation. Comparison of white and Hispanic subjects, *Contact Dermatitis,* 19, 136, 1988.

27. Lammintausta, K., Maibach, H. I., and Wilson, D., Irritant reactivity in males and females, *Contact Dermatitis,* 17, 276, 1987.

28. Agner, T., Basal transepidermal water loss, skin thickness, skin blood flow and skin color in relation to sodium-lauryl-sulphate-induced irritation in normal skin, *Contact Dermatitis,* 25, 108, 1991.

29. Agner, T., Damm, P., and Skouby, S. O., Menstrual cycle and skin reactivity, *J. Am. Acad. Dermatol.,* 24, 566, 1991.

30. Agner, T., Skin susceptibility in uninvolved skin of hand eczema patients and healthy controls, *Br. J. Dermatol.,* 125, 140, 1991.

31. Hannuksela, A. and Kinnunen T., Moisturizers prevent irritant dermatitis, *Acta Derm. Venereol.,* 72, 42, 1992.

32. Agner, T., An experimental study of irritant effects of urea in different vehicles, *Acta Derm. Venereol. Suppl.,* 177, 44, 1992.
33. Johansson, J. and Lahti, A., Topical non-steroidal anti-inflammatory drugs inhibit non-immunologic immediate contact reactions, *Contact Dermatitis,* 19, 161, 1989.
34. Duteil, L., Queille, C., Poncet, M., and Czernielewski, J., Processing and statistical analysis of laser Doppler data applied to the assessment of systemic anti-inflammatory drugs, *J. Dermatol. Sci.,* 2, 376, 1991.
35. Queille-Roussel, C., Duteil, L., Padilla, J., Poncet, M., and Czernielewski, J., Objective assessment of topical anti-inflammatory drug activity on experimentally induced nickel contact dermatitis: comparison between visual scoring, colorimetry, laser Doppler velocimetry and transepidermal water loss, *Skin Pharmacol.,* 3, 248, 1990.
36. Frödin T. and Anderson, C., Multiple parameter assessment of skin irritancy, *Contact Dermatitis,* 17, 92, 1987.
37. Brasch, J., Burgard, J., and Sterry, W., Common pathogenetic pathways in allergic and irritant contact dermatitis, *J. Invest. Dermatol.,* 98, 166, 1992.
38. Larmi, E., PUVA treatment inhibits nonimmunologic immediate contact reactions to benzoic acid and methyl nicotinate, *Int. J. Dermatol.,* 28, 609, 1989.
39. Larmi, E., Lahti, A., and Hannuksela, M., Effect of ultraviolet B on nonimmunologic contact reactions induced by dimethyl sulfoxide, phenol and sodium lauryl sulphate, *Photodermatology,* 6, 258, 1989.
40. Larmi, E., Lahti, A., and Hannuksela, M., Ultraviolet light inhibits nonimmunologic immediate contact reactions to benzoic acid, *Arch. Dermatol. Res.,* 280, 420, 1988.
41. Eun, H. C., Lee, H. G., and Paik, N. W., Patch test responses to rock wool of different diameters evaluated by cutaneous blood flow measurements, *Contact Dermatitis,* 24, 270, 1991.
42. de Boer, E. M., Scholten, J. P. M., van Ketel, W. G., and Bruynzeel, D. P., The irritancy of metal working fluids: a laser Doppler flowmeter study, *Contact Dermatitis,* 22, 86, 1990.
43. Quinn, A. G., McLalland, J., Essex, T., and Farr, P. M., Measurement of cutaneous inflammatory reactions using a scanning laser Doppler velocimeter, *Br. J. Dermatol.,* 125, 30, 1991.

Chapter 6

Skin Pharmacology

Andreas J. Bircher

CONTENTS

I. INTRODUCTION

Blood flow has been measured on virtually all human organ surfaces such as the nasal and gastrointestinal mucosa, in the retina, and on bone and muscular tissue; the skin, however, remains the most readily accessible organ. The skin microcirculation plays an important role in physiologic processes and is affected in a variety of skin and other disorders. There is, therefore, considerable interest in measuring cutaneous blood flow (CBF) objectively under physiologic conditions and in pathologic disorders, and an extensive literature on many aspects of CBF has been published in recent years. Laser Doppler flowmetry (LDF) has been used extensively in basic research and clinical investigations. It permits noninvasive, continuous determination of the superficial CBF and is therefore also particularly suited to observe the influence of topically or systemically applied pharmacological agents on CBF or to study the pharmacodynamics of vasoactive drugs and compounds. This chapter, based on an earlier review,[1] presents some newer aspects concerning skin pharmacology.

II. SUBJECT-RELATED VARIABLES

Some individual and environmental factors may influence baseline and stimulated CBF. This has been reviewed recently and recommendations to circumvent such problems have been given.[2] Of particular interest with regard to skin pharmacology are age, gender, and possibly ethnic background, and several studies have addressed these issues.

0-8493-8371-4/95/$0.00+$.50
© 1995 by CRC Press, Inc.

The CBF response to topically applied methyl nicotinate (MN), a potent vasodilator, was evaluated by laser Doppler velocimetry (LDV) and photopulse plethysmography (PPG) in 20- to 30-year-old black and white individuals and in 63- to 80-year-old white individuals.[3] Only the magnitude of the peak response revealed some significant differences; the other parameters evaluated gave similar results. In a further study on 20- to 30-year-old Caucasian, Oriental, and black subjects, the response to different topical doses of MN was measured by visual scoring and by LDV. Again only one parameter (area under the response-time curve) was significantly different when black and Oriental subjects were compared with the white individuals.[4] On the contrary, methyl and hexyl nicotinate applied under different conditions (untreated, occlusion, delipidized) to the skin of black and white subjects[5] gave variable results. Especially in untreated and preoccluded sites a significantly higher CBF response in whites was seen.

Human skin undergoes several changes with age. Several histological and biochemical changes within this tissue have been identified. There are controversial results with regard to the age-related changes in baseline and stimulated CBF in aging skin.[6] For example, in young white and black and in old white individuals, 100 mM MN was applied for a short period and the response followed by LDF. Except for the magnitude of the peak response, a remarkable similarity of the response in the different subject populations was found.[3] In an investigation of the dose-response behavior to topical MN in a young and an old age group, few significant differences in the area under the response-time curve were observed. This was due at least in part to the high interindividual variability of CBF responses to the vasodilator. The results indicate a trend that "older" individuals show an attenuated response as determined by the response onset, the peak reaction, the magnitude of the reaction, and possibly also the clearance of the compound.[7]

III. PERCUTANEOUS ABSORPTION

Nicotinic acid and its derivatives have served as model substances for the investigation of the route and kinetics of percutaneous absorption.[8] In particular, methyl nicotinate is the most widely used agent; it has been used to address questions with regard to skin pharmacokinetics and penetration enhancement and also to evaluate the influence of topical and systemic drugs on MN-induced vasodilation. The mechanism of the vasodilative action of nicotinic acid derivatives is not completely clear. In a study using the hamster cheek pouch a direct drug action on vascular smooth muscle cells and mediation by local nerve conduction was suggested.[9] However, there is also evidence for involvement of second messenger substances. For example, pretreatment of facial flushing induced by nicotinic acid with oral indomethacin[10] and aspirin[11] assessed by skin temperature changes reduced the flushing significantly. The cutaneous vascular response to topical application of several MN concentrations determined by LDF was substantially suppressed by oral prostaglandin synthetase inhibitors, but not by histamine antagonists or a placebo.[12] The erythematous reaction to topical MN, benzoic acid, and other agents inducing a nonimmunologic contact urticaria was also suppressed by acetylsalicylic acid,[13] but not by terfenadine.[14] These findings imply that prostaglandins are involved, at least in part, in nicotinate-induced vasodilation.

LDF has been extensively employed to study the pharmacodynamic effects of vasodilating agents such as methyl nicotinate on the cutaneous microcirculation.[15] The percutaneous absorption was investigated by measuring the CBF response to increasing topical doses of aqueous solutions of MN. With increasing concentration the microcirculatory response was saturable such that the duration of the effect, rather than the peak response, was enlarged.[16] In a LDF study, no local difference in CBF changes, stimulated by three concentrations of hexyl nicotinate, at two skin sites, although in some individuals an insignificant dose response was found.[17]

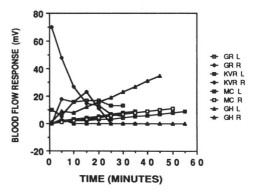

Figure 1 Cutaneous blood flow response to topically applied water. Legend markers are subjects' initials and indicate left (L) or right (R) arm. (From Bircher, A., Roskos, K., Maibach, H., and Guy, R., *Int. J. Pharm.*, 45, 263, 1988. With permission.)

IV. TOPICAL DRUGS AND AGENTS

A. VEHICLE EFFECTS

Supposedly ''inactive'' vehicle agents such as water or ethanol may themselves influence the CBF. In a study the selection of an appropriate vehicle for studies with nicotinates was emphasized. Water and lower aliphatic alcohols, but not kerosene, propylene, or polyethylene glycol, elicited an inconsistent and irreproducible vasoresponse.[18] Water[19] and ethanol[20] may induce inconsistent changes in CBF, but these effects are usually of a short duration and small extent (Figure 1). In other investigations a significant stimulation of CBF by water was found,[21,22] while in yet other studies no influence of water on CBF was found.[23,24] Probably different skin and vehicle temperatures and occlusive conditions account for the observed variations.

B. PENETRATION ENHANCERS

Methyl nicotinate has been used as a potent vasodilating agent to investigate the influence of a variety of systemic and topical drugs and substances with the potential to enhance percutaneous absorption of CBF. In one study the effects of two topical penetration enhancers, dimethyl sulfoxide and 2-pyrrolidone, were studied by the application of hexyl nicotinate incorporated into propylene glycol and isopropanol as vehicles. Only 2-pyrrolidone showed a significant penetration increase as measured by the nicotinate effect.[25] Another study involved pretreatment of the skin site with lauracapram or 2-pyrrolidone[26] followed by challenge with a hexyl nicotinate solution. The CBF response was significantly increased by simple occlusion alone and was further elevated by pretreatment with either penetration enhancer.

C. MINOXIDIL®

Vasoactive drugs have been evaluated by LDF with variable results. In one study, 1 to 5% concentrations of minoxidil in a propylene glycol/ethanol/water vehicle increased CBF in balding scalps in a dose-dependent manner.[19] On the other hand, in another investigation 3% minoxidil in an ethanol/propylene glycol/water base did not increase CBF, whereas 0.1% hexyl nicotinate used as a control in the same vehicle triggered vasodilation.[27]

D. NITROGLYCERIN

Inconsistent results also have been reported with nitroglycerin. Topical application of a 2% nitroglycerin ointment increased CBF significantly over baseline.[28] The peak response, however, did not correlate with the application time, and even repetitive applications did not further increase the CBF response. It has been suggested that nitroglycerin is slowly delivered into the dermis. Subsequently, CBF was monitored for 10 h at the application site of a

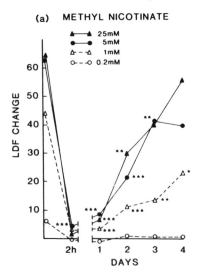

Figure 2 Effect of 2 g aspirin on cutaneous blood flow induced by four concentrations of methyl nicotinate, as measured by laser Doppler flowmetry (LDF), $*p < 0.05$, $**p < 0.01$, $***p <$ 0.001. (From Kujala, T. and Lahti, A., *Contact Dermatitis*, 21, 60, 1989. With permission.)

transdermal nitroglycerin delivery system. A rapid local increase of CBF to a sustained, relatively constant level was found. Despite the occurrence of nitroglycerin-related headaches, CBF on the contralateral arm remained at baseline levels; therefore, a systemically mediated action on CBF was not found.[28] On the contrary, during a 24-h administration via a nitroglycerin patch on the chest, a 25% increase of CBF on the forehead was measured throughout the dosing period. The majority of subjects experienced mild to severe headache, indicating a systemic drug action.[29]

E.   LOCAL ANESTHETICS

The effects of histamine on the microvasculature were evaluated by applying a topical eutectic mixture of prilocaine and lidocaine (EMLA®). The local anesthetic greatly diminished the CBF reaction and accelerated the decay of the response, whereas the weal was not affected. This implies that the flare is mediated more by neurogenic mechanisms than directly by histamine.[30] The action of amethocaine (another local anesthetic which can be administered topically) on CBF was investigated. Some local anesthetics induce vasodilation themselves; therefore, amethocaine was studied in varying concentrations. The higher the concentration the more pronounced was the CBF response. The CBF, however, did not parallel the magnitude of the local anesthesia.[31] Upon intradermal injection a new local anesthetic, ropivacaine, induced a relatively weak vasodilation as measured by LDF. When ropivacaine was combined with epinephrine the vasoconstrictive effect of epinephrine was diminished.[32]

V.   SYSTEMIC DRUGS AND AGENTS

Methyl nicotinate and benzoic acid have been used particularly to study CBF measured by LDF in nonimmunologic contact urticaria reactions. The influence of several systemic compounds has been tested in this model.

A.   NONSTEROIDAL ANTI-INFLAMMATORY DRUGS

The influence of 2 g acetylsalicylic acid on contact urticarial reaction elicited by methyl nicotinate and a series of other compounds was studied in healthy subjects. A significant decrease of the CBF in the reactions was measured with all test substances, indicating that arachidonic acid metabolites might play a common role in such reactions.[13] In a later study it was shown that the action of a single 2-g dose of acetylsalicylic acid decreased CBF stimulated by methyl nicotinate and benzoic acid up to 4 days previously[33] (Figure 2). There-fore, it was suggested that administration of nonsteroidal anti-inflammatory drugs (NSAIDs) should be stopped several days, or possibly longer, prior to experiments involving such agents. This was confirmed for acetylsalicylic acid in another investigation with methyl nicotinate. Treatment with indomethacin and tiaprofenic acid, two potent prostaglandin syn-thesis inhibitors, gave intermediate results.[34]

B. MISCELLANEOUS SYSTEMIC AGENTS

A 1-mg dose of peroral prazosin caused a sudden increase of forehead blood flow for ap-proximately 1 h, while 6 mg norepinephrine induced a significant decrease lasting up to 6 h.[29] On the other hand, 5 mg sublingual nitroglycerin[35] increased CBF transiently on the forearm, with a rapid onset within minutes and a return to baseline levels after 9 to 10 min.

Infusion of arginine vasopressin[36] caused facial pallor, reduction of facial temperature, and a significant decrease of interdigital CBF as measured by LDF; forearm and finger blood flow rose slightly, whereas heart rate did not change. This is in agreement with other find-ings[37] where elevated vasopressin levels and decreased CBF after smoking were observed. The same group,[38] on the other hand, found a prolonged exogenous vasopressin-induced slowing of the heart rate and a significant decrease of CBF in the forearm of healthy subjects. The administration of a vasopressin antagonist did not change any hemodynamic parameters, indicating that unstimulated endogenous vasopressin has no sustaining action on blood pressure.

In several studies it has been shown that smoking decreases finger CBF significantly, probably by inducing vasopressin. Transdermally administered nicotine, on the other hand, did not change the CBF, heart rate, or cardiac stroke volume.[39]

Facial flushing is a physiologic sign of heat exposure and of emotional distress, or it can be a reaction due to pathologic enzyme activity or a hormonal disorder. In some individuals, particularly those of Oriental origin, flushing after alcohol consumption is common due to a low activity of alcohol dehydrogenase and/or aldehyde dehydrogenase. Other substances causing flushing reactions are nicotinic acid and possibly monosodium glutamate. In patients with an underlying enzyme abnormality the flushing response was increased after oral prov-ocation with ethanol.[40] LDF was found to be superior to thermometry in quantification of alcohol-provoked flushing with regard to sensitivity and specificity; the methods, however, were significantly correlated with each other and with clinically apparent flushing.[41] A non-selective β-antagonist, nadolol, was used to treat experimentally induced flushing caused by hot water, ethanol, and nicotinic acid. Although nadolol had a clear influence on the systemic vasculature, no effects were observed on the flushing.[42] The flushing as one symptom of the "Chinese restaurant syndrome" was examined with increasing doses of monosodium glu-tamate, the agent thought to cause the reaction. No flushing reactions were observed in 18 individuals with a positive history with doses up to 3 g monosodium glutamate. In a pro-spective study, up to 18.5 g monosodium glutamate was tolerated without flushing, although some subjects experienced other symptoms. It was concluded that, in contrast to earlier beliefs, monosodium glutamate-induced flushing is rare.[43]

VI. MEDIATORS

All of the mediators discussed have a considerable influence on CBF, mostly by inducing a strong vasodilation. This results either in an isolated erythema or in a weal and flare response. Visual scoring and planimetry are still the standard methods used to determine the reaction of the cutaneous vasculature to topically applied vasodilating mediators. Methods such as LDF and others allow a semiquantitative determination of such reactions and a more sensitive evaluation of the dose-response relationship. The injection trauma itself, however, induces a considerable although short-lived LDF response and has to be taken into consideration. This has been circumvented in some experiments by epicutaneous or iontophoretically applied substances.

A. HISTAMINE

In an early LDF study with histamine, CBF increase at the injection site was higher for the needle stick than for saline and histamine.[44] In another study the discrimination between different concentrations of histamine was possible only when measuring sites were chosen at different distances from the injection point. Directly at the injection site, discrimination between saline and histamine was not possible; only measurements which were temporally and spatially separated permitted deconvolution of the traumatic and mediator-induced effect. The best quantification of the response was achieved by determinations 10 min postinjection, 5 mm from the weal center.[45] In another LDF investigation examining three intracutaneous histamine concentrations (6.5×10^{-5} $M, 10^{-4}$ M, and 10^{-3} M), measurement of CBF with LDF was found to have a good reproducibility based on repeat measurements.[46] Here the LDF response was quantified by integrating the CBF values measured at different distances from the injection site. With this approach it was possible to differentiate between the three histamine concentrations and the saline control for up to 60 min after administration. Skin reactivity to histamine was determined and shown to be significantly dependent on age, but not on sex and atopic diatheses. An increase in the response from early childhood to adulthood that then decreased to a plateau after the age of 60 was found.[47]

The immediate CBF response to histamine and allergens was examined in allergic patients. Skin prick tests with the negative control alone induced a short CBF increase. Histamine induced an immediate weal and flare response which lasted for 45 min, and the allergens resulted in a prolonged CBF increase for over 60 min.[48] In an extension of this investigation the immediate and the late cutaneous reaction to histamine and birch or grass pollen was studied in allergic individuals. The control, histamine, and allergens induced in all an initial short-lived CBF response. While the control and the histamine-stimulated CBF returned to baseline within 2 h, the allergen-stimulated blood flow remained significantly elevated above baseline for 6 and even 24 h. This indicates that in allergen-induced skin responses mediators other than histamine play a role in the late-phase reaction.[49] In the evaluation of the influence of several drugs, such as a topical corticosteroid (clobetasol-17-propionate), an H_1 antagonist (Loratadine®, or an α-adrenoceptor agonist (pseudoephedrine), on histamine- and allergen-induced CBF, variable results were obtained. The local pretreatment with the corticosteroid affected neither the visually scored weal and flare reaction nor the initial and late CBF response. As expected, the H_1 antagonist significantly decreased the weal and flare to histamine and allergen. The immediate and late CBF response to histamine, however, was not affected by pretreatment with the antihistamine or pseudoephedrine[50] (Figure 3). Evaluation of CBF response to histamine and compound 48/80, a histamine releaser, was performed in the finger of patients with Raynaud's disease. In this area, rich with arterio-venous anastomoses, no clear differences were found.[51] In a comparison of two laser systems, a helium-neon and an infrared laser, intradermally injected histamine increased CBF significantly at a distance of 5 to 10 cm from the injection site. There was also a clear dose-response relationship for histamine doses of 25, 27, 250, and 750 pmol per site.[52] In patients with acute atopic dermatitis and in controls the influence of iontophoretically administered histamine was measured. At higher currents the CBF response of the atopic individuals was significantly reduced compared to the controls, indicating an altered histamine response in atopics.[53]

B. EICOSANOIDS

The leucotrienes (LT) are particularly potent inflammatory mediators and may cause a sustained vascular reaction. The effects of intradermal equimolar concentrations of LTC_4, LTD_4, and histamine were studied in humans.[54] The measurements were made twice at a single site over 15 to 30 min; in that experimental setup all agents maximally dilated the vascular bed.

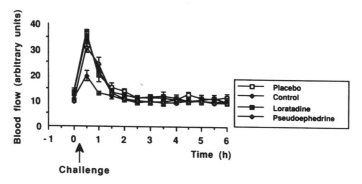

Figure 3 Cutaneous blood flow values (mean ± SEM) of histamine prick test after pretreatment with Loratadine® (an H1-receptor antagonist), pseudoephedrine, and a control. (From Hammarlund, A., Olsson, P., and Pipkorn, U., *Allergy,* 45, 64, 1990. With permission.)

In a dose-response study of LTD_4, planimetric measurement of erythema, LDF, and ^{133}Xe washout were compared. The latter two techniques were found to be superior to planimetric grading; ^{133}Xe washout was better than LDF in discriminating interindividual differences in CBF.[55] In an extensive investigation,[56] intradermal LTD_4 increased CBF equipotently to histamine, with a peak response within 5 min and a duration of 60 min. Late reactions were not observed. Saline always provoked lower CBF responses of a shorter duration than LTD_4. Pretreatment with H_1 and H_2 antagonists did not alter the CBF response to LTD_4.[56]

Similar investigations have been performed in animals; there are, however, apparent species differences in the response to LT. Intradermal LTB_4, LTF_4, prostaglandin (PG) E_2, and histamine were tested in pigs.[57] LTC_4 and LTD_4 induced vasodilation at doses of 1 ng and caused vasoconstriction at 1 μg. Indomethacin did not influence the CBF response to any agent, while saline induced a short-term 200 to 300% increase from the baseline CBF. Compared to human skin, the reaction in the pig skin is shorter; 1-μg doses of LTC_4 and LTD_4 induce vasoconstriction in the pig, whereas at that dose vasodilation occurs in human skin. Still the pig seems to be the model most relevant to human skin in which to study the effects of leucotrienes.

In type I allergic skin reactions PGD_2 and histamine are released into the local tissue and into the venous blood in cold and heat urticaria. PGD_2 increased CBF and the area of erythema in a dose-dependent manner. The amount of PGD_2 necessary to elicit a response was higher than its blood levels measured in urticaria, and its injection in combination with histamine did not significantly enhance the respective effect of either mediator alone on CBF and erythema.[58]

C. NEUROPEPTIDES

The action of the neuropeptide calcitonin-gene-related peptide (CGRP) on the cutaneous microvasculature was also monitored by LDF.[59] Compared to saline, a significant increase in CBF which gradually decreased over 4 h was found. Whereas CGRP induced the largest and most prolonged erythema, no weal and flare reactions were elicited by histamine, PGE_2, PGI_2, substance P, or vasoactive intestinal peptide. In healthy controls and in patients with primary and secondary Raynaud's phenomenon, intradermal injection of CGRP and substance P resulted in a significant and sustained vasodilation as measured by LDF.[51] The clinical use of CGRP in Raynaud's phenomenon was evaluated after intradermal and intravenous injection in controls and patients. In both groups intradermal injection resulted in significantly increased immediate and delayed CBF responses. Four patients were treated

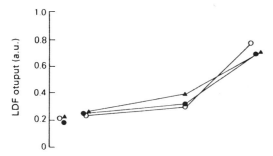

Figure 4 Cutaneous blood flow values following intradermal injection of bradykinin (0,1, 2.5, and 5µg) at 24 h after pretreatment with placebo (o), captopril (●) and enalapril (▲). LDF= Laser Doppler Flowmetry. (From Li Kam Wa, T. C., Cooke, E. D., and Turner, P., *Br. J. Clin. Pharmacol.*, 35, 8, 1993. With permission.)

intravenously with α-CGRP, which increased CBF considerably. Adverse effects of this short-term therapy were not observed.[60]

Again in patients with atopic dermatitis (AD) the symptoms and signs due to intradermally injected substance P and the topically applied substance P-releaser mustard oil were studied. In both groups a clear dose-response of the substance P doses (10^{-9} to 10^{-12}mol) was present, but in atopics the itch, the flare, and the CBF reaction were significantly reduced compared to healthy controls,[61] indicating an attenuated response in individuals with AD.

D. MISCELLANEOUS MEDIATORS

To circumvent the problem of CBF changes induced by the injection trauma or the alterations of CBF by the systemic action of intravenously administered agents, iontophoresis has been used to deliver norepinephrine into finger skin.[62] Additionally, this technique makes possible a continuous recording of CBF during the administration of the drug. With the same technique α-1 and α-2 adrenoceptors in finger skin were identified by using specific agonists and antagonists.[63]

The effects of iontophoretically applied acetylcholine and sodium nitrite were examined before and after several days of topical capsaicin pretreatment. Whereas neurovascular response to the electrically elicited axon reflex and the acute vasodilating effect of capsaicin were significantly reduced, no CBF changes due to acetylcholine and sodium nitrite were observed.[64]

Bradykinin causes a dose-dependent increase in CBF and weal formation after intradermal injection. Bradykinin also may increase prostaglandin synthesis by activating phospholipase A_2. Angiotensin converting enzyme (ACE) or kininase II is involved in the degradation of bradykinin. ACE inhibitors are widely used antihypertensive drugs, and some of their adverse effects, such as angioedema, rashes, and chronic cough, may be mediated by bradykinin. In several studies it has been demonstrated that ACE inhibitors such as captopril and enalapril potentiate the cutaneous effects of bradykinin.[65] Both drugs alone had no influence on CBF, but potentiated the weal and CBF changes induced by bradykinin over time periods up to 24 h [66] (Figure 4). Addition of indomethacin to captopril did not yield results different from those obtained by captopril treatment alone, indicating that cyclooxygenase products do not contribute to the cutaneous effects of bradykinin.[67]

VII. CONCLUSIONS

LDF is an interesting tool to study particularly changes in the cutaneous microcirculation. Its advantages are that it is noninvasive and can be used on virtually any skin surface. It also allows long-term, continuous surveillance of the action of vasoactive compounds on CBF, i.e., *in vivo* measurements in real time, which cannot be achieved with other methods. When some precautions with regard to the instrumental, individual, and environmental variables are taken and a well-defined stimulus, such as an exact dose of a vasoactive compound, is used, precise and reproducible measurement of its effects on skin vessels is possible. For some investigations, e.g., the study of irritation, LDF must be used in combination with other techniques, such as ultrasound, transepidermal water loss, and other measuring techniques.

REFERENCES

1. Bircher, A., Guy, R., and Maibach, H., Skin pharmacology and dermatology, in *Laser-Doppler Blood Flowmetry,* Shepherd, A. and Öberg, P., Eds., Kluwer Academic Publishers, Boston, 1990, 141.

2. Bircher, A., de Boer, E., Agner, T., Wahlberg, J., and Serup, J., Guidelines for the measurement of cutaneous blood flow by laser Doppler flowmetry, *Contact Dermatitis,* 30, 65, 1994.

3. Guy, R., Tur, E., Bjerke, S., and Maibach, H., Are there age and racial differences to methyl nicotinate-induced vasodilation in human skin?, *J. Am. Acad. Dermatol.,* 12, 1001, 1985.

4. Gean, C., Tur, E., Maibach, H., and Guy, R., Cutaneous responses to topical methyl nicotinate in black, Oriental and Caucasian skin, *Arch. Dermatol. Res.,* 281, 95, 1989.

5. Berardesca, E. and Maibach, H., Racial differences in pharmacodynamic response to nicotinates in vivo in human skin: black and white, *Acta Derm. Venereol.,* 70, 63, 1990.

6. Bircher, A., Roskos, K., Maibach, H., and Guy, R., Laser Doppler-measured cutaneous blood flow: effects with age, in *Aging Skin: Properties and Functional Changes;* Leveque, J. and Agache, P., Eds., Marcel Dekker, New York, 1993, 105.

7. Roskos, K., Bircher, A., Maibach, H., and Guy, R., Pharmacodynamic measurements of methyl nicotinate percutaneous absorption: the effect of aging on microcirculation, *Br. J. Dermatol.,* 122, 165, 1990

8. Albery, W. and Hadgraft, J., Percutaneous absorption: in vivo experiments, *J. Pharm. Pharmacol.,* 31, 140, 1979.

9. Fulton, G., Farber, E., and Moreci, A., The mechanism of action of rubefacients, *J. Invest. Dermatol.,* 33, 317, 1959.

10. Svedmyr, N., Heggelund, A., and Åberg, G., Influence of indomethacin on flush induced by nicotinic acid in man, *Acta Pharmacol. Toxicol.,* 41, 397, 1977.

11. Wilkin, J., Wilkin, O., Kapp, R., Donachie, R., Chernovsky, M., and Buckner, J., Aspirin blocks nicotinic acid-induced flushing, *Clin. Pharmacol. Ther.,* 31, 478, 1982.

12. Wilkin, J., Fortner, G., Reinhardt, L., Flowers, O., Kilpatrick, S., and Streeter, W., Prostaglandins and nicotinate-provoked increase in cutaneous blood flow, *Clin. Pharmacol. Ther.,* 38, 273, 1985.

13. Lahti, A., Väänänen, A., Kokkonen, E., and Hannuksela, M., Acetylsalicylic acid inhibits non-immunologic contact urticaria, *Contact Dermatitis,* 16, 133, 1987.

14. Lahti, A., Terfenadine does not inhibit non-immunologic contact urticaria, *Contact Dermatitis,* 16, 220, 1987.

15. Guy, R., Tur, E., Wester, R., and Maibach, H., Blood flow studies and percutaneous absorption, in *Percutaneous Absorption,* Bronaugh, R. L. and Maibach, H. I., Eds., Marcel Dekker, New York, 1985, 393.

16. Guy, R., Bugatto, B., Gaebel, C., Sheiner, L., and Maibach, H., Pharmacodynamic measurements of methyl nicotinate percutaneous absorption, *Pharm. Res.,* 1, 76, 1984.

17. Dowd, P., Whitefield, M., and Greaves, M., Hexyl nicotinate-induced vasodilation in normal human skin, *Dermatologica,* 174, 239, 1987.

18. Kohli, R., Archer, W., and Li Wan Po, A., Laser velocimetry for the non-invasive assessment of the percutaneous absorption, *Int. J. Pharm.,* 36, 91, 1987.

19. Bircher, A., Roskos, K., Maibach, H., and Guy, R., Laser Doppler velocimetric measurement of skin blood flow: a reply, *Int. J. Pharm.,* 45, 263, 1988.

20. Wester, R., Maibach, H., Guy, R., and Novak, E., Minoxidil stimulates cutaneous blood flow in human balding scalps: pharmacodynamics measured by laser Doppler velocimetry and photopulse plethysmography, *J. Invest. Dermatol.*, 82, 515, 1984.

21. Nilsson, G., Otto, U., and Wahlberg, J., Assessment of skin irritancy in man by laser Doppler flowmetry, *Contact Dermatitis*, 8, 401, 1982.

22. Blanken, R., van der Valk, P., and Nater, J., Laser-Doppler flowmetry in the investigation of irritant compounds on human skin, *Derm. Beruf Umwelt*, 34, 5, 1986.

23. Wahlberg, J., Erythema-inducing effects of solvents following epicutaneous administration to man-studied by laser Doppler flowmetry, *Scand. J. Work Environ. Health*, 10, 159, 1984.

24. Wahlberg, J. and Nilsson, G., Skin irritancy from propylene glycol, *Acta Derm. Venereol.*, 64, 286, 1984.

25. Guy, R., Carlstrøm, E., Bucks, D., Hinz, R., and Maibach, H., Percutaneous penetration of nicotinates: in vivo and in vitro measurements, *J. Pharm. Sci.*, 75, 968, 1986.

26. Ryatt, K., Stevenson, J., Maibach, H., and Guy, R., Pharmacodynamic measurement of percutaneous penetration enhancement in vivo, *J. Pharm. Sci.*, 75, 374, 1986.

27. Bunker, C. and Dowd, P., Alterations in scalp blood flow after the epicutaneous application of 3% minoxidil and 0.1% hexyl nicotinate in alopecia, *Br. J. Dermatol.*, 117, 668, 1987.

28. Stevenson, J., Maibach, H., and Guy, R., Laser Doppler and photoplethysmographic assessment of cutaneous microvasculature, in *Models in Dermatology*, Vol.3, Maibach, H. I. and Lowe, N., Eds., S. Karger, Basel, 1987, 121.

29. Sundberg, S. and Castrén, M., Drug- and temperature-induced changes in peripheral circulation measured by laser-Doppler flowmetry and digital-pulse plethysmography, *Scand. J. Clin. Lab. Invest.*, 46, 359, 1986.

30. Harper, E., Swanson Beck, J., and Spence, V., Effect of topically applied local anaesthesia on histamine flare in man measured by laser Doppler velocimetry, *Agents Actions*, 28, 192, 1989.

31. Woolfson, A., McCafferty, D., McGowan, K., and Boston, V., Non-invasive monitoring of percutaneous anaesthesia using laser Doppler velocimetry, *Int. J. Pharm.*, 51, 183, 1989.

32. Cederholm, I., Evers, H., and Lofstrom, J. B., Skin blood flow after intradermal injection of ropivacaine in various concentrations with and without epinephrine evaluated by laser Doppler flowmetry, *Reg. Anesth.*, 17, 322, 1992.

33. Kujala, T. and Lahti, A., Duration of non-immunologic immediate contact reactions by acetylsalicylic acid, *Contact Dermatitis*, 21, 60, 1989.

34. Queille-Roussel, C., Duteil, L., and Czernielewski, J., Evaluation par laser Doppler de l'effet des anti-inflammatoires non stéroidiens systémiques sur l'urticaire de contact induite par le nicotinate de méthyle, *Ann. Derm. Venereol.*, 116, 285, 1989.

35. Sundberg, S., Acute effects and long-term variations in skin blood flow measured with laser Doppler flowmetry, *Scand. J. Clin. Lab. Invest.*, 44, 341, 1984.

36. Wiles, P., Grant, P., and Davies, J., The differential effect of arginine vasopressin on skin blood flow in man, *Clin. Sci.*, 71, 633, 1986.

37. Waeber, B., Schaller, M., Nussberger, J., Bussien, J., Hofbauer, K., and Brunner, H., Skin blood flow reduction induced by cigarette smoking: role of vasopressin, *Am. J. Physiol.*, 247, H895, 1984.

38. Bussien, J., Waeber, B., Nussberger, J., Schaller, M., Gavras, H., Hofbauer, K., and Brunner, H., Does vasopressin sustain blood pressure of normally hydrated healthy volunteers?, *Am. J. Physiol.*, 246, H143, 1984.

39. Müller, P., Imhof, P., Mauli, D., and Milanovic, D., Human pharmacological investigations of a transdermal nicotine system, *Methods Find. Clin. Pharmacol.*, 11, 197, 1989.

40. Wilkin, J. and Fortner, G., Cutaneous vascular sensitivity to lower aliphatic alcohols and aldehydes in Orientals, *Alcohol Clin. Exp. Res.*, 9, 522, 1985.

41. Wilkin, J., Quantitative assessment of alcohol-provoked flushing, *Arch. Dermatol.*, 122, 63, 1986.

42. Wilkin, J., Effect of nadolol on flushing reactions in rosacea, *J. Am. Acad. Dermatol.*, 20, 202, 1989.

43. Wilkin, J., Does monosodium glutamate cause flushing (or merely "glutamania")?, *J. Am. Acad. Dermatol.*, 15, 225, 1986.

44. Holloway, G., Cutaneous blood flow responses to injection trauma measured by laser Doppler velocimetry, *J. Invest. Dermatol.*, 74, 1, 1980.

45. Serup, J. and Staberg, B., Quantification of weal reactions with laser Doppler flowmetry, *Allergy*, 40, 233, 1985.

46. Hovell, C., Beasley, C., Mani, R., and Holgate, S., Laser Doppler flowmetry for determining changes in cutaneous blood flow following intradermal injection of histamine, *Clin. Allergy*, 17, 469, 1987.

47. Skassa-Brociek, W., Manderscheid, J., Michel, F., and Bousquet, J., Skin test reactivity to histamine from infancy to old age, *J. Allergy Clin. Immunol.*, 80, 711, 1987.

48. Olsson, P., Hammerlund, A., and Pipkorn, U., Weal-and-flare reactions induced by allergen and histamine: evaluation of blood flow with laser Doppler flowmetry, *J. Allergy Clin. Immunol.*, 82, 291, 1988.

49. Hammarlund, A., Olsson, P., and Pipkorn, U., Blood flow in dermal allergen-induced immediate and late-phase reactions, *Clin. Exp. Allergy*, 19, 197, 1989.

50. Hammarlund, A., Olsson, P., and Pipkorn, U., Blood flow in histamine- and allergen-induced weal and flare responses, effects of an H1 antagonist, a-adrenoceptor agonist and a topical glucocorticoid, *Allergy*, 45, 64, 1990.

51. Bunker, C., Foreman, J., and Dowd, P., Digital cutaneous vascular responses to histamine and neuropeptides in Raynaud's phenomenon, *J. Invest. Dermatol.*, 96, 314, 1991.

52. Coulsen, M., Hayes, N., and Foreman, J., Comparison of infrared and helium-neon lasers in the measurement of blood flow in human skin by the laser Doppler technique, *Skin Pharmacol.*, 5, 81, 1992.

53. Heyer, G., Hornstein, O., and Handwerker, H., Skin reactions and itch sensation induced by epicutaneous histamine application in atopic dermatitis and controls, *J. Invest. Dermatol.*, 93, 492, 1989.

54. Bisgaard, H., Kristensen, J., and Søndergaard, J., The effect of leucotrienes C4 and D4 on cutaneous blood flow in humans, *Prostaglandins*, 23, 797, 1982.

55. Bisgaard, H. and Kristensen, J., Quantitation of microcirculatory blood flow changes in human cutaneous tissue induced by inflammatory mediators, *J. Invest. Dermatol.*, 83, 184, 1984.

56. Bisgaard, H., Vascular effects of leukotriene D4 in human skin, *J. Invest. Dermatol.*, 88, 109, 1987.

57. Chan, C. and Ford-Hutchinson, A., Effects of synthetic leukotrienes on local skin blood flow and vascular permeability in porcine skin, *J. Invest. Dermatol.*, 84, 154, 1985.

58. Maurice, P., Barr, R., Koro, O., and Greaves, M., The effect of prostaglandin D2 on the response of human skin to histamine, *J. Invest. Dermatol.*, 89, 245, 1987.

59. Brain, S., Tippins, J., Morris, H., MacIntyre, I., and Williams, T., Potent vasodilator activity of calcitonin gene-related peptide in human skin, *J. Invest. Dermatol.*, 87, 533, 1986.

60. Bunker, C., Foreman, J., O'Shaughnessy, D., Reavley, C., and Dowd, P., Calcitonin-gene-related peptide in the treatment of severe Raynaud's phenomenon, *Br. J. Dermatol.*, 121, 43, 1989.

61. Heyer, G., Hornstein, O., and Handwerker, H., Reactions to intradermally injected substance P and topically applied mustard oil in atopic dermatitis patients, *Acta Derm. Venereol.*, 71, 291, 1991.

62. Lindblad, L., Ekenvall, L., Ancker, K., Rohman, H., and Öberg, P., Laser Doppler flow-meter assessment of iontophoretically applied norepinephrine on human finger skin circulation, *J. Invest. Dermatol.,* 87, 634, 1986.

63. Lindblad, L. and Ekenvall, L., Alpha-adrenoceptors in the vessels of human finger skin, *Acta Physiol. Scand.,* 128, 219, 1986.

64. Roberts, R. G., Westerman, R. A., Widdop, R. E., Kotzmann, R. R., and Payne, R., Effects of capsaicin on cutaneous vasodilator responses in humans, *Agents Actions,* 37, 53, 1992.

65. Li Kam Wa, T., Almond, N., Cooke, E., and Turner, P., Effect of captopril on skin blood flow following intradermal bradykinin measured by laser Doppler flowmetry, *Eur. J. Clin. Pharmacol.,* 37, 471, 1989.

66. Li Kam Wa, T. C., Cooke, E. D., and Turner, P., A comparison of the effects of captopril and enalapril on skin responses to intradermal bradykinin and skin blood flow in the human forearm, *Br. J. Clin. Pharmacol.,* 35, 8, 1993.

67. Li Kam Wa, T. C., Cooke, E. D., and Turner, P., Cutaneous blood flow changes and weal induced by intradermal bradykinin following pretreatment with indomethacin and captopril, *Eur. J. Clin. Pharmacol.,* 44, 41, 1993.

Chapter 7

Scleroderma

Dorothea Hiller and Hans-Peter Albrecht

CONTENTS

ABSTRACT: This chapter reviews the aspects of cutaneous blood flow and erythema evaluated by laser Doppler velocimetry (LDV) and photoplethysmography (PP) in scleroderma or systemic sclerosis. This condition presents itself as a chronic generalized connective tissue disease characterized by fibrosis. Although it is a rather rare disorder, it can serve as a model for the study of pathophysiological events leading to fibrosis and sclerosis with several animal models as well as known human scleroderma-like and scleroderma-associated syndromes. The main characteristics—such as the immunological pathway, the role of connective tissue, and especially the vascular factors—are discussed in relation to the topic of this volume. Furthermore, the various therapeutic approaches relating to the three pathogenetic complexes are reported.

Emphasis is put on reviewing the microcirculatory data base in scleroderma during the last decade. Two tables present an extract of the available publications in scleroderma involving LDV and PP. Where appropriate, reports using other microcirculatory methods are included for comparison. One table lists studies concerned with general pathophysiological aspects, while the other table shows microcirculatory studies of therapeutic regimens used in scleroderma, with a total number of 22 and 20 separate references evaluated, respectively.

The third part of the chapter summarizes some guidelines to follow when applying the above methods in scleroderma, particularly concentrating on ambient conditions and patient- and disease-related factors. An exemplary study protocol is presented together with dynamic parameters of functional microcirculation which describe valuable information to be drawn from such experiments.

Keeping the restrictions due to the methods and pathophysiological limits of the disease in mind, the conclusion shows that both methods provide good non-invasive measuring medical research tools for the continuous *in vivo* assessment of the vascular aspects and a more objective and quantitative therapeutic control of this interesting but also confusing disease. Further monitoring of the cutaneous blood flow with these and other methods might elucidate more of the physiological and pathological conditions prevailing in scleroderma, which seems especially touched by a disturbed microcirculation, sometimes also termed "flow of life."

I. INTRODUCTION

A medical thesaurus defines scleroderma as a "chronic, progressive dermatosis characterized by boardlike hardening and immobility of the affected skin, with visceral involvement, especially of esophagus, lungs, kidneys, and heart. It may be accompanied by calcinosis, Raynaud's phenomenon, and teleangiectasis. It includes acrosclerosis and sclerodactyly."

This definition shows that "scleroderma" includes a heterogeneous group of limited and systemic conditions which cause hardening of the skin. Although the systemic nature of the disease was recognized early, it was only defined and put into a disease concept in 1945 by Goetz,[1] who proposed the name "progressive systemic sclerosis," thus implying an inevitable advancement of the disease; this is not always the case. Another—maybe more appropriate—title is "systemic sclerosis,"pointing out the importance of visceral features in this disorder. In contrast, the term "scleroderma"(Scl)—which will be used in the paragraphs to follow—emphasizes the skin involvement which is of main interest in this book on skin bioengineering. However, one has to keep in mind that regardless of the name used, various disease subsets exist with established criteria for diagnosis[2] and different classifications for disease activity.

The separate disorder entities for the term "scleroderma" range from localized scleroderma (morphea) to systemic scleroderma—which will be the focus of this chapter—to overlap syndromes. The latter, also showing a highly variable severity and course, can again be divided into acroscleroderma, acroscleroderma with ascending sclerosis, and diffuse scleroderma.[3] Another attempt at classification is based exclusively on the involvement of the skin with the wrist as a "borderline" for type I, ascending sclerosis including the forearm for type II, and the beginning of sclerosis at the trunk in type III.[4] Only two forms of Scl, the "limited cutaneous systemic sclerosis (lSSc)" and the "diffuse cutaneous systemic sclerosis (dSSc)," are differentiated by an international group of internists and rheumatologists.[5] However, to assure that clinicians and researchers are "talking the same language" it is important to use some sort of standardized classification, regardless of the type, especially when comparing results from different patient groups. Additionally, it provides a framework for further research leading to advances in understanding and treatment of this interesting disease.

II. THE DISEASE AND WHY MEASUREMENTS OF CUTANEOUS BLOOD FLOW ARE OF INTEREST

A. EPIDEMIOLOGY

Scleroderma is a rare disease with a morbidity of approximately 105 per million inhabitants (U.S.), an incidence of about 3 to 12 per million annually,[6] and an age-adjusted mortality for the total population (U.S.) of 3.08 per million person-years.[7] No climatic or geographic differences are known, a relatively greater proportion exists in non-whites vs. whites, and the female-to-male ratio is 3:1 to 5:1, the latter being even more marked in cases with disease onset before age 45. Whereas in general scleroderma is rare within the juvenile age range, incidence as well as sex ratio and prognosis change impressively according to the age of onset (puberty, postmenopause), suggesting that hormonal-metabolic factors are influencing the target tissues.[7] In addition, there are reports of significantly lower capillary pressure (finger nailfold) together with a slightly lower capillary pressure pulse amplitude (CPPA) and a significant relationship between temperature and CPPA in healthy women compared to healthy men.[8] These microcirculatory findings show a cold-induced elevation in precapillary resistance that is particularly pronounced in women and could be a further explanation for their predominance in scleroderma. Furthermore, a study[9] of peripherally mediated vascular response to temperature showed that normal women had an intermediate response between normal men and a group of patients with primary Raynaud's phenomenon.

Relating to etiology, the disease shows a genetic predisposition and increased incidence of chromosome abnormalities in affected families with a HLA-B8 linkage with regard to the disease course and impaired cell-mediated immunity.[7] In addition, the epidemiology indicates some similarities with other connective tissue diseases, e.g., systemic lupus erythematosus.

B. SCLERODERMA AS A MODEL DISEASE

Even though scleroderma is a rather rare disorder, it can serve as a model for the study of pathophysiological events leading to fibrosis and sclerosis. This is supported by findings[10,11] that hypoxia in sclerodermatous skin may play a role in the modulation of dermal fibroblast proliferation and activity. Moreover, there are experimental animal models for scleroderma available, such as graft-vs.-host disease in rats, genetically transmitted avian disease[12] in a mutant line of white leghorn chickens, and the tight-skin mouse.[13]

One hypothesis on development of sclerosis is the formation of free radicals,[14] which can be influenced by ischemia and certain environmental triggers[15] such as vinyl chloride, silica dust, toxic oil, organic solvents, epoxy resins, bleomycin, L-5-hydroxytryptophan, pentazocine, and breast augmentation. In the last few decades, scleroderma-like syndromes[16] caused by above-mentioned agents, vibration, eosinophilic fasciitis, scleredema of Buschke, scleromyxedema Arndt-Gottron, porphyria, and amyloid and graft-vs.-host reaction and scleroderma-associated syndromes[16] such as overlap syndromes, Werner's syndrome, phenylketonuria, carcinoid syndrome, and bronchoalveolar carcinoma have been found to share some but not all features of the idiopathic disease.[15] In general, scleroderma seems to be a model for a variety of other disorders characterized by fibrosis and sclerosis. Additionally, the account of the distinguishing peculiarities in this disease makes it interesting to measure them by several bioengineering techniques—for example, to quantify skin thickness (e.g., by ultrasound[17–19]), skin stiffness (e.g., by a suction device[20]), cutaneous oxygen supply (e.g., by polarography[10,21]), vascular morphology (e.g., by capillaroscopy[22]), and cutaneous blood flow (e.g., by laser Doppler velocimetry, videocapillaroscopy,[23] photoplethysmography,[24–26] or thermography[27–29]).

C. CHARACTERISTIC FEATURES OF SCLERODERMA

Although the etiology of scleroderma is still unknown, its three main features—diffuse vascular damage, fibrosclerosis of cutaneous and visceral connective tissues, and immunological abnormalities—give rise to different schools of pathophysiological concepts. Thus, primary damage at the vascular and probably the endothelial level is in one view involved in the pathogenesis of scleroderma,[30,31] whereas others[3,32,33] hypothesize an alteration in collagen metabolism, possibly due to abnormal release of cytokines from mononuclear and/or endothelial cells.[34]

For a better understanding of the connection between the topic "cutaneous blood flow and erythema" and scleroderma, a short review of the major characteristics and therapeutic approaches of the disease in relation to the above theme seems appropriate.

1. The Immunological Pathway

The immunological pathway—which will not be covered in detail—is characterized by circulating autoantibodies to recently identified nuclear antigens,[3] similar to other connective tissue diseases. Some of the antibodies are highly specific and, therefore, important for diagnosis and defining subsets of scleroderma, but the antibody titers do not correlate in general with the disease activity. For that reason their pathogenetic significance is questionable, probably representing secondary phenomena. However, the clinically scleroderma-like features of graft-vs.-host reactions could hint at an immune trigger mechanism[16] leading to recurrent endothelial injury and repair, thus initiating fibroblast proliferation and, consequently, overproduction of collagen.

2. The Role of Connective Tissue

Fibrosclerosis as the final result of activating the connective tissue metabolism can involve all such tissue-containing organs, with the skin, the gastrointestinal tract, the kidneys, the lungs, and the heart being the main targets. Alterations of the internal organs often determine the prognosis of scleroderma. However, their characteristics will not be discussed further, in contrast to the skin involvement.

The uncontrolled and irreversible proliferation of connective tissue together with the striking vascular changes considerably impairs the normal physiological functions of the above organs, resulting in increased morbidity and mortality.[35] Mainly type I, III, and VI collagens and, to a lesser extent, proteoglycans, fibronectin, and tenascin are the extracellular matrix proteins known to be excessively produced in scleroderma. Their interrelationship as well as the stimuli which activate the process are still unknown. The thickened and indurated tissues in Scl represent the ultimate irreversible phase in the development of a disorder that progresses through numerous stages and usually requires many months or years to evolve. Especially the early, often-missed developments, particularly the presclerodermatous events such as swollen edematous skin and puffy stiff hands, seem to be of great importance. The stage of the disease also correlates with different types of collagen produced, with type III collagen prevalent in the early lesions and type I predominant at a later stage. Cutaneous involvement can also be divided into an early edematous, a sclerotic, and a late atrophic phase.

Mostly starting at acral areas (finger, hand, face), nonpitting edema and often slight erythema, indicating inflammation, gradually changes into sclerosis. This leads to the typical cutaneous signs such as a beaked nose, radial furrowing around the lips, constricted opening of the mouth, shortening of the frenulum of the tongue, inability to fully extend the fingers and hands, slow healing, sclerodactyly, and development of painful ulcers around the fingertips, in areas with underlying calcinosis, and over the knuckles. Hair loss and anhidrosis in affected skin reflect involvement of the appendages by the surrounding fibrosis. Hyper- or hypopigmentation, probably due to postinflammatory changes, can develop along with the sclerotic skin signs.

3. Vascular Factors

Early in the course of the disease three related groups of vascular changes can be seen: vasomotor instability, intravascular abnormalities, and small-vessel structural abnormalities.[36] This supports the hypothesis that vascular factors are a primary event in the pathogenesis of Scl. However, it is still uncertain whether the vascular disorder and tissue fibrosis occur simultaneously or whether one is the consequence of the other. Therefore, their relationship to each other must be discussed (for review see Reference 36).

Telangiectasia, mostly periungual and reflecting dilatated nailfold capillaries, and Raynaud's phenomenon (RP) are early observed vascular symptoms in Scl. The characteristic episodic vascular spasm on cold- or stress-induced vasoconstriction with the typical three-stage sequence of painful ischemia, local cyanosis, and arterial hyperemia occurs in 60 to 90% of patients suffering from Scl.[6] Because RP is for the majority of patients suffering from Scl one of the most disturbing clinical symptoms, various therapies make a major effort to improve this debilitating condition (see Table 2). Since the distinguishing phenomenon is due to the underlying connective tissue disease (CTD) it is termed secondary RP. Scl is the leading CTD causing RP; however, other CTD can also induce these striking attacks.

The histopathology of early lesions reveals a usually mild cellular infiltrate consisting of lymphocytes, monocytes, and histiocytes. This inflammatory infiltration is mainly present in the perivascular region and in the interface between dermis and subcutaneous tissue.[30] It precedes fibrosis and often disappears completely in later stages of the disease. There are several additional signs of a close relationship between the hallmarks of Scl, tissue fibrosis and vascular disorder: thickening and hyalinization of dermal vessel walls, the beginning of connective tissue deposition in the reticular dermis around the blood vessels, and accumulation of newly synthesized procollagen in the same area.[37] Capillaries in scleroderma skin show numerous alterations, like gaps between endothelial cells (EC), granular degeneration of the nucleus and multiple vacuoles in the cells, and reduplication of the basal lamina reflecting endothelial cell injury and subsequent attempts at repair. This leads to disruption of ECs and often obstruction of the vessel lumina. In addition to the capillary abnormalities, involution of arteriovenous anastomoses has been shown in morphological studies of digital pulp in Scl.[38]

Proliferative intimal arterial lesions—conforming to the clinical finding of luminal narrowing of 75% or more in 80% of the digital arteries[39]—and the obliterative microvascular process both can be explained by the hypothesis of EC injury and dysfunction as the central event.[36] Normal ECs participate in vasodilatation, anticoagulation, fibrinolysis, and a variety of immunological reactions. Thus, injury of ECs occurring in small arteries leads to platelet aggregation, myointimal cell proliferation and migration into the injured region, and deposition of connective tissue, followed by a reduced lumen diameter and changed vessel distensibility. In the capillary bed, without a population of myointimal cells, the same endothelial injury leads to little proliferative response but does cause changes in permeability which may activate fibroblasts to produce more collagen.[16] Thus, inhibition of new capillary growth or physical exclusion of capillaries from areas of scarring results in devascularization and atrophy. The devascularization of crucial viscera and the extent and severity of vascular dysfunction lead to a state of underperfusion and chronic ischemia of organ systems and are extremely important for the prognosis and eventual outcome of scleroderma. Another clinical characteristic, the visible telangiectasia known from nailfold capillaroscopy,[22] ultimately develops through condensation of existing vascular channels and expansion into the dilated capillaries and normally shows a brisk resting flow.[16] However, cold exposure leads to exaggerated vasoconstriction with a resulting standstill of capillary flow. For the pathogenetic role of the vasculature in scleroderma the following "cascade" is proposed.[3] Endothelial cell damage—representing a key event in the development of the fibrotic reaction—leads to platelet aggregation and enhanced adhesion to collagen. Platelets contain several active mediators, such as platelet-derived growth factor, transforming growth factor-β, epidermal growth factor, and basic fibroblast growth factor, which when released into the tissue are potent modulators of fibroblast function. Additionally, the increased permeability of the vessel walls allows mononuclear cells access to the tissue as well as formation of the already described perivascular infiltrates in early stages of Scl. Furthermore, the above-mentioned potent mediators can easily diffuse through the defective basement membrane with subsequent induction of fibroblasts in the surrounding tissue.

The cause of the continuing endothelial injury is still unknown, with conflicting reports regarding the mechanism. One cause is thought to be an EC cytotoxic factor with proteolytic activity, which has been found in the sera of over 50% of patients with scleroderma.[40] Another proposed mechanism is an antibody-dependent cellular cytotoxicity.[41] Both causes fit the suggestion of a circulating factor involved in EC injury. This suggestion is supported by reports of improvement of Scl following plasmapheresis and immunosuppressive therapy associated with decreased serum EC cytotoxicity *in vitro.*[42]

4. Therapeutic Approaches

Since the etiopathogenesis of the disease is still insufficiently known, at present there is no specific drug or therapeutic regimen to cure scleroderma. However, therapy to reduce the clinical symptoms is directed at three pathogenetic complexes: influence on the vascular changes, influence on the metabolism of connective tissue, and influence on the inflammatory signs. Prostacyclin,[43,44] prostaglandin E_1,[45,46] calcitonin,[21] plasma expanders,[47] calcium channel blockers[48] (especially nifedipine), angiotensin-converting enzyme inhibitor[49] (captopril; for treatment of renal crises), serotonin-receptor blocker[50,51,53,62] (ketanserin), and formerly sympathectomy[54-57] were used as vasoactive substances/regimens to inhibit platelet hyperaggregation and/or to induce vasodilatation, reducing the number and severity of Raynaud's attacks. Yet a significant effect on the further progression of Scl could not be shown.

Inhibition of collagen synthesis is the second therapeutic approach influencing the metabolism of connective tissue. D-Penicillamine is still the most important agent.[58] This drug given as a long-term low-dose therapy can reduce skin involvement, leading to a greater 5-year cumulative survival rate, but also leads to considerable side effects. Other antifibrotic agents[3] are gamma-interferon,[59] potassium aminobenzoate,[60-62] proline analogs, cyclofenil,[63,64] procollagen peptides, and prolyl hydroxylase inhibitor. However, for some of these substances studies on the positive effects on inhibition of collagen production are still in the *in vitro* phase, or they have proved to be highly toxic *in vivo.* The third type of treatment is by anti-inflammatory agents/regimens[3] like corticosteroids,[65] azathioprine, cyclophosphamide, cyclosporine,[66] plasmapheresis,[42,67,68] and photopheresis.[69] The anti-inflammatory action is especially indicated in patients showing inflammatory, immunologically active forms of the disease with, e.g., myositis or interstitial lung involvement. Symptomatic response to this therapy is often good, but unfortunately the prognosis of scleroderma cannot be altered by the immunosuppressive agents.[70]

In addition to the above-discussed substances, general treatment with intensive physiotherapy, psychological guidance, special diet, and avoidance of all risks of cutaneous trauma, infections, cold, smoking, or other vasospasm-inducing agents is very important.

III. A REVIEW OF THE MICROCIRCULATORY DATA BASE IN SCLERODERMA

The striking vascular changes in scleroderma stress the necessity for quantitative measurements of the circulatory aspects of the disease. Although the topic of this chapter is laser Doppler velocimetry (LDV) and photoplethysmography (PP), the evaluation of results by these methods also requires comparison and/or combination with other microcirculatory methods such as the radioactive tracer clearance technique, oxygen tension (pO_2) measurement, thermography, capillaroscopy, strain-gauge plethysmography, and skin temperature measurements. However, the measuring principles and hardware of these methods will not be covered (for more details see the reference list, respectively Chapter 3 of this book). No review can ever be complete, therefore, only data from publications in the English language (MEDLINE index) and mostly only that within the last decade have been considered, although total coverage cannot be achieved. Due to this temporal restriction and the fact that photophlethysmography[71,72] is a method developed much earlier,[73] the data base—with the most recent publications at the beginning—will mainly concentrate on LDV measurements. It will be presented in a tabular form, with Table 1 concerning general pathophysiological aspects of the disease and Table 2 (grouped by the substances used) showing microcirculatory studies of therapeutic regiments in scleroderma. Of course, the conditions, results, and conclusions of each report could only be pointed out in a very abbreviated form, thus, for further details the respective references must be consulted. In the column "Disease", information is given only on the various patient groups, not on the control groups of approximately the same size. The latter always consisted of healthy, mostly age- and sex-matched subjects without any vascular alterations. In the column "Result(s)/Conclusion(s)" an attempt was made to quote the authors as closely as possible within the available space. Results from LDV and PP measurements in scleroderma sometimes vary considerably depending on the localization used (with or without existence of arteriovenous anastomoses), skin temperature, and the patient group. Therefore, a "Note" column concerning these three restrictions was included in the tables. In comparing different groups of patients, often persons affected by the so-called "primary Raynaud's phenomenon" (pr.RP) were used. This is said to be present if there is at least a 2-year history of RP in the absence of any clinical or laboratory evidence of an underlying CTD—such as scleroderma—or structural vasculopathy.[74] Furthermore, there has to be a normal nailfold capillaroscopy, absence of characteristic antibodies, and *in vivo* platelet activation. Due to the extent of the tables and the topic "scleroderma", most reports focusing only on patients with primary Raynaud's phenomenon have not been included in the tabular review, although this could be of interest since some of these patients eventually progress to scleroderma after several years.

Of course, this data base is only an excerpt of studies on Scl. For comparison of above-described results the early work of Kristensen and co-workers[116-121] seems especially valuable. The ^{133}Xe washout technique either with epicutaneous labeling or with intra- or subcutaneous injections was used to determine the capillary blood flow in the various skin compartments. Applying these methods in scleroderma Kristensen[116] found, as did Albrecht et al.[21] that the morphological changes of vascular and extravascular tissue in Scl did not affect the distensibility of the vascular bed in subcutaneous tissue, whereas in the fingertips a lowered skin compliance was suggested.[118] In further experiments Kristensen used arterial occlusion,[120] a stepwise release of arterial occlusion[119] to study blood flow in cutaneous tissue at the dorsum of the hand (via epicutaneous and intradermal tracer application), and local perfusion pressure of the fingers (via epicutaneous labeling) in Scl. Surprisingly, he found at the first location a 100% increase in washout rate in Scl compared to normal. Using a different disease model Serup and Kristensen[94] also found an increased blood flow by LDV and ^{133}Xe washout technique in the sclerotic white-colored center and the perilesional areas (lilac ring) of localized scleroderma (morphea). From this it becomes very clear that those bioengineering techniques are more objective criteria by which to describe and evaluate skin status and function than morphology, which deceives one by suggesting a reduced perfusion in the whitish and lilac areas of morphea plaques. Serup and Kristensen suggested that inflammatory and vascular changes seen in the early stages of morphea are the cause of the increased LDV. Kristensen and Wadskov explained their findings[120] as a result of a decreased resistance to flow offered by defective dilated arterioles and concluded that the mechanisms of local regulation of cutaneous blood flow were defective in Scl. The experiments on local perfusion pressure of the fingers[119] showed that in normals finger blood pressure (BP) did not differ from mean arm BP and, furthermore, that local perfusion BP was significantly decreased compared to mean arm BP in Scl and to local perfusion pressure of normal fingertips. This corresponds to the results of multiple other authors (Tables 1 and 2) which showed significantly reduced LDV in Scl at the fingertip location. It was thought by Kristensen to be consistent with the changes in digital arteries in Scl found by arteriography and an increased resistance to blood flow in digital arteries. These seemingly contrasting findings at the dorsum of the hand and the finger emphasize the importance of the measuring location. The majority

Table 1   **Microcirculatory studies in scleroderma (mainly from the last decade and using for the most part laser Doppler velocimetry and photoplethysmography) concerning general pathophysiological aspects of the disease**

Method(s)	Disease {Number of Patients} [Measuring Site]	1) Physiological Aim(s) 2) Stress Test(s)
LDV, pO_2, skin temp.	Scl (type I, II) {19} [heated dor. midhand, prox. to 3rd digit] [unheated wrist, dor. midhand, dor. surface of 3rd digit, fingertip]	1) Microcirculatory functions in Scl, static and functional parameters relevant for the microcirculatory response to defined stimuli; are they additional parameters of therapeutic concepts? 2) Art. occlusion (3 min), local skin heating (42° C), 10-day i.v. calcitonin infusion (100 IU/day)
LDV, capillaroscopy, immunology laboratory parameters	Isolated RP {30} [nailfold, digit]	1) Correlation among capillaroscopic abnormalities, digital flow, and immunologic findings 2) Standardized cold and rewarming test
LDV, capillaroscopy	RP in Scl {6} [nailfold, adjacent areas]	1) Differences in signal pattern between LDV and RBV, ability to detect early, if treatment is successful? 2) Cold stress test
LDV, planimetry	RP in Scl {8} pr.RP {12} [dor. prox. digital phalanx]	1) Cut. vasc. responses to histamine and neuropeptides to see if there is a defect in local histamine vasodilator mechanism as a local fault for developing RP 2) Intradermal injection of saline, histamine, and histamine-releasing agent compounds 48/80, substance P, calcitonin-gene-related-peptide (CGRP)
LDV, $tcpO_2$	Scl {24} [dor. aspect of hand, interscapular region]	1) Correlation between $tcpO_2$ and LDV at different temp. and between them and clinical aspects of the patient 2) Temp. 36°–37° C vs. 44° C, oxygen administration.

Result(s)/Conclusion(s)	Note[a]	Ref. (Year)
Scl shows initially hyperemic hypoxia with $\uparrow$ LDV and $\downarrow$ pcuO$_2$. LDV and pcuO$_2$ response to hyperemic stimuli are $\downarrow$ due to exhausted functional reserves and yield evidence of severely disturbed reactivity of the dermal vessels in Scl. During and after calcitonin infusion pcuO$_2$ $\uparrow$ and LDV $\downarrow$. Testing of microcirculatory functions provides additional parameters for evaluation of therapeutic agents.		21 (1993)
A significant correlation between digital BF and capillaroscopic score was found, but capillaroscopic pattern did not correlate well with immunologic findings. In ANA-positive patients intensity and length of flow stop during cold test correlated with ANA titers. LDV may detect potentially secondary RP patients during a standardized thermic test even if a definite Scl-like capillaroscopic pattern is absent.	−A−	75 (1993)
RBV seems to indicate improvement after start of therapy more clearly than LDV. Discrepancies between RBV and LDV are due to LDV detecting other vessels in addition to the superficial nutritional capillaries.		76 (1992)
Response to histamine or 48/80 was not consistently different between the groups. Response to CGRP was equally great at 60 min. In Scl no specific pharmocological deficit was seen. The long-enduring erythema after CGRP points to the possibility that this may be a useful treatment in RP.	−A− −B−	77 (1991)
At 44° C: LDV and tcpO$_2$ were $\downarrow$ compared to normals in sclerotic and nonsclerotic areas. At 37° C: LDV was $\uparrow$ only in sclerotic areas; for tcpO$_2$ no difference from the normals was found. The $\uparrow$ LDV at physiological temperatures balances partly the $\downarrow$ tissue pO$_2$; however, under maximal hyperemia-inducing temperatures Scl vessels are unable to increase their BF enough to ensure normal tcpO$_2$.		78 (1991)

Table 1 *Continued*

Method(s)	Disease {Number of Patients} [Measuring Site]	1) Physiological Aim(s) 2) Stress Test(s)
[133]Xe washout	pr.RP {9} RP in Scl {7} [subcut. tissue at dor. hand and dor. midphalanx]	1) To evaluate if a generally ↑ sympathetic response in Scl exists by investigating orthostatic induced local and central sympathetic reflexes 2) Hand lowering, head tilted up
LDV	pr.Rp {9} RP in Scl {7} [dor. foot skin]	1) To determine if an abnormal vasc. response to temp. change independent of central sympathetic control exists in RP 2) Localized heating of dor. foot skin
LDV, strain gauge plethysmography (SPG, skin temp.)	pr.RP {8} RP in Scl {8} UCTD {8} [3rd fingertip—LDV; 2nd prox. finger—SGP]	1) Is physiological separation of Scl patient from other forms of RP possible? 2) Clinical maneuvers to evoke sympathetic tone, handwarming cycles, hand cooling with and without central cooling
LDV	pr.RP {13} RP in Scl {21} [warmed 4th fingertip]	1) Postischemic hyperemic response 2) Art. occlusion (5 min.), infusion of a potent vasodilator (iloprost)
LDV, tcpO2, skin thickness, skin extensibility	loc. Scl {16} Scl {11} [Scl: ventral forearm, loc. Scl: N/A]	1) Which are the most sensitive parameters in the study of the progression of loc. Scl and Scl? 2) None
LDV	pr.RP {29} RP in Scl {38} AOD [fingertip]	1) Postocclusive RH of fingertips in the diagnosis of RP for discrimination of RP types 2) Art. occlusion (5 min.)
LDV	RP in Scl {26} [fingertip]	1) Hyperemic responses 2) Art. occlusion (2 min)

Result(s)/Conclusion(s)	Note[a]	Ref. (Year)
When supine pr.RP showed ↑ finger and hand washout rates compared to controls and Scl. A ↓ response to lowering of the hand was observed in pr.RP and was normal in Scl. Head-up tilt showed parallel responses in all three groups. The not ↑ responses on local or central sympathetic reflexes make a generally ↑ sympathetic activity as a pathophysiological background for vasospastic attacks not likely, but the hyperemic state in pr.RP might have influenced orthostatic sympathetic responses.	–A– –B–	79 (1991)
pr.RP showed early vasodilatation compared to normals. Responses in RP in Scl widely over lapped the other groups. Therefore, an abnormal peripheral vascular response to temperature change exists in pr.RP.	–A–	80 (1990)
Reflex sympathetic influences on peripheral perfusion are maintained in Scl and at the level of arterial circulation did not differ from normals, pr.RP, or UCTD. Loss of arterial flow, whether medicated by local temp. or sympathetic tone, is associated with loss of nutritive BF in Scl. Therefore, therapy would be best directed at maintenance of arterial inflow (digital preload).	–A– –B– –C–	74 (1990)
Pr.RP showed normal postischemic and ↑ baseline BF. Compared to normals and pr.RP, SCL showed ↓ baseline BF and a blunted and delayed HR. Local skin warming and iloprost infusion did not improve the abnormal response, suggesting that this is a consequence of a fixed structural vascular bed.	–A– –C–	81 (1990)
Skin in loc. Scl showed ↓ extensibility, ↑ thickness, ↑ LDV (cut. microcirculation), ↓ tcpO$_2$ compared to normals, whereas in Scl ↓ extensibility, ↑ thickness, but no change in cut. microcirculation and tcpO$_2$ (44° C) was found. Therefore, skin thickness and extensibility are the most useful parameters in the study of the progression in loc. Scl and Scl.	–A– –B– –C–	82 (1990)
Of all the indices analyzed, none showed an optimal discrimination among the several subtypes of finger ischemia, with parameters representing the ↑ LDV after occlusion differentiated best between pr.RP and PR due to Scl. The time variable after peak proved more suitable to differentiate Scl from AOD.	–A– –B– –C–	83 (1989)
Without handwarming, SCL patient showed a very low resting BF and in 50% an absent HR. After vasodilation HR was comparable to that of controls but had an ↓ slope. RH is a mechanical phenomenon; vessel wall reactivity is abnormal in Scl, producing delayed RH. The magnitude of RH depends on the initial flow rate, and lack of response in Scl is a result of their low but reversible resting BF.	–B– –C–	84 (1989)

Table 1   *Continued*

Method(s)	Disease {Number of Patients} [Measuring Site]	1) Physiological Aim(s) 2) Stress Test(s)
LDV	RP in Scl {17} pr.RP {29} AOD {8} [fingertip]	1) Is it possible to separate different forms of RP? 2) Local warming (40° C) for 5 min
LDV	RP in Scl {12} [fingertip]	1) Duration and number of RA 2) Hot hand bath (5 min) every 4 h throughout the day during alternate weeks of a 6-week study
LDV	RP in Scl {15} [3rd fingertip]	1) The effects of simple warming procedures on finger BF in Scl 2) In an ambient temp. of 28° C: 10–min warm water bath (35° C) followed by a cold stress (18° C) for 10 min
TcpO₂	Scl {33} [dor. forearm as involved skin, medial aspect of arm as uninvolved skin]	1) Transcutaneous oxygen tension (at 44° C probe temp.) in patient with Scl 2) Breathing room air vs. 31% oxygen
LDV	RP in Scl {10} [fingertip]	1) BF response to increased temperature 2) Hand warming (10 min) in 35° C with subsequent cooling
LDV, skin temp.	RP in Scl {15} pr.RP {16} [fingertip]	1) Comparison of different patient groups with RP 2) Indirect (body) cooling
Photoplethysmography (PP)	pr.RP {32} RD with documented art. obstructions {30} [finger]	1) Measurement of systolic finger blood pressure to differentiate between pr.RP and RD 2) Cooling

Result(s)/Conclusion(s)	Note[a]	Ref. (Year)
Pr.RP could be distinguished from normals by heating. Scl and AOD showed ↓ response to heating compared to normals and pr.RP. Therefore, measurements of hyperemia during heating may play a role in determining the degree of obstructive vascular involvement in RP.	–A– –C–	85 (1988)
A hot water hand bath ↓ number and duration of RA with a ↑ of digital BF accompanied by clinical improvement. This simple hand warming appears to be effective in the management of RP in Scl.	–C–	86 (1988)
Resting LDV was significantly ↓ in Scl compared to controls. After hand warming both groups did not differ and response to a cold stress thereafter neither differed in level nor in time course. Additionally, resting LDV 20 min after this showed no difference. The induced vasodilation persisted for at least 2 h. These results indicate that local and central thermoregulatory reflexes in Scl are intact. Cold-induced symptoms are related to low resting flow, not to cold sensitivity. Therefore, simple warming may provide a useful treatment.	–B– –C–	87 (1988)
Sclerodermatous skin is hypoxic compared to uninvolved skin in Scl or in normals. TcpO$_2$ is directly related to skin thickness; no correlation with pulmonary functions of arterial pO$_2$ in the Scl patient was found. The administration of oxygen ↑ tcpO$_2$ in Scl to normal values. Therefore, it is concluded that thickened skin in Scl is hypoxic, tcpO$_2$ measurements may be helpful in objectively assessing the degree of skin thickness, and hypoxia may play a role in the modulation of dermal fibroblast proliferation and synthetic activity.		11 (1988)
Resting BF with a complex relationship to skin temp was ↓ in Scl but could be returned to normal by simple warming. For up to 3 h after warming the response to cooling is similar to controls. These patients should thoroughly warm up before entering cold environments.	–B– –C–	88 (1987)
Patients with RP showed ↓ resting BF compared to controls. 2 min indirect cooling showed no differences in the relative BF ↓ between the groups. 10 min cooling: BF in Scl ↓ more pronounced than in pr.RP. Vasomotion was preserved in the patients with observed influences of sympathetic vasomotor fibers during cooling. Finger temp. changes during cooling proved significant only between normals and pr.RP.	–A– –B– –C–	89 (1986)
In patients with RP and art. obstructions in 91% of fingers a ↓ digital pressure was found, with a ↓ of finger pressure of 12.3% during cooling. Pr.RP showed normal digital pressure in 97% of fingers but a 59.2% ↓ of pressure during cooling.	–A–	90 (1986)

Table 1 *Continued*

Method(s)	Disease {Number of Patients} [Measuring Site]	1) Physiological Aim(s) 2) Stress Test(s)
PP, capillaroscopy	pr.RP {16} RP in Scl {30} UCTD {40} [finger, nailfold]	1) Digital BF and number of enlarged capillary loops according to different patient groups 2) None
LDV, SGP, ^{133}Xe washout	pr.RP {7} RP in Scl {15} [finger, fingertip]	1) Comparison of LDV with other techniques during RA 2) Combined body and finger cooling
Videocapillaroscopy, -densitometry	Patient with peripheral vasc. disease, in particular RD and acrocyanosis. REVIEW	1) Blood flow pattern 2) Cold stimulus
LDV, ^{133}Xe washout	loc. Scl {15} [N/A, various sites]	1) BF measurements in various areas of loc. Scl. 2) None
LDV	RP in Scl {10} pr.RP {1} [fingertip]	1) To elucidate the function of sympathetic nervous system and digital BF in Scl; pathophysiological classification of RP on the basis of function tests 2) Direct and indirect cooling, deep breath, venous stastis, Valsalva maneuver

Note: Abbreviations—↑ = increase/increased; ↓ = decrease/decreased; AOD = arterial occlusive disease; ANA = antinuclear antibody; art. = arterial; BF = blood flow; BP = blood pressure; CTD = connective tissue disease; cu(t). = cutaneous; dor. = dorsal; gen. = generalized; HR = hyperemic response; i.v. = intravenous; LDV = laser Doppler velocimetry; loc. = localized; p0$_2$ = oxygen tension; N/A = information not available; PP = photoplethysmography; pr. = primary; prox. = proximal; RA = Raynaud's Attack; RBC = red blood cell(s); RBV = red blood cell velocity; RD = Raynaud's Disease; RH = reactive hyperemia; RP = Raynaud's phenomenon; Scl = scleroderma; sec. = secondary; tc = transcutaneous; temp. = temperature; UCTD = undifferentiated connective tissue disease; vasc. = vascular.

Result(s)/Conclusion(s)	Note[a]	Ref. (Year)
Capillaroscopy was the most sensitive and specific test, with a positive predictive value of 90% for Scl. Only capillaroscopy was able to differentiate between pr.RP and undifferentiated CTD.	–A– –B–	91 (1986)
LDV is more sensitive at extremely low BF values. Elicitation and duration of RP can be observed by LDV, and zero BF by LDV could be confirmed by ^{133}Xe washout	–A– –B– –C–	92 (1985)
In acrocyanosis different flow patterns in adjacent capillaries were observed together with delayed fillings of single capillaries or capillary groups.	–A–	93 (1985)
In the sclerotic white-colored center and in perilesional areas of plaques LDV was ↑^{133}Xe washout showed an analog tendency. Scars of other origin did not show ↑ BF. The ↑ BF in plaques seems to become normal with time. Inflammatory and vascular changes seen in the early stages of morphea are the suggested cause of those findings.	–A– –B– –C–	94 (1984)
In sec. RP fingertip BF reacted like normal skin without shunt vessels with ↓ resting BF and prolonged rewarming period. In pr.RP an even more prolonged rewarming period was observed after direct or indirect cooling. A pathophysiological classification of RP may be possible on the basis of function tests.	–A– –B– –C–	95 (1983)

[a]**A:** Attention has to be given to the patient groups used; either they are nonhomogeneous (i.e., just designated RP without an exact underlying cause being given) or patients were also used other than those suffering from Scleroderma. **B:** Notice that either uneven skin temperature prevailed in the different groups measured (this is often the case when comparing patients and controls without using a heated probe holder, especially in locations like the fingertip, where significant temperature differences can be observed: see also Table 7–3) or no information on skin temperature was given. **C:** The measuring location (e.g., fingertip) needs special consideration because the anatomical site contains a variety of vessels (additional arteriovenous anastomoses) and has a different innervation control; both could have a major influence on the results. In addition, temperature differences (see note B) are more likely.

Table 2   **Microcirculatory studies of therapeutic regiments used in scleroderma (mainly from the last decade and using for the most part laser Doppler velocimetry and photoplethysmography)**

Method(s)	Disease {Number of Patients} [Measuring Site]	1) Physiological Aim/ Therapeutic Regimen 2) Possibly Combined with Stress Test(s)
Prostaglandin Derivatives		
Capillaroscopy, capillary pressure gradient	RP in Scl {8} [nailfold]	1) Prostaglandin E_1 infusion
Photoplethysmography (PP)	RP in Scl {13} [finger]	1) Response of digital BF and pulse amplitude to $3\times$ 8 h infusions of a synthetic prostacyclin analogue (iloprost) on consecutive days, with a 10-week follow-up period
PP	RP in Scl {23} [finger]	1) BF measurements before and after 4, 8, 12, and 16 weeks of treatment with iloprost infusion vs. oral nifedipine in a randomized, double-blind study
LDV, skin temp.	RP in Scl {8} pr.RP {12} [dor. prox. phalanx of finger]	1) Responsiveness of patient with severe RP to intradermal injections of 25 µl of $1\mu M$ calcitonin-gene-related peptide or saline; later i.v.-therapy for 15 min on 5 successive days in four Scl patients
PP	RP in Scl {5} [finger]	1) Effect of long-term iloprost therapy on RP in Scl patients to 6–h iloprost infusions on 6 successive days followed by weekly infusions during the winter months

Result(s)/Conclusion(s)	Note[a]	Ref. (Year)
↑ of transcapillary pressure gradient, ↑ of capillary RBC velocity. This therapy improves nutritive capillary circulation by lowering precapillary resistance.		96 (1982)
An ↑ of digital BF and pulse amplitude was seen but did not reach statistical significance. However, there was a significant ↓ of digital peripheral vasc. resistance and a significant ↑ of the dicrotic notch proportion of pulse amplitude.		97 (1987)
Microcirculatory flow was consistently ↑ by iloprost but not by nifedipine. However, ↓ of clinical symptoms was equal in both drugs. This is explained by the different mode of actions of the two drugs. Therefore, it may be beneficial to use both drugs in Scl.	–N/A–	98 (1989)
All subjects produced significantly greater immediate and delayed response to CGRP than to saline. All patients treated showed ↑ of LDV and/or finger temp. CGRP may be an useful treatment for severe RP with its long-term efficacy under evaluation.	–B–	99 (1989)
Prominent ↓ of number, duration, and severity of Raynaud's attacks. Improvement or normalization of digital PP combined with clinical improvement and healing of finger ulcerations.		100 (1993)

Table 2   *Continued*

Method(s)	Disease {Number of Patients} [Measuring Site]	1) Physiological Aim/ Therapeutic Regimen 2) Possibly Combined with Stress Test(s)
LDV, strain-gauge photoplethysmography, skin temp.	RP in Scl {35} [fingertip, finger]	1) Physiological effects and clinical usefulness of i.v. iloprost in a double-blind, placebo-controlled study of iloprost infusion over 6 h for 5 successive days 2) Cold challenge, finger warming to 38° C, arterial occlusion
Topical Vasodilators		
PP	RP {12} [finger]	1) Efficacy in vasospastic RP during winter of topical ointment of 200 mg isosorbide dinitrate in a double-blind, placebo-controlled crossover study 2) Local cooling (3 min) in ice water
LDV	pr.RP {13} RP in Scl {12} [ventral forearm, dor. hand, dor. middle phalanx]	1) Efficacy on cut. BF in patients with RP of topically applied 0.1% and 1.0% hexyl nicotinate lotion on 2 cm² skin in various sites
Nifedipine		
Venous occlusion air plethsmography	pr.RP {3} RP in CTD {5} [fingertip]	1) Effects on clinical symptoms and cut. BF of sublingual nifedipine administration in a double-blind, placebo-controlled crossover study
Strain-gauge plethysmography, PP	RP in Scl {10} [forearm, index finger, fingertips 3rd digits]	1) To evaluate the efficacy of nifedipine on the number, severity, and duration of RA in a double-blind, placebo-controlled crossover trial of 10 mg nifedipine, 3× daily for 3 weeks
LDV, thermography, bolometry, ultrasonic Doppler	pr.RP {9}	1) To determine the acute effects of 20 mg sublingual nifedipine

Result(s)/Conclusion(s)	Note[a]	Ref. (Year)
Improved healing of cut. lesions. ↓ of critical ischemic temp (21.3– 16.1° C). Improvement of skin temp. recovery after cold test. No changes in digital BF, digital skin ambient temp, or platelet activation. Positive effects were seen up to 10 weeks after therapy. The mechanism of this remains unclear.	–B– –C–	101 (1993)
Treatment was ineffective compared to placebo.	–A–	102 (1983)
Both concentrations produced a 100% response with ↑ LDV at the forearm. In Scl the response rate at more distal sites ranged between 33% (finger) to 55% (hand), only rarely showing erythema. The 1% lotion induced somewhat higher response rates, while no negative side effects were observed. This treatment may be of value for patients with RP who are intolerant of systemic vasodilators.	–A– –B–	103 (1988)
Frequency and severity of RA were ↓ with partial healing of finger ulcers. Total BF ↑. Therefore nifedipine may be a useful agent for treatment of digital vasospasm.	–A– –B– –C–	104 (1983)
Duration of RA ↓ significantly; however, ↓ of number and severity of RA together with ↑ digital BF was not significant. Clinically, shortening of the duration of RA is a more valuable therapy outcome than ↓ in number of RA, making this a useful therapy.		105 (1987)
There was some evidence of protection against reduction in BF. Therefore, self-administration before cold exposure may be effective for preventing RA, in contrast with the conventional form of regular oral administration.	–A–	106 (1987)

Table 2 *Continued*

Method(s)	Disease {Number of Patients} [Measuring Site]	1) Physiological Aim/ Therapeutic Regimen 2) Possibly Combined with Stress Test(s)
Ketanserin		
LDV, PP, skin temp.	RP in CTD {10} [LDV: volar middle phalanx; PP: 4th fingertip; temp. 5th fingertip]	1) Effectiveness in RP due to CTD of 20 mg ketanserin 3× daily for 10 days followed by 40 mg 3× daily for a total treatment period of 3 weeks and 3 periods in a double-blind, placebo-controlled crossover study 2) Cooling test (2 min, 15° C water bath)
LDV, systolic finger blood pressure	RP in Scl {9} [fingertip]	1) Efficacy in treatment of RP of 20 mg ketanserin 3× for the 1st week and 40 mg 3× daily for 2nd to 5th week 2) Body and finger warming (15 min, 30° C), cooling test (15 min on a 15° C blanket)
LDV, thermography, bolometry	RP in Scl {11} [N/A]	1) Therapeutic effect of ketanserin (serotonin antagonist)— 10 mg i.v. as a bolus followed by infusion (2 mg/h) over 72 h, then oral therapy with 40 mg 3× daily for 4 days
Prazosin		
LDV, skin temp.	pr. and sec. RP {24} [finger]	1) Effects of prazosin (1 mg 3× daily in a randomized, double-blind, placebo controlled crossover study 2) Standard finger cooling test
LDV, skin temp.	RP {24} [finger]	1) Dose-response study of prazosin: clinical effectiveness vs. side effects of prazosin (3 mg, 6 mg, and 12 mg daily)

Result(s)/Conclusion(s)	Note[a]	Ref. (Year)
LDV is the most useful method of estimating variations in digital BF in Scl. ↓ of recovery time after cold provocation combined with an unchanged resting BF and healed ulcers. The number and duration of RA ↓. It is suggested that ketanserin acts by blocking the platelet aggregation and vasoconstriction induced by serotonin. Therefore, it may be an effective and well-tolerated treatment for RP, especially in Scl.	–A– –B–	107 (1984)
No significant ↑ in LDV or finger BP during cold challenge could be found. An ↓ in duration of induced vasospasm was observed, while number of RA did not ↓. The dosage used was not effective in treatment of RP, but interference of the concomitant drug (penicillamine) was discussed as a possible explanation for this result.	–C–	108 (1988)
Skin BF as measured by thermography, LDV, bolometry, and ultrasound-Doppler pulses showed significant improvement with clinical healing of ulcers. Combined i.v. and oral regimen is effective in treatment of critical digital ischemia in Scl.	–B– –C–	109 (1989)
During the cooling test beneficial effects of prazosin for finger skin temp. and LDV were seen. No difference in reaction to therapy between pr. RP and sec. RP was found. In ⅔ of patients with RP this agent is useful in treating digital vasospastic disease.	–A–	110 (1986)
Number and duration of RA ↓ significantly. Skin BF and skin temp. ↑ evaluated by the cooling test. Skin BF before, during, and after cold challenge showed no significant difference between the three regimens. Side effects ↑ with ↑ dosage. Therefore, the dosage of 3 mg daily gave best balance between clinical effectiveness and side effects.	–A– –B–	111 (1988)

Table 2 **Continued**

Method(s)	Disease {Number of Patients} [Measuring Site]	1) Physiological Aim/ Therapeutic Regimen 2) Possibly Combined with Stress Test(s)
Various Regimens		
PP, mercury gauge pulsed plethysmography	pr. RP {12} [finger]	1) Comparison of the two plethysmographic techniques for measurement of the vasodilator effect of 8 mg dihydroergokryptine daily for 4 weeks
PP, various clotting parameters	pr. RP {11} [finger]	1) Effectiveness of 400 mg pentoxifylline 3× daily per os for 2 months 2) Exposure to cold conditions
Capillaroscopy	Rp in Scl {18}	1) Microcirculatory and hemorrheological parameters before and after plasmapheresis once a week over 4 weeks
LDV, strain-gauge plethysmography	RP in Scl {6} [LDV: hand, plethysmography: finger, forearm]	1) Effect of reserpine administered by local retrograde transvenous application once

Note: Abbreviations—↑ = increase, increased; ↓ = decrease, decreased; AOD = arterial occlusive disease; art. = arterial; BF = blood flow; BP = blood pressure; CTD = connective tissue disease; cu(t.) = cutaneous; dor. = dorsal; gen. = generalized; HR = hyperemic response; i.v. = intravenous; LDV = laser Doppler velocimetry; loc. = localized; pO_2 = oxygen tension; N/A = information not available; PP = photoplethysmography; pr. = primary; prox. = proximal; RA = Raynaud's attack; RBC = red blood cell(s); RBV = red blood cell velocity; RH = reactive hyperemia; RP = Raynaud's phenomenon; Scl = scleroderma; sec. = secondary; tc = transcutaneous; temp. = temperature; UCTD = undifferentiated connective tissue disease; vasc. vascular.

Result(s)/Conclusion(s)	Note[a]	Ref. (Year)
Mercury gauge plethysmography measures global digital BF and appears to be more sensitive than PP, which measures dermal and hypodermal BF, for follow-up of effects of vasodilator treatment on RP.	–A–	112 (1987)
7 of 11 patients treatment showed distinct improvement of peripheral BF and clinical symptoms under basal conditions as well as after exposure to cold. Positive changes were also found in the various clotting variables.	–A– –B–	113 (1987)
RBC velocity ↑; RBC aggregation and plasma viscosity ↓. Skin ulcers healed and did not return up to 3 years after treatment. Hemorrheologic factors returned to initial values, but skin capillary BF remained ↑ for 24 months. Disturbed hemorrheology plays a role in the diminished skin BF.		114 (1991)
Total finger BF and LDV ↑, persisting for 3 days. Reserpine initiated mainly ↑ of BF to skin, including microcirculation (LD). This combined with subjective improvement. Therefore, local sympathetic denervation using reserpine in Bier's technique might be clinically useful.	–A– –B–	115 (1992)

[a]**A:** Attention has to be given to the patients group used; they are either nonhomogeneous (i.e., just designated RP without an exact underlying cause being given) or patients were also used other than those suffering from scleroderma. **B:** Notice that either uneven skin temperature prevailed in the different groups measured (this is often the case when comparing patients and controls without using a heated probe holder, especially in locations like the fingertip, where significant temperature differences can be observed; see also Table 7–3) or no information on skin temperature was given. **C:** The measuring location (e.g. fingertip) needs special consideration because the anatomical site contains a variety of vessels (additional arteriovenous anastomoses) and has a different innervation control; both could have a major influence on the results. In addition, temperature differences (see note B) are more likely.

of studies have used the fingertip or finger location, whereas Albrecht et al.[21] also performed LDV and pO_2 measurements at the dorsal midhand, finding as initial state of hyperemic hypoxia in Scl. Decreased capillary density in Scl, impaired oxygen diffusion, perivascular fibrosis and partial thickening of the dermis, and a compensatory increased perfusion of the precapillary vasculature in order to maintain a minimal circulation in the nutritive-capillary area were the reasons discussed for this.

Again, studies on the local regulation of blood flow (BF) in cutaneous tissue[117] at the back of the hand, and of digital BF,[121] were performed by Kristensen and colleagues using either orthostatic changes in vascular transmural pressure caused by lowering and elevating the hand in relation to heart level or local heating, arterial occlusion, and local cooling as stress tests. They found an abolished vasoconstrictor response to an increase in venous trans-mural pressure and a defective autoregulation to changes in the latter. Thus, the normal edema protecting mechanisms are defective, and BF in cutaneous tissue on the dorsum of the hand appears to depend mainly on the "arterial perfusion pressure head". The experiment on regulation of digital BF used different measuring locations, such as the dorsum of the hand, dorsal proximal phalanx, the area immediately proximal to the nailfold, and the apex of the finger. There[121] it was shown that in Scl from hand to fingers and from subcutis to cutis there was decreased reactivity of the vascular bed upon changes in transmural pressure and in the arterial perfusion pressure head. This was seen as an expression of deteriorating vascular smooth muscle function. In subcutaneous as well as in cutaneous tissue, reactive hyperemia to corresponding stimuli was lowered, indicating an increased digital artery resistance and demonstrating that a passive vascular bed is an important factor in RP. Digital vascular resistance was found to be greater than in normals even during warming. RP in Scl may be dependent on a low-pressure vascular bed where a local factor in blood flow control is more or less defective. Thus, it seems that vasodilator drugs which lower systemic blood pressure are contraindicated while warm environments and an increase in systemic blood pressure would be beneficial.[121]

IV. GUIDELINES TO FOLLOW WHEN APPLYING LASER DOPPLER VELOCIMETRY IN SCLERODERMA

A. AMBIENT CONDITIONS, PATIENT- AND DISEASE-RELATED FACTORS

The general guidelines for measuring LDV and PP have already been outlined in Chapter 4 and in a very recent review from the standardization group of the European Society of Contact Dermatitis.[122] Of course, they should also be adhered to in patients suffering from Sclero-derma. Therefore, only a few points especially important in Scl will be raised.

Due to the special sensitivity of scleroderma patient to cold or cold-inducing stress it is even more important than in normals to keep the ambient conditions stable and comfortable in order to not provoke Raynaud attacks involuntarily. This means that the room temperature normally suggested (20 to 25° C) for LDV measurements should be above 22° C rather than below. Furthermore, the environment has to be draught-free since this could also lead to cooling effects.[123,124] In addition, room humidity should be registered and kept constant in the normal range of 40 to 50%. A quiet room with dimmed lights is as important as avoiding mental stress[125] and unnecessary talking, which modifies respiration.

An adaptation time of 20 to 30 min to room conditions is thought to be adequate for the usual subject. However, scleroderma patients seem to need a longer time to "equilibrate", especially when they come from outdoors during the cold season. Usually the patient himself can determine when he feels comfortable in the room, which often takes longer than 20 to 30 min. Therefore, an adaptation time of 30 to 40 min seems mandatory, in the wintertime even longer. Since the sensitivity of LDV to motion artifacts requires that the patients lie still, they tend to get stiff and later painful joints. Therefore, the duration of the experimental

Table 3 **Flux and skin surface temperature at different sites on the unheated skin (hand) in patients with Scleroderma (Scl) (*n* = 19) and controls (*n* = 15)[a]**

Site	Controls	Scl
	Flux Measurements (rU)	
Wrist	9.70 ± 2.40	19.95 ± 9.58[b]
Dorsal hand	14.73 ± 5.45	29.85 ± 17.46[c]
Digitus 3	19.73 ± 9.22	26.25 ± 18.63[e]
Fingertip	124.40 ± 62.90	77.45 ± 47.20[d]
	Temperature Measurement (° C)	
Wrist	29.21 ± 1.96	29.87 ± 1.64[e]
Dorsal hand	30.02 ± 1.86	29.98 ± 1.72[e]
Digitus 3	30.80 ± 2.36	28.76 ± 2.90[e]
Fingertip	32.07 ± 2.24	26.98 ± 3.40[d]

[a]Mean ± SD, [b]$p < 0.01$, [c]$p < 0.05$, [d]$p < 0.001$, [e]$p > 0.05$, not significant.
From Albrecht, H.-P., Hiller, D., Hornstein, O. P., Bühler–Singer, S., Mück, M. and Gruschwitz, M., *J. Invest. Dermatol.*, 101 No. 2, 211–215, 1993. With permission.

procedure should not be longer than 1 to 2 h. Measurements involving cooling tests or provocations of Raynaud attacks have to be performed within the same season[101] of the year, and the conditions should be stated in the report.

When comparing groups of patients with each other or normals, an exact classification of the disease must be given, regardless of the classification scheme used. There should be no patient groups classified as just having Raynaud's phenomenon, since this condition may be due to a variety of different underlying causes like scleroderma, lupus erythematosus, or other connective tissue diseases as well as the so-called primary RP.

Of course, the anatomical site of measurement is of utmost importance.[122] Aside from insuring the same measuring location in each subject/patient as close as possible, special consideration must be given to the selection of the most appropriate measuring site. The fingertip (one location chosen very often) has—in the authors' eyes—some serious drawbacks. First of all, it is not possible to place more than one measuring probe on it, something that should be encouraged very much. Second, this location has a different innervation control and variety of vessels present compared to the hand and is usually afflicted by proliferative intimal arterial lesions of the digital arteries typical in Scl. Thus, this macrocirculatory component can superimpose on the microcirculatory alterations in Scl.[94] When using stimulus-response experiments, measurements at multiple sites with subsequent averaging of the single LDV values are not possible, therefore, the site of measurement should have the highest possible LDV values to control for the well-known problem of spatial heterogeneity. This was shown to be the case for the dorsal midhand (Table 3), which is usually affected by sclerosis. The local temperature of the measuring site is also a very important variable for microcirculatory methods such as LDV and PP. Skin and probe temperature should always be given. As seen in Table 3, flux in unheated skin correlated with the skin temperature at the fingertip location and was significantly different from the other locations. Kristensen[118] even reported a 3° C difference at the dorsum of the finger just proximal to the nailfold. Therefore, ensuring the same local temperature (e.g., by a heated probe holder) seems appropriate. The reduced blood flow in the fingertip of Scl patients found with unheated probes in a variety of studies (see Table 1) might be due in part to the significant temperature difference at this location. Goodfield et al.[87] reported significantly lower resting flood flow (LDV values) for unheated skin of the fingertip than in normals. However, after hand warming in 35° C water, LDV in Scl and controls did not differ significantly. He stated: "The warmed hand appears to be a useful starting point for comparative investigations." In a model of local arterial insufficiency, something which also can occur in the fingers of Scl patients, an investigation of LDV and transcutaneous pO_2 as a function of temperature was performed. The results strongly indicate that, at least in this model, LDV and transcutaneous pO_2 measurements reflect changes in the local arteriovenous gradient only when measured in warmed skin.[126] Studying the effects of duration of ischemia and local heating on reactive hyperemia in skin of the human foot measured by LDV, Wamsley and Wiles[127] showed that the duration of ischemia required to produce maximal postischemic peak flow was longer for foot skin than for the whole limb and that these results were consistent with temperature differences being the cause of this. Thus, a standard skin temperature of 32 to 33° C was suggested as useful for studying cutaneous blood flow responses since acclimatization to room conditions alone appeared to be inadequate. Furthermore, different temperatures could affect the biological zero in LDV, possibly due to the involvement of Brownian molecular motion within the tissue[128] and/or the various states of temperature-dependent vasodilatation, which was shown by Wahlberg et al.[129] to increase the biological zero markedly.

Another point not often addressed is the probe pressure. The effect of pressure loading on the underlying skin blood flow can be of considerable clinical interest when interpreting blood flow measurements.[130,131] It was shown by Obeid et al.[132] that LDV values could be dramatically reduced by relatively small pressures (<15 mmHg) in the range of systemic venous pressure, with 5 mmHg already producing approximately a 50% reduction of LDV values. Newson and Pearcy[133] found a minimal "cutoff pressure" of 15.2 kPa=114 mmHg in human skin above the prominence of bones and showed that transcutaneous measurement (45° C) of oxygen tension was not seriously affected by leads of up to 10% of the cutoff pressure. The usual commercially available oxygen electrodes together with their probe holders produce an average pressure of 1 to 2% of the cutoff load (=1.1 to 2.3 mmHg). The large-area LDV probes with their flexible leads cause a pressure of approximately 1 mmHg. However, smaller cylindrical LDV probes without a large probe holder can produce pressure loads in the range of 3 to 5 mmHg, thus affecting the underlying microcirculation. Although skin in scleroderma is sclerotic and the dermis is partially thickened, one should not conclude that microcirculation in Scl is less sensitive to probe pressure than in normal individuals. Thus, it seems important to ensure that the pressure produced does not exceed the above values, especially when using cylindrical pencil probes with very small contact surfaces.

B. EXEMPLARY STUDY PROTOCOL

The methods used (LDV, PP) have physiological limits due to the inter- and intraindividual spatial and temporal heterogeneity of the microcirculation.[134-136] These restrictions have to be kept in mind even when minimizing external and internal influences on the microcirculation by adhering to the guidelines described above and in Chapter 4.[122] Always using the same anatomically defined position and exact marking of the skin position of the probes, the probe holders, and their position relative to each other can partly control spatial heterogeneity. Through this the spatial interindividual mistake is then evenly distributed statistically within all measurements and thus is attenuated by group data evaluation, which usually shows great standard deviations of the mean values (LDV). Therefore, individual microcirculatory data should not be simply compared with each other. However, comparison of groups showing statistically significant differences is usually regarded as valid by most researchers, despite their great standard deviation. The problem of the great variations in blood flow at adjacent sites could also be overcome by a proposed contour mapping of cutaneous microvasculature[137] with a specially designed computerized probe holder, but this cannot be applied

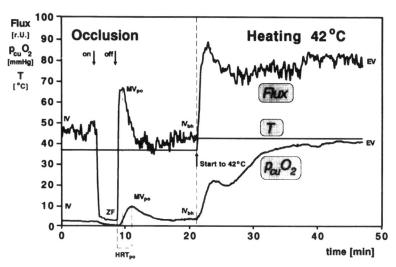

Figure 1 Examplary study protocol using 3-min arterial occlusion and local heating to 42°C as two subsequent stimulus-response tests during the experimental procedure. Filtered original recordings of LDV (flux), LDV probe temperature (T), and cutaneous pO_2 ($p_{cu}O_2$) in a 54-year-old female patient suffering from scleroderma IV, initial values; ZF, zero flux; HRT_{po}, postocclusive hyperemic recovery time; MV_{po}, maximum postocclusive values; IV_{bh}, initial values before heating; EV, end values after heating.

for fast stimulus-response tests. There are conflicting reports on the effect of temporal heterogeneity on LDV measurements. Temporal day-to-day variation may be significant,[136] whereas temporal diurnal factors are said to be minor.[122] Generally, it is proposed that long-term studies be performed with repetitive measurements at the same time of day, even though no major changes in cutaneous blood flow during the day have been observed. However, a recent study[138] on day-time variations in peripheral micro- and macrocirculation in man using capillaroscopy , LDV, and venous occlusion plethysmography found that forearm (predominantly muscle) blood flow increased during the day, thermoregulatory skin blood flow (LDV) decreased, while nutritive skin blood flow (capillaroscopy) showed no significant changes. Thus, different regulatory mechanisms for skin nutritive, skin thermoregulatory, and forearm muscle blood flow seem to be present. Since thermoregulatory skin blood flow (LDV) is of special interest in Scl and the three compartments of blood flow can be affected differently in scleroderma, measurements at the same time of day, preferably in the morning hours, are suggested.

Further lessening of the interindividual variations can be achieved by stimulus-response experiments like arterial occlusion and local heating, where reliable reproducibility had been shown in earlier studies.[139,140] The comparison of time-responses in the dynamic tests together with assuring the same local measuring temperature by using a heated probe or probe holder also reduces part of the interindividual heterogeneity in microcirculation. Therefore, the following study protocol for LDV measurements in scleroderma (Figure 1) is suggested:

1. If at all possible, combine LDV with other microcirculatory methods such as cutaneous oxygen tension measurement (pO_2) or capillaroscopy, ideally in combined electrodes,[139] or apply them on directly adjacent spots.

2. When using the fingertip or finger as a measuring location, use a second spot, also sclerotic (e.g., the back of the hand), which is not influenced by the alterations of the digital arteries typical in Scl. Additionally, measuring a third nonsclerotic area would give even more information.

3. Measure baseline values of LDV and local skin temperature on unheated skin, then determine and control an identical local measuring temperature in each subject/Scl patient by using a heated probe/probe holder. Choose a temperature which still enables satisfactory responses to stimulatory tests and is compatible with the other methods used simultaneously, such as 36 to 37° C for combination with pO_2 measurements.

4. After obtaining stable initial values with the moderately heated probes, produce an abrupt arterial occlusion by immediate automatic inflation of a correctly sized blood pressure cuff to 40 mmHg above systolic pressure. Maintain and record zero perfusion together with the individual biological zero in LDV ("zero flux") for 3 min, then deflate the cuff immediately.

5. A recovery time of 10 min after arterial occlusion is sufficient for subjects. However, after this time scleroderma patients showed[21] a still increased cutaneous pO_2 and a slight increase in LDV. Therefore, a longer recovery time (sometimes up to 40 min) appears to be better. In order not to prolong the procedure too much one probably has to compromise on this point, e.g., with a 15-min recovery time. In any case, the same experimental set up has to be used in the control group for valid comparison.

6. Induce a local heating stimulus by abruptly heating the electrodes to, e.g., 42° C using a built-in electrical heating system with identical time pattern in all probes. From the beginning of point 3, continuous recording is essential throughout the whole experimental procedure and has to last up to at least 25 min after start of heating.

Concerning photoplethysmography, it must be mentioned that sometimes this method cannot be used in scleroderma patients, possibly because of the tightness of the skin and low digital blood flow in Scl, with a resulting low pulsatile flow.[95,107]

C. VALUABLE INFORMATION TO BE DRAWN FROM SUCH EXPERIMENTS

Using the above-presented study protocol the data for LDV and, e.g., $p_{cu}O_2$ can be evaluated for each stimulus separately with regard to their levels, time courses, and relation to each other. By these means the following parameters have been specified by Albrecht et al.,[21] as relevant to the microcirculatory response to defined stimuli.

The parameters to study in occlusion are: IV=initial values with a slightly heated probe, e.g., at 36° C, ZF=biological zero flux during occluded perfusion; MVpo=maximum postocclusive values; HRVpo=postocclusive hyperemic response values (MVpo−IV); HRTpo=postocclusive hyperemic recovery time up to MVpo, and OE=occlusion effect (HRVpo×10/HRTpo). The parameters to study in the heating (e.g. 42° C) test are: IVbh=initial values before heating, EVh=end values after 25 min of heating (42° C), and RW=regulation width (EVh − IVbh). All the above-defined static and dynamic parameters have to be corrected for the individual biological zero flux (ZF). This is necessary to improve the interindividual comparability of the data[129] since the amount of ZF is, within a certain range, dependent[130] on the individual measuring location, the system used, and the temperature. Under good perfusion conditions this phenomenon was found to be negligible. However, in a low perfusion situation (e.g., peripheral arterial occlusive disease) ZF could amount to up to 80% of the total LDV signal, therefore, it must be taken into account, especially when measuring changes following provocation tests in Scl. With the device used [21] at the back of the hand in Scl patients the ZF was found to range between 10 and 20% of the initial value (IV).

Short-term arterial occlusion and local thermal stimulation both cause cutaneous hyperemia, the measurement of which enables the testing of the capacity of poststimulatory vasodilation. Using the presented study protocol pronounced differences in the mean initial values for LDV (and $pcuO_2$) were found between Scl patients and controls by Albrecht et al.[21] Furthermore, the dynamic parameters, occlusion effect (OE) and regulation width (RW), were significantly lower in Scl and were combined with a significant reduction of the postocclusive hyperemic response value (HRVpo) and the postocclusive hyperemic recovery time (HRTpo), which both reflect the microcirculatory functional performance.

The RW is a partial measure for what has been termed cutaneous vascular reserve.[139] The OE indicates the ability for postocclusive compensation through the peripheral vessels. HRVpo and HRTpo yield appropriate measures for the reactivity of the dermal vasculature to hypoxia and ischemia. The IV and maximum postocclusive values (MVpo), which determine HRVpo, give information if reductions of HRVpo are due to either a restriction of MVpo or an elevation of IV. Therefore, the distensibility of the vascular bed under study can be evaluated. At the dorsal midhand in Scl the latter was not changed because limitation of HRVpo was due to an increased IV and not to a restricted MVpo. The maximum postocclusive value was significantly higher in Scl than in controls.[21] The prolongation of HRTpo in Scl pointed to an altered compliance of the dermal vessels due to their disturbed reactivity and/or fibrosclerotic perivascular tissue.

Thus, the presented static and especially the dynamic parameters contain valuable information on microcirculatory functions in scleroderma and also can be used to evaluate therapeutic concepts if they can positively influence microcirculation in Scl.

The importance of such dynamic parameters for diagnosis, assessment of the course, and therapy of Scl can be seen from the data of Tables 1 and 2. Stimulatory testing was applied four/one times using occlusion, eight/two times using heating, nine/six times using cooling, and three/zero times using sympathetic stresses such as tilting, Valsalva maneuver, etc., respectively. Such protocols were also used[75,79–81,83,85,90,91] to further evaluate whether the clinical sign—Raynaud's attack—was due to a secondary cause such as Scl or other connective tissue diseases or whether it was instead related to the so-called primary Raynaud's phenomenon. Except for Serup and Kristensen[94] who showed that in localized Scl the increased LDV values became normal with time, there have been to our knowledge up to now no data available using the above microcirculatory testing to assess the course of scleroderma. However, the authors of this chapter will be involved in a prospective study of calcitonin in Scl using microcirculatory testing. Therapeutic evaluations in Scl are a stronghold using LDV and photoplethysmography to study the effect on cutaneous blood flow (see Table 2).

V. COMMENTS

In summary, the above paragraphs show that laser Doppler velocimetry and, to a lesser extent, photoplethysmography provide good noninvasive measuring medical research tools for the continuous *in vivo* assessment of the vascular aspects and a more objective and quantitative control of therapeutic regimens in scleroderma, which is a very interesting but also confusing disease due to the broad range of clinical manifestations and characteristic features. For the future, such methods should be used even more often as objective criteria to describe and evaluate skin status and function instead of morphology alone, which can be deceptive. With more objective quantification of disturbing symptoms in skin diseases—such as Raynaud's phenomenon in Scl—it is hoped that a better understanding of the underlying pathophysiology and, ultimately, more effective patient management will be achieved. New developments in microcirculatory methods such as laser Doppler imaging will probably further enhance the information on scleroderma already available due to the above procedures. Yet measurements using laser Doppler imaging in scleroderma have not been published to our knowledge.

ACKNOWLEDGMENT

The authors are very much indebted to Professor Otto Paul Hornstein, whose genuine interest in skin microcirculation and scleroderma "catalyzed" several studies on these subjects and thus enabled this contribution.

REFERENCES

1. Goetz, R. H., Pathology of progressive systemic sclerosis (generalized scleroderma) with special reference to changes in the viscera, *Clin. Proc.,* 4, 337, 1945.
2. Subcommittee for Scleroderma Criteria of the American Rheumatism Association Diagnostic and Therapeutic Criteria Committee, Preliminary criteria for the classification of systemic sclerosis (scleroderma). *Arthritis Rheum.,* 23, 581, 1980.
3. Krieg, T. and Meurer, M., Systemic scleroderma, *J. Am. Acad. Dermatol.,* 18(3), 457, 1988.
4. Arbeitsgruppe Sklerodermie der Arbeitsgemeinschaft Dermatologische Forschung (ADF), Klinik der progressiven systemischen Sklerodermie (PSS), *Hautarzt,* 37, 320, 1986.
5. LeRoy, E. C., Black, C., Fleishmayer, R., Jablonska, S., Krieg, T., Medsger, T. A., Rowell, N., and Wollheim, F., Scleroderma (systemic sclerosis). Classification, subsets and pathogenesis, *J. Rheumatol.,* 15, 202, 1988.
6. Krieg, T. and Meurer, M., Diseases of connective tissue, in *Dermatology,* Braun-Falco, O., Plewig, G., Wolff, H. H., and Winkelmann, R. K., Eds., Springer-Verlag, Berlin, 1991, chap.18.
7. Masi, A. T., Clinical-epidemiological perspective of systemic sclerosis (scleroderma), in *Systemic Sclerosis: Scleroderma,* Jayson, M. I. V., and Black, C. M., Eds., John Wiley & Sons, Chichester, 1988, chap. 2.
8. Shore, A. C., Sandeman, D. D., Mawson, D. M., and Tooke, J. E., The relationship between skin temperature and capillary pressure pulse amplitude (CPPA) in healthy men and women, *Int. J. Microcirc. Clin. Exp.,* 11 (Suppl. 1), 220, 1992.
9. Wamsley, D. and Goodfield, M. J. D., Evidence for an abnormal peripherally mediated vascular response to temperature in Raynaud's phenomenon, *Br. J. Rheumatol,* 29(3), 181, 1990.
10. Hiller, D., Kessler, M., Hornstein, O. P., Vergleichende kutane Sauerstoffdruckmessung ($p_{cu}O_2$) bei Gesunden und bei Patienten mit progressiver Sklerodermie, *Hautarzt,* 37, 83, 1986.
11. Silverstein, J. L., Steen, V. D., Medsger, T. A., and Falanga, V., Cutaneous hypoxia in patients with systemic sclerosis (scleroderma), *Arch. Dermatol.,* 124, 1379, 1988.
12. Gruschwitz, M., Moormann, S., Kroemer, G., Sgonz, R., Dietrich, H., Böck, G., Gershwin, M. E., and Wick, G., Phenotypic analysis of skin infiltrates in comparison with peripheral blood lymphocytes, spleen cells and thymocytes in early avian scleroderma, *J. Autoimmun.,* 4, 577, 1991.
13. Jimenez, S. A., Experimental models of scleroderma, in *Systemic Sclerosis: Scleroderma,* Jayson, M. I. V. and Black, C. M. Eds., John Wiley & Sons, Chichester, 1988, chap. 8.
14. Murell, D. F., A radical proposal for the pathogenesis of scleroderma, *J. Am. Acad. Dermatol.,* 28(1), 78, 1993.
15. Welsh, K. I. and Black, C. M., Environmental and genetic factors in scleroderma, in *Systemic Sclerosis: Scleroderma,* Jayson, M. I. V. and Black, C. M., Eds., John Wiley & Sons, Chichester, 1988, chap. 3.
16. LeRoy, E. C., Scleroderma (systemic sclerosis), in *Textbook of Rheumatology,* Vol. 2, Kelley, W. N., Harris, E. D., Jr., Ruddy, S., and Sledge, C. B., Eds., W. B. Saunders, Philadelphia, 1985, chap. 76.

17. Serup, J., Localized scleroderma (morphoea): thickness of sclerotic plaques as measured by 15 MHz pulsed ultrasound, *Acta Derm. Venereol.*, 64(3), 214, 1984.

18. Serup, J., Quantification of acrosclerosis: measurement of skin thickness and skin-phalanx distance in females with 15 MHz pulsed ultrasound, *Acta Derm. Venereol.*, 64(1), 35, 1984.

19. Hoffmann, K., Gerbaulet, U., el-Gammal, S., and Altmeyer, P., 20-MHz B-mode ultrasound in monitoring the course of localized scleroderma (morphea), *Acta Derm. Venereol. Suppl.*, 164, 3, 1991.

20. Serup, J. and Northeved, A., Skin elasticity in localized scleroderma (morphoea). Introduction of a biaxial in vivo method, and measurement of tensile distensibility, hysteresis and resilient distension of diseased and normal skin, *J. Dermatol. (Tokyo)*,12(1), 52, 1985.

21. Albrecht, H.-P., Hiller, D., Hornstein, O. P., Bühler-Singer, S., Mück, M., and Gruschwitz, M., Microcirculatory functions in systemic sclerosis: additional parameters for therapeutic concepts?, *J. Invest. Dermatol.*, 101, 211–215, 1993.

22. Maricq, H. R., Widefield capillary microscopy: technique and rating scale for abnormalities seen in scleroderma and related disorders, *Arthritis Rheum.*, 24, 1159, 1981.

23. Carpentier, P. H. and Maricq, H. R., Microvasculature in systemic sclerosis, *Rheum. Dis. Clin. North Am.*, 16(1), 75, 1990.

24. Roald, O. K. and Seem, E., Treatment of Raynaud's phenomenon with ketanserin in patients with connective tissue disorders, *Br. Med. J. Clin. Res. Ed.*, 289(6445), 577, 1984.

25. Lee, P. Sarkozi, J., Bookman, A A., Keystone, E. C., and Armstrong, S. K., Digital blood flow and nailfold capillary microscopy in Raynaud's phenomenon, *J. Rheumatol.*, 13(3), 564, 1986.

26. Tordoir, J. H., Haeck, L. B., Winterkamp, H., and Dekkers, W., Multifinger photoplethysmography and digital blood flow pressure measurement in patients with Raynaud's phenomenon of the hand, *J. Vasc. Surg.*, 3(3), 456, 1986.

27. Keller, J., Huben, H., and Hornstein, O. P., Telethermometrische Messungen zur Beurteilung des Raynaud-Phänomens bei Patienten mit Kollagenosen, *Hautarzt*, 34, (Suppl. 6), 397, 1983.

28. McHugh, N. J., Csuka, M., Watson, H., Belcher, G., Amadi, A., Ring, E. F., Black, C. M., and Maddison, P. J., Infusion of iloprost, a prostacyclin analogue, for treatment of Raynaud's phenomenon in systemic sclerosis, *Ann. Rheum. Dis.*, 47(1), 43, 1988.

29. O'Reilly, D., Taylor, L., el-Hadidy, K., and Jayson, M. I., Measurement of cold challenge responses in primary Raynaud's phenomenon and Raynaud's phenomenon associated with systemic sclerosis, *Ann. Rheum. Dis.*, 51(11), 1193, 1992.

30. Campell, P. M. and LeRoy, E. C., Pathogenesis of systemic sclerosis: a vascular hypothesis, *Semin. Arthritis Rheum.*, 4, 351, 1975.

31. Jayson, M. I. V., Systemic sclerosis—a microvascular disorder?, *J. Ro. Soc. Med.*, 76, 635, 1983.

32. Fleischmajer, R., The pathophysiology of scleroderma, *Int. J. Dermatol.*, 16, 310, 1977.

33. Gruschwitz, M., Müller, P. U., Sepp, N., Hofer, E., Fontana, A., and Wick, G., Transcription and expression of transforming growth factor type beta in the skin of progressive sclerosis: a mediator of fibrosis?, *J. Invest. Dermatol.*, 94, 197, 1990.

34. Kähäri, V. M., Sandberg, M., Kalimo, H., Vuorio, T., and Vuorio, E. Identification of fibroblasts responsible for increased collagen production in localized scleroderma by in situ hybridization, *J. Invest. Dermatol.*, 90, 664, 1988.

35. Bailey, A. J., and Back, C. M., The role of connective tissue in the pathogenesis of scleroderma, in *Systemic Sclerosis: Scleroderma*, Jayson, M. I. V. and Black, C. M., Eds., John Wiley & Sons, Chichester, 1988, chap. 6.

36. Kahaleh, M. B. and LeRoy, E. C., Vascular factors in the pathogenesis of systemic sclerosis, in *Systemic Sclerosis: Scleroderma*, Jayson, M. I. V. and Black, C. M., Eds., John Wiley & Sons, Chichester, 1988, chap. 7.

37. Fleischmajer, R., Dessau, W., Timpl, R., Krieg, T., Luderschmidt, C., and Wiestner, M., Immunofluorescence analysis of collagen, fibronectin and basement membrane protein in scleroderma skin, *J. Invest. Dermatol.*, 75, 270, 1980.

38. Burch, G. E., Harb, J. M., and Sun, C. S., Fine structure of digital vascular lesions in Raynaud's phenomenon and disease, *Angiology*, 21, 361, 1981.

39. Rodnan, G. P., Myerowitz, R. L., and Justh, G. O., Morphologic chances in the digital arteries of patients with progressive systemic sclerosis (scleroderma) and Raynaud's phenomenon, *Medicine*, 59, 393, 1980.

40. Kahaleh, M. B., Sherer, G. K., and LeRoy, E. C., Endothelial injury in scleroderma, *J. Exp. Med.*, 149, 1326, 1979.

41. Penning, C. A., Cunningham, J., French, M. A. H., Harrison, G., Rowell, N. R., and Hughes, P., Antibody-dependent cellular cytotoxicity of human vascular endothelium in systemic sclerosis, *Cancer Res.*, 46, 891, 1984.

42. Dau, P. C., Kahaleh, M. B., and Sagebiel, R. W., Plasmapheresis and immunosuppressive drug therapy in scleroderma, *Arthritis Rheum.*, 24, 1128, 1981.

43. Belch, J. J. F., Newmann, P., Drury, J. K., McKenzie, F., Capell, H., Lieberman, P., Forbes, C. D., and Prentice, C. R. M., Intermittent epoprosterol (prostacyclin) infusion in patients with Raynaud's syndrome, *Lancet*, 1, 313, 1983.

44. Dowd, P. M., Martin, M. F. R., Cooke, E. D., Bow-Cock, S. A., Jones, R., Dieppe, P. A., and Kirby, J. D. T., Treatment of Raynaud's phenomenon by intravenous infusion of prostacyclin (PGI$_2$), *Br. J. Dermatol.*, 106, 81, 1982.

45. Martin, M. F. R., Dowd, P. M., Ring, E. F. J., Cooke, E. D., Dieppe, P. A., and Kirby, J. D. T., Prostaglandin E$_1$ infusion for vascular insufficiency in progressive systemic sclerosis, *Ann. Rheum. Dis.*, 41, 350, 1981.

46. Mohrland, J. S., Porter, J. M., Smith, E. A., Belch, J., and Simms, M. H. A., Multi-clinic, placebo-controlled, double-blind study of prostaglandin E$_1$ in Raynaud's syndrome, *Ann. Rheum. Dis.*, 44, 754, 1985.

47. Holti, G., The effect of intermittent low molecular dextran infusions upon digital circulation in systemic sclerosis, *Br. J. Dermatol.*, 77, 560, 1965.

48. Roedeheffer, R. J., Rommer, J. A., and Wigley, F., Controlled double-blind trial of nifedipine in the treatment of Raynaud's phenomenon, *N. Engl. J. Med.*, 308, 880, 1983.

49. Lopez-Overjero, J. A., Saal, S. D., D'Angeelo, W. A., Cheigh, J. S., Stenzel, K. H., and Lavagh, J. H., Reversal of vascular and renal crisis of scleroderma by oral angiotensin-converting-enzyme blockade, *N. Engl. J. Med.*, 300, 1417, 1979.

50. Strandern, E., Roald, O. K., and Krogh, K., Treatment of Raynaud's phenomenon with 5-HT2-receptor antagonist ketanserin, *Br. Med. J.*, 4, 1069, 1982.

51. Roald, O. K. and Seem, E., Treatment of Raynaud's phenomenon with ketanserin in patients with connective tissue disorders, *Br. Med. J. Clin. Res. Ed.*, 289(6445), 577, 1984.

52. Klimiuk, P. S., Kay, E. A., Mitchell, W. S., Taylor, L., Gush, R., Gould, S., and Jayson, M. I., Ketanserin: an effective treatment regimen for digital ischaemia in systemic sclerosis, *Scand. J. Rheumatol.*, 18(2), 107, 1989.

53. Ortonne, J. P., Torzuoli, C., Dujardin, P., and Fraitag, B., Ketanserin in the treatment of systemic sclerosis: a double-blind controlled trial, *Br. J. Dermatol.*, 120, 261, 1989.

54. Gahhos, F., Ariyan, S., Frazier, W. H., and Cuono, C. B., Management of scleroderma finger ulcers, *J. Hand Surg. Am.*, 9(3), 320, 1984.

55. De-Trafford, J. C., Lafferty, K., Potter, C. E., Roberts, V. C., and Cotton, L. T., An epidemiological survey of Raynaud's phenomenon, *Eur. J. Vasc. Surg.*, 2(3), 167, 1988.

56. Carpentier, P., Magne, J. L., Guidicelli, H., and Franco, A., Sympathectomie and microcirculation: physiopathologic reflections, *J. Mal. Vasc.*, 10 (Suppl. A), 62, 1985.

57. Kristensen, J., Blood flow and blood pressure in fingers in generalized scleroderma, *Int. J. Dermatol.*, 21(7), 404, 1982.

58. Steen, V. D., Medsger, T. A., and Rodnan, G. P., D-penicillamine therapy in progressive systemic sclerosis: a retrospective analysis, *Ann. Intern. Med.* 97, 652, 1082.

59. Hein, R., Behr, J., Hündgen, M., Hunzelmann, N., Meurer, M., Braun-Falco, O., Urbanski, A., and Krieg, T., Treatment of systemic sclerosis with gamma-interferon, *Br. J. Dermatol.*, 126, 496, 1992.

60. Priestley, G. C. and Brown, J. C., Effects of potassium para-aminobenzoate on growth and macromolecule synthesis in fibroblasts cultured from normal and sclerodermatous human skin and human synovial cells, *J. Invest. Dermatol.*, 72, 161, 1979.

61. Zarafonetis, C. J. D., Antifibrotic therapy with Potaba, *Am. J. Med. Sci.*, 248, 550, 1964.

62. Zarafonetis, C. J. D., Retrospective studies in scleroderma. Effect of potassium para-aminobenzoate on survival, *J. Clin. Epidemiol.*, 41, 193, 1988.

63. Bloom-Boelow, B., Oberg, K., Wollheim, F. A., Persson, B., Jonson, B., Malmerg, P., Bostrom, H., Herbai, G., Cyclofenil versus placebo in progressive systemic sclerosis, *Acta Med. Scand.*, 210, 419, 1981.

64. Gibbson, T. and Grahame, R., Cyclofenil treatment of scleroderma—a controlled study, *Br. J. Rheumatol.*, 22, 218, 1983.

65. Rodnan, G. P., Black, R. L., Bollett, A. J., and Bunim, J. J., Observations on the use of prednisolone in patients with progressive systemic sclerosis (diffuse scleroderma), *Ann. Intern. Med.*, 44, 16, 1956.

66. Clements, P. J., Lachenbruch, P. A., Sterz, M., Danovitch, G., Hawkins, R., Ippoliti, A., and Paulus, H. E., Cyclosporine in systemic sclerosis, *Arthritis Rheum.*, 36(1), 75, 1993.

67. Mascaro, G., Cardario, G., Bordini, G., Tarditi, M., and Ferraris, G., Plasma exchange in the treatment of nonadvanced stages of progressive systemic sclerosis, *J. Clin. Apheresis,* 3, 219, 1987.

68. Jacobs, M. J., Jorning, P. J., Van Rhede-van der Kloot, E. J., Kitslaar, P. J., Lemmens, H. A., Slaaf, D. W., and Reneman, R. S., Plasmapheresis in Raynaud's phenomenon in systemic sclerosis: a microcirculatory study, *Int. J. Microcirc. Clin. Exp.*, 10(1), 1, 1991.

69. Rook, A. H., Freundlich, B., Jegasothy, B. V., Perez, M. I., Barr, W. G., Jimenez, S. A., Rietschel, R. L., Wintroub, B., Kahaleh, B., Varga, J., Heald, P. W., Steen, V., Massa, M. C., Murphy, G. F., Perniciaro, C., Istfan, M., Ballas, S. K., and Edelson, R. L., Treatment of systemic sclerosis with extracorporeal photochemotherapy, *Arch. Dermatol.*, 128, 337, 1992.

70. Steigerwald, J. C., Progressive systemic sclerosis. Management. III. Immunosuppressive agents, *Clin. Rheum. Dis.*, 5, 284, 1979.

71. Challoner, A. V. J., Photoelectric plethysmography for estimating cutaneous blood flow, in *Non-Invasive Physiological Measurements,* Vol. 1, Rolfe, R., Ed., Academic Press, London, 1979, 125.

72. Guy, R. H., Tur, E., and Maibach, H. I., Optical techniques for monitoring cutaneous microcirculation, *Int. J. Dermatol.*, 24(2), 88, 1985.

73. Hertzmann, A. B., Randall, W. C., and Jochim, K. E., The estimation of cutaneous blood flow with the photoelectric phlethysmograph, *Am. J. Physiol.*, 145, 716, 1946.

74. Engelhart, M. and Seibold, J. R., The effect of local temperature versus sympathetic tone in digital perfusion in Raynaud's phenomenon, *Angiology,* 41, 715, 1990.

75. Binaghi, F., Cannas, F., Mathieu, A., and Pitzus, F., Correlation among capillaroscopic abnormalities, digital flow and immunologic findings in patients with isolated Raynaud's phenomenon. Can laser Doppler flowmetry help identify a secondary Raynaud phenomenon?, *Int. Angiol.*, 11(3), 186, 1992.

76. Hahn, M., Jünger, M., Geiger, H., and May, M., Synchronous measurement of laser Doppler flux and red blood cell velocity in healthy controls and in patient with Raynaud's phenomenon during local cold exposure, *Int. J., Microcirc. Clin. Exp.*, 11 (Suppl. 1), 52, 1992.

77. Bunker, C. B., Foreman, J. C., and Dowd, P. M., Digital cutaneous vascular response to histamine and neuropeptides in Raynaud's phenomenon, *J. Invest. Dermatol.*, 96(3), 134, 1991.

78. Valentini, G., Leonardo, G., Moles, D. A., Apaia, M. R., Maselli, R., Tirri, G., and Del Guercio, R., Transcutaneous oxygen pressure in systemic sclerosis: evaluation at different sensor temperatures and relationship to skin perfusion, *Arch. Dermotol. Res.*, 283, 285, 1991.

79. Engelhart, M. and Kristensen, J. K., Local and central orthostatic sympathetic reflexes in Raynaud's phenomenon, *Scand. J. Clin. Lab. Invest.*, 51, 191, 1991.

80. Wamsley, D. and Goodfield, M. J., Evidence for an abnormal peripherally mediated vascular response to temperature in Raynaud's phenomenon, *Br. J. Rheumatol.*, 29(3), 181, 1990.

81. Wigley, F. M., Wise, R. A., Mikdashi, J., Schaefer, S., and Spence, R. J., The post-occlusive hyperemic response in patients with systemic sclerosis, *Arthritis Rheum.*, 33(11), 1620, 1990.

82. Kalis, B., De Rigal, J., Leonard, F., le Leveque, J. L., De Riche, O., Corre, Y. L., and Lacharriere, O. D., In vivo study of scleroderma by non-invasive techniques, *Br. J. Dermatol.*, 122(6), 785, 1990.

83. Wollersheim, H., Reyenga, J., and Thien, T., Postocclusive reactive hyperemia of fingertips, monitored by laser Doppler velocimetry in the diagnosis of Raynaud's phenomenon, *Microvasc. Res.*, 38(3), 286, 1989.

84. Goodfield, M., Hume, A., and Rowell, N., Reactive hyperemic responses in systemic sclerosis patients and healthy controls, *J. Invest. Dermatol.*, 93(3), 368, 1989.

85. Wollersheim, H., Reyenga, J., and Thien, T., Laser Doppler velocimetry of fingertips during heat provocation in normals and in patients with Raynaud's phenomenon, *Scand. J. Clin. Lab. Invest.*, 49(1), 91, 1988.

86. Goodfield, M. J. and Rowell, N. R., Hand warming as a treatment for Raynaud's phenomenon in systemic sclerosis, *Br. J. Dermatol.*, 119(5), 643, 1988.

87. Goodfield, M. D. J., Hume, A., and Rowell, N. R., The effect of simple warming procedures on finger blood flow in systemic sclerosis, *Br. J. Dermatol.*, 118, 661, 1988.

88. Goodfield, M. J. and Rowell, N. R., The effect of changes in temperature on peripheral blood flow in systemic sclerosis, *Br. J. Dermatol.*, 117 (Suppl. 32), 21, 1987.

89. Engelhart, M. and Kristensen, J. K., Raynaud's phenomenon: blood supply to fingers during indirect cooling, evaluated by laser Doppler flowmetry, *Clin. Physiol.*, 6(6), 481, 1986.

90. Tordoir, J. H., Haeck, L. B., Winterkamp, H., and Dekkers, W., Multifinger photoplethysmography and digital blood pressure measurement in patients with Raynaud's phenomenon of the hand, *J. Vasc. Surg.*, 3(3), 456, 1986.

91. Lee, P., Sarkozi, J., Bookman, A. A., Keystone, E. C., and Armstrong, S. K., Digital blood flow and nailfold capillary microscopy in Raynaud's phenomenon, *J. Rheumatol.*, 13(3), 564, 1986.

92. Engelhart, M., Nielsen, H. V., and Kristensen, J. K., The blood supply to fingers during Raynaud's attack: a comparison of laser Doppler flowmetry with other techniques, *Clin. Physiol.*, 5(5), 447, 1985.

93. Bollinger, A., Function of the precapillary vessels in peripheral vascular disease, *J. Cardiovasc. Pharmacol.*, 7 (Suppl. 3), 147, 1985.

94. Serup, J. and Kristensen, J. K., Blood flow of morphoea plaques as measured by laser Doppler flowmetry, *Arch. Dermatol. Res.*, 276(5), 322, 1984.

95. Kristensen, J. K., Engelhart, M., and Nielsen, T., Laser Doppler measurement of digital blood flow regulation in normals and in patients with Raynaud's phenomenon, *Acta Derm. Venereol.*, 63(1), 43, 1983.

96. Martin, M. F. and Tooke, J. E., Effects of prostaglandin E1 on microvascular haemodynamics in progressive systemic sclerosis, *Br. Med. J. Clin. Res. Ed.*, 285(6356), 1688, 1982.

97. Rademaker, M., Thomas, R. H., Provost, G., Beacham, J. A., Cooke, E. D., and Kirby, J. D., Prolonged increase in digital blood flow following iloprost infusion in patients with systemic sclerosis, *Postgrad. Med. J.*, 63(742), 617, 1987.

98. Rademaker, M., Beacham, J. A., Cooke, E. D., Meyrick Thomas, R. H., Man, T. G., and Kirby, J. D., Is intravenous iloprost better than oral nifedipine in systemic sclerosis?, *Br. J. Dermatol.*, 121 (Suppl. 34), 43, 1989.

99. Bunker, C. B., Foreman, J., O'Shaugnessy, D., Reavley, C., and Dowd, P. M., Calcitonin-gene-related peptide in the treatment of severe Raynaud's phenomenon, *Br. J. Dermatol.*, 121 (Suppl. 34), 43, 1989.

100. Cordioli, E., Virgilio, S., Ghirardi, R., and Martinelli, M., Effects of long-term iloprost therapy on Raynaud's phenomenon in progressive systemic sclerosis, *Minerva Med.*, 83(11), 739, 1992.

101. Wigley, F. M., Seibold, J. R., Wise, R. A., McCloskey, D. A., and Dole, W. P., Intravenous iloprost treatment of Raynaud's phenomenon and ischemic ulcers secondary to systemic sclerosis, *J. Rheumatol.*, 19(9), 1407, 1992.

102. Diehm, C., Muller-Buhl, U., and Morl, H., Isosorbide dinitrate ointment in Raynaud's disease, *Z. Kardiol.*, 72 (Suppl. 3), 185, 1983.

103. Bunker, C. B., Lanigan, S., Rustin, M. H., and Dowd, P. M., The effects of topically applied hexyl nicotinate lotion on the cutaneous blood flow in patients with Raynaud's phenomenon, *Br. J. Dermatol.*, 119, 771, 1988.

104. Winston, E. L., Pariser, K. M., Miller, K. B., Salem, D. N., and Creager, M. A., Nifedipine as a therapeutic modality for Raynaud's phenomenon, *Arthritis Rheum.*, 26(10), 1177, 1983.

105. Meyrick Thomas, R. H., Rademaker, M., Grimes, S. M., MacKay, A., Kovacs, I. B., Cooke, E. D., Bowcock, S. M., and Kirby, J. D. T., Nifedipine in the treatment of Raynaud's phenomenon in patients with systemic sclerosis, *Br. J. Dermatol.*, 117, 237, 1987.

106. Gush, R. J., Taylor, L. J., and Jayson, M. I., Acute effects of sublingual nifedipine in patients with Raynaud's phenomenon, *J. Cardiovasc. Pharmacol.*, 9(5), 628, 1987.

107. Roald, O. K. and Seem, E., Treatment of Raynaud's phenomenon with ketanserin in patients with connective tissue disorders, *Br. Med. J. Clin. Res. Ed.*, 289(6445), 577, 1984.

108. Engelhart, M., Ketanserin in the treatment of Raynaud's phenomenon associated with generalized scleroderma, *Br. J. Dermatol.*, 119, 751, 1988.

109. Klimiuk, P. S., Kay, E. A., Mitchell, W. S., Taylor, L., Gush, R., Gould, S., and Jayson, M. I., Ketanserin: an effective treatment regimen for digital ischaemia in systemic sclerosis, *Scand. J. Rheumatol.*, 18(2), 107, 1989.

110. Wollersheim, H., Thien, T., Fennis, J., Van Elteren, P., and van't Laar, A., Double-blind, placebo-controlled study of prazosin in Raynaud's phenomenon, *Clin. Pharmacol. Ther.*, 40(2), 219, 1986.

111. Wollersheim, H. and Thien, T., Dose-response study of prazosin in Raynaud's phenomenon clinical effectiveness versus side effects, *J. Clin. Pharmacol.*, 28(12), 1089, 1988.

112. Lasfargues, G., Royere, D., Perrotin, D., Titon, J. P., Soutif, D., and Guilmot, J. L., A study of the plethysmographic amplification factor during vasodilator treatment. Healthy subject versus primary Raynaud phenomenon, *J. Mal. Vasc.*, 12(4), 319, 1987.

113. Neirotti, M., Longo, F., Molaschi, M., Macchione, C., and Pernigotti, L., Functional vascular disorders: treatment with pentoxifylline, *Angiology*, 38(8), 575, 1987.

114. Jacobs, M. J., Jorning, P. J., Van Rhede-van der Kloot, E. D., Kitslaar, P. J., Lemmens, H. A., Slaaf, D. W., and Reneman, R. S., Plasmapheresis in Raynaud's phenomenon in systemic sclerosis: a microcirculatory study, *Int. J. Microcirc. Clin. Exp.*, 10(1), 1, 1991.

115. Stümpflen, A., Atteneder, M., Kaiser, S., Koppensteiner, R., Larch, E., Minar, E., and Ehringer, H., Local retrograde transvenous application of reserpine: effects on finger blood flow and laser-Doppler flux in patients with secondary Raynaud's syndrome, *Int. J. Microcirc. Clin. Exp.*, 11(1), 54, 1992.

116. Kristensen, J. K. and Henriksen, O., Distensibility of the vascular bed in subcutaneous tissue in generalized scleroderma, *J. Invest. Dermatol.*, 70(3), 156, 1978.

117. Kristensen, J. K. and Henriksen, O., Local regulation of blood flow in cutaneous tissue in generalized scleroderma, *J. Invest. Dermatol.*, 70(5), 260, 1978.

118. Kristensen, J. K., Reactive hyperemia in cutaneous tissue in generalized scleroderma, *J. Invest. Dermatol.* 71(4), 269, 1978.

119. Kristensen, J. K., Finger skin perfusion pressure in generalized scleroderma, *Acta Derm. Venereol.*, 58, 491, 1978.

120. Kristensen, J. K. and Wadskov, S., Increased ^{133}Xenon washout from cutaneous tissue in generalized scleroderma indicates increased blood flow, *Acta Derm. Venereol.*, 58, 313, 1978.

121. Kristensen, J. K., Local regulation of digital blood flow in generalized scleroderma, *J. Invest. Dermatol.*, 72(5), 235, 1979.

122. Bircher, A., De Boer, E. M., Agner, T., Wahlberg, J. E., and Serup, J., Guidelines for measurement of cutaneous blood flow by laser Doppler flowmetry, *Contact Dermatitis*, 30(2), 65, 1994.

123. Nilsson, A. L., Blood flow, temperature and heat loss of skin exposed to local radiative and convective cooling, *J. Invest. Dermatol.*, 88, 586, 1987.

124. Nilsson, A. L., Eriksson, L. E., and Nilsson, G. E., Effects of local convective cooling and rewarming on skin blood flow, *Int. J. Microcirc. Clin. Exp.* 5, 11, 1986.

125. Catalano, M., Ninno, D., Pasquino, M., Ninno, V., Milanese, F., and Libretti, A., Catecholamines and microcirculatory response by laser Doppler during mental stress, *Int. J. Microcirc. Clin. Exp.*, 9 (Suppl 1), 192, 1990.

126. Bircher, A. J. and Maibach, H. I., The assessment of the cutaneous microcirculation by laser Doppler and photoplethysmographic techniques: an update, in *Models in Dermatology*, Vol 4, Maibach, H. I. and Lowe, N. J., Eds., S. Karger, Basel, 1989, 209.

127. Wamsley, D. and Wiles, P. G., Reactive hyperaemia in skin of the human foot measured by laser Doppler flowmetry: effects of duration of ischaemia and local heating, *Int. J. Microcirc. Clin. Exp.*, 9, 345, 1990.

128. Caspary, L., Creutzig, A., and Alexander, K., Biological zero in laser Doppler fluxmetry, *Int. J. Microcirc. Clin. Exp.*, 7, 367, 1988.

129. Wahlberg, E., Olofsson, P., Swedenborg, J., and Fagrell, B., The effects of locally induced hyperemia and edema on the biological zero in laser Doppler fluxmetry (LDF), *Int. J. Microcirc. Clin. Exp.*, 9 (Suppl. 1), 187, 1990.

130. Daley, C. H., Chimoskey, J. E., Holloway, G. A., and Kennedy, D., The effect of pressure loading on the blood flow rate in the human skin, in *Bedsore Biomechanics*, Kenedi, R. M. and Cowden, J. M., Eds., Macmillan, London, 1976, 69.

131. Castronuovo, J. J., Pabst, T. S., Flanigan, D. P., and Forster, L. G., Non-invasive determination of skin perfusion pressure using a laser Doppler, *J. Cardiovasc. Surg.*, 28, 253, 1987.

132. Obeid, A. N., Barnett, N. J., Dougherty, G., and Ward, G., A critical review of laser Doppler flowmetry, *J. Med. Eng. Technol.*, 14(5), 178, 1990.

133. Newson, T. P. and Pearcy, M. J., Skin surface pO_2 measurement and the effect of externally applied pressure, *Arch. Phys. Med. Rehabil.*, 62(8), 390, 1981.

134. Nilsson, G. E., Tenland, T., and Öberg, P. A., Evaluation of a laser-Doppler flowmeter for measuring tissue blood flow, *IEEE Trans. Biomed. Eng.*, 27, 12, 1980.

135. Braverman, I. M., Keh, A., and Goldminz, D., Correlation of laser Doppler wave patterns with underlying microvascular anatomy, *J. Invest. Dermatol.*, 95, 238, 1990.

136. Tenland, J. E., Salerud, E. G., Nilsson, G. E., and Öberg, P. A., Spatial and temporal variations in human skin blood flow, *Int. J. Microcir. Clin. Exp.,* 2, 81, 1983.
137. Braverman, I. M. and Schechner, J. S., Contour mapping of the cutaneous microvasculature by computerized laser Doppler velocimetry, *J. Invest. Dermatol.,* 97(6), 1013, 1991.
138. Houben, A. J. H. M., Schaper, N. C., Slaaf, D. W., and Nieuwenhuijzen Kruseman, A. C., Day-time variations in peripheral micro and macrocirculation in man, *Int. J. Microcirc. Clin. Exp.,* 11 (Suppl. 1), 59, 1992.
139. Franzeck, U. K., Stengele, B., Panradl, U., Wahl, P., and Tillmanns, H., Cutaneous reactive hyperemia in short-term and long-term type I diabetes—continuous monitoring by a combined laser Doppler and transcutaneous oxygen probe, *Vasa,* 19, 8, 1990.
140. Oestergren, J. and Fagrell, B., Skin capillary blood cell velocity in man, characteristics and reproducibility of the reactive hyperemic response, *Int. J. Microcirc. Clin. Exp.,* 5, 37, 1986.

Chapter 8

Psoriasis

Ethel Tur, Anat Tamir, Gavriel Tamir, and Sarah Brenner

CONTENTS

I. INTRODUCTION

This chapter describes the use of laser Doppler flowmetry (LDF) to investigate the pathogenesis of psoriasis and the follow-up of its treatment. It starts with a brief review of the link between psoriasis and cutaneous microvasculature, followed by some of the relevant advantages and disadvantages of LDF. Then two types of studies probing its pathogenesis are outlined, namely morphological studies of the microvasculature and blood flow measurements. The use of blood flow measurements for the assessment of the disease process, its severity, and the response to therapy is discussed next. New developments of LDF as used in the study of psoriasis conclude the chapter. Whenever of importance, other blood flow measurement techniques that add to the understanding of the disease processes are mentioned as well.

II. PSORIASIS AND ITS LINK TO THE CUTANEOUS MICROVASCULATURE

Psoriasis is a common, chronic skin disease characterized by circumscribed, erythematous, dry patches of various sizes, covered by fine, silvery scales.[1] As the scales are removed by gentle scraping, fine bleeding points appear (Auspitz's sign). This bleeding is caused by the proximity of the upper dermal capillaries to the skin surface. The superficial vessels are tortuous, elongated, and dilated. The clinical picture results from increased epidermal cell turnover. Psoriasis is a multifactorial disease, and the exact nature of its initial changes is still debated. Those changes relevant to the topic of this chapter are endothelial cell swelling and separation[2] and microvascular changes preceding other events, including vascular dilation and tortuosity,[3] as well as increased blood flow.[4] The increase in blood flow is probably associated with a diffusible, possibly humoral, initiating factor stimulating transformation to a psoriatic plaque.[4] Whatever the initial event in psoriasis is, microvascular abnormalities play an essential role in its pathogenesis. Various studies of the microvascular changes focus on the histopathogenesis of the disease, as well as on treatment evaluation and follow-up.

III. ADVANTAGES AND DISADVANTAGES OF LASER DOPPLER FLOWMETRY FOR THE INVESTIGATION OF PSORIASIS

Since skin blood flow is increased in psoriasis and varies during the lifetime of the disease, blood flow measurement techniques should be useful for the assessment of the severity of the disease and for the evaluation of treatment. In particular, LDF, which is a noninvasive, easy-to-handle, relatively fast and inexpensive technique, appears to be a very attractive practical tool to objectively monitor the disease severity and the effect of treatment. Further developments in instrumentation are constantly being introduced, offering more advantages. The most recent one is the scanning laser Doppler, which is able to measure blood flow over a large area without contact with the skin surface, thus not disturbing the blood flow to the measured area. For psoriasis it offers the possibility to measure simultaneously the area of increased blood flow and the severity of the psoriasis plaque in terms of mean blood flow. Objective evaluation of therapeutic response may be obtained by serial scans.

LDF is less sensitive than the ^{133}Xe washout technique in giving quantitative data concerning blood flow rate, which seems to be a drawback of the LDF technique. However, when comparing LDF to other techniques, it is important to note that different methods measure different sections of the microvasculature. It is likely that the flux signal shown by LDF represents the large volume of red blood cells moving within the larger blood vessels, particularly the subpapillary plexus, rather than the much smaller volume of red blood cells residing within the nutritive capillaries. The depth of laser penetration in the wavelengths used is approximately 1 mm. Therefore, in normal skin it is likely to include the subpapillary plexus as well as the capillaries in the subpapillary dermis. In psoriasis the epidermal ridges are elongated, and this may alter the relative contribution of superficial and deep blood flow to the laser Doppler signal. Another alteration may follow when thick psoriatic scales produce some reduction in the blood flow reading. This factor exhibits a disadvantage of the method, for it may create inaccuracies in LDF evaluation of plaques with varying degrees of scaling. Another disadvantage of LDF is that results obtained by various instruments or by the same instrument in various subjects cannot be compared.

These criticisms aside, laser Doppler flowmetry adds another measurement dimension to the quantitative characterization of cutaneous blood flow in general and to a better understanding of psoriasis in particular.

IV. PATHOGENESIS

A. MORPHOLOGICAL STUDIES OF THE MICROVASCULATURE

Microvasculature morphology in psoriatic skin has been investigated by light microscopy, electron microscopy, and *in vivo* capillaroscopic techniques.[1,5–14] These studies have shown the abnormalities of the upper dermal blood vessels to consist of dilated, elongated, coiled, and tortuous capillaries in the dermal papillae, without an increase in the number of capillaries. These changes are especially prominent in the fully developed chronic psoriatic plaque. In a study of the ultrastructure of the psoriatic capillary loops, Braverman and Yen have demonstrated that their irregular coiling results from marked elongation of the venous portion, and these microvascular changes parallel the disease activity.[5] The capillary loops gradually revert to normal during the healing process.[2,10,11] This transformation may start within 24 to 72 h after initiation of treatment.[5,6,11] Yet recently healed skin may still display abnormal microvascular findings, and these may persist long after clinical resolution.[5,13,14] Moreover, microvascular changes, though less abundant, can be observed in nonlesional psoriatic skin.[1,7] Ross[7] showed that morphological abnormalities in dermal blood vessels might be detected in the clinically unaffected skin of psoriatic patients.

Barton et al.[9] tried to quantify the microvascular changes in biopsies taken from psoriatic skin lesions in terms of the endothelial and luminal volume with respect to the volume of dermal components of the skin. They concluded that vascular mass and vessel dilation were greatly increased in psoriatic lesions and, to a lesser extent, in the uninvolved skin of psoriatic patients as compared to normal controls. It is still debated whether the pathogenetic process is primarily vascular and inflammatory with secondary epidermal alteration or consists of epidermal proliferation with secondary dermal involvement. However, there is a direct correlation between the vascular changes and those in the overlying epidermis, and epidermal hyperplasia cannot occur without vascular proliferation.[10]

B. BLOOD FLOW MEASUREMENTS

Recognizing the important role played by blood vessels in psoriasis, various qualitative and quantitative methods of measuring blood flow have been utilized for its objective assessment.[15] Among these are photopulse and piezoelectric plethysmographic measurements, venous occlusion, plethysmography, spectrophotometry, and skin temperature measurements, allowing indirect evaluation of the cutaneous vasculature. The radioactive indicator washout technique (e.g., local ^{133}Xe washout) and LDF have also served to assess vascular reactivity. Although the results obtained by the various methods are not always in agreement, they all show a correlation between blood flow and disease activity and indicate a decrease in blood flow of psoriatic lesions during successful treatment.

Klemp and Staberg[16] tried to quantify blood flow changes using the ^{133}Xe washout technique. They found that the mean cutaneous blood flow in involved psoriatic skin on the forearm was about 63.5 ml/100 g/min, while the value in normal subjects was 6.3 ml/100 g/min. Thus, in untreated lesional skin of patients with active psoriasis, the cutaneous blood flow is approximately ten times higher than normal cutaneous blood flow. They also found that cutaneous blood flow in nonlesional skin of psoriatic patients with active disease was twice as high as normal skin blood flow. The same authors[17] also used the ^{133}Xe washout technique to measure cutaneous blood flow in nonlesional skin of patients with only minimal psoriatic skin manifestations. These measurements yielded no significant difference in comparison to normal individuals. They thus reason that cutaneous blood flow in both lesional and nonlesional skin relates directly to the severity of psoriasis, so the activity of the disease can be monitored by cutaneous blood flow measurements in nonlesional skin. They postulate that cutaneous blood flow in non-lesional skin is elevated with increasing psoriatic activity due to changing levels of certain humoral factors.

Using LDF, Hull et al.[4] measured skin blood flow in the active advancing edges of psoriatic plaques. Over a 3-week period they serially traced 82 untreated plaques from 15 patients and 38 treated plaques from 6 patients; 57% of the former and 65% of the latter showed consistent asymmetrical movement, thus allowing identification of an active and inactive edge of each plaque. The active edge of two or more plaques in each of ten patients was detected using this technique. Blood flow, as measured by LDF, was 2.5 to 4.5 times higher at the active edge as compared to the inactive edge of the plaque, while no detectable histologic changes were yet apparent. Routine histologic examination of punch biopsies taken from the sites investigated by LDF did not reveal any epidermal changes. Moreover, using monoclonal antibody immunohistology, no T-lymphocytic infiltration of the skin at this site was found. The increased microcirculatory blood flow clearly predates the inflammatory cell infiltration as well as the epidermal changes. In addition, since no excess of T-cells was present at the active edge of the psoriatic plaques, they suggested that diffusible and possibly humoral vasodilators might accumulate there, thus initiating the expansion of the lesions.

Both ^{133}Xe washout technique and LDF serve as indirect methods to assess blood flow. Neither of these techniques offers a direct measurement; they only allow a semiquantitative assessment of blood flow changes. As mentioned above, it is likely that the flux signal shown by LDF represents the large volume of red blood cells moving within the larger blood vessels, particularly the subpapillary plexus, rather than the much smaller volume of red blood cells moving within the nutritive capillaries.[18]

Bull et al.[8] investigated the psoriatic microcirculation by means of intravital capillaroscopy using a videomicroscopy system with and without contrast dyes. Native capillaroscopy allows direct visualization of the nutritive or exchange capillaries in the upper dermis. Their results are in agreement with the other methods exhibiting differences between normal and psoriatic skin. In comparison to healthy control skin, both plaque and uninvolved psoriatic skin showed higher perfusion. In psoriatic plaques the capillaries were much larger than in normal skin. The density of capillaries was not increased in plaque or uninvolved psoriatic skin, indicating expansion of existing vessels rather than new vessel formation. Fluorescein angiography indicated greater vessel transcapillary diffusion in plaque and uninvolved psoriatic skin than in normal skin. The authors hold that this direct noninvasive technique gives more representative answers than indirect means of measuring blood flow such as LDF and isotope washout.

Psoriatic arthritis was investigated by Korotaeva et al.,[19] who, in addition to using the LDF and ^{133}Xe washout techniques, measured the viscosity of the plasma. Patients who exhibited microvascular abnormalities also had an elevated viscosity and a lowered hematocrit. They suggested that these abnormalities might be involved in the pathogenesis of psoriatic arthritis.

V. FOLLOWING THE DISEASE PROCESS AND RESPONSE TO THERAPY

Since, as shown above, skin blood flow is increased in psoriasis, blood flow measurement techniques may be useful in the assessment of the disease severity and treatment evaluation. Existing techniques for assessment of the area of involved skin in psoriasis are unsatisfactory. Visual assessment of the extent of psoriasis shows wide variations among different observers, and there is a tendency to overestimate the area of involvement.[20] Tracing of plaques is time-consuming and difficult, especially when measuring small plaques. Given the advantages of LDF in this context, attempts have been made to use it as a practical tool to monitor the effect of treatment objectively.

Many treatment modalities are used in psoriasis, including topical and systemic. Each of these affects the epidermis and vasculature in different ways. Their mechanisms of action are not yet completely understood, in part because the pathogenesis of the disease is still obscure. Investigators have tried to assess the efficacy of some of these modalities objectively by using various laboratory methods, correlating them to the clinical course. In addition, changes in laboratory parameters during treatment may shed some light on the pathogenesis of the disease. Some of these studies have concentrated on the cutaneous microvasculature and blood flow during and following treatment.

As already noted, microvascular anatomical changes are seen within the first few days of treatment with various antipsoriatic agents.[12,13] Transformation of abnormal psoriatic capillaries to a normal morphology has been reported to precede both clinical improvement and the decrease in labeling index of the basal cell layer.[10] Staberg and Klemp used LDF to follow psoriatic patients during Goeckerman or beech tar therapy.[21] They measured cutaneous blood flow in lesional and uninvolved skin of eight psoriatic patients during these treatments. Measurements were performed before treatment and at weekly intervals following its initiation. The results were compared to a clinical psoriasis index composed of the sum of the individual scores of visual assessment of infiltration, erythema, and scaling. The pretreatment value of blood flow in lesional skin was significantly (nine-fold) higher than in uninvolved skin. During treatment a rapid decrease in cutaneous blood flow was noticed, approaching that of uninvolved skin after 3 to 4 weeks. Furthermore, clinical improvement was linearly correlated with decreased blood flow rates. These results imply that the LDF technique could be used to obtain a quantitative measure of the disease activity during evaluation of antipsoriatic treatment.

The same authors also compared LDF to ^{133}Xe washout before and during treatment with tar.[22] They measured cutaneous blood flow in involved and uninvolved psoriatic skin before and on the 3rd, 7th, 14th, and 28th day of treatment. Control experiments were performed in normal individuals. Again, they confirmed their earlier finding of a significantly higher blood flow rate in involved vs. uninvolved psoriatic skin. Their results demonstrated a significant decrease in cutaneous blood flow of the involved psoriatic skin 3 days following onset of treatment. The decline in blood flow continued, although at a more moderate rate, until the 28th day of treatment, reaching a value that was not significantly different from uninvolved skin in the same patients. Cutaneous blood flow of uninvolved skin had decreased at the end of treatment to values similar to those observed in the skin of normal individuals, but was still significantly higher. Clinical improvement, estimated by the same clinical score index, showed a parallel trend, although no conclusions were drawn as to whether it preceded or succeeded the observed decrease in blood flow. LDF measurements did not reflect changes in the uninvolved skin in psoriatic patients during treatment. Compared to the ^{133}Xe washout results, LDF was not sensitive enough for the evaluation of uninvolved psoriatic skin or for normal skin. Therefore, they deduced that the LDF method was less sensitive and could only be used to give rough estimates of treatment-induced changes of blood flow in skin at high perfusion rates—namely, in involved skin.

Khan et al.[23] have also attempted to follow the effect of treatment objectively, utilizing LDF. They followed the resolution of two psoriatic plaques in seven patients during 1 month of therapy. Six of them received the Goeckerman regimen, while one was treated with psoralen-ultraviolet A (PUVA) therapy. LDF readings taken from uninvolved skin during that period served as controls. Daily LDF readings showed a decrease in cutaneous blood flow in the psoriatic lesions of the Goeckerman-treated patients, reaching levels comparable to those of uninvolved skin early in the course of treatment. The decrease in LDF readings and the clinical improvement were in good agreement, but a striking difference was noted: the blood flow decreased much faster than the clinical grading. This decline significantly preceded the observed clinical resolution from 4 to 8 days after initiation of treatment. They thus proposed that LDF measurements could provide advance knowledge of disease improvement, which might prove valuable for the design and follow-up of new treatments. LDF readings closely followed the course of the disease: transient elevation of LDF readings accompanied the visible flare-ups which sometimes appeared at intervals when patients were untreated. As to the PUVA-treated patient, perfusion of his lesions remained higher than that of uninvolved skin, despite clinical healing. In a separate series of experiments, the same authors[23] measured blood flow to the extensor surface of the forearm and compared measurements of normal subjects to uninvolved psoriatic skin and to untreated lesional skin of psoriatic patients. Significantly higher perfusion rates were measured in lesional skin compared to uninvolved psoriatic or normal control skin, yet no significant difference was found between blood flow to uninvolved psoriatic skin and to the skin of healthy controls. These findings support those of Klemp and Staberg,[22] showing that LDF is less sensitive than the ^{133}Xe washout technique in giving quantitative data concerning blood flow rate. Yet it may be incorrect to compare numerically LDF measurements found by different groups of investigators, since discord between different studies may originate from the biological variation within the disease state. Using LDF, numerical comparison of data is reliable only when the same subject groups are studied at the same time. Khan et al.[23] point out that, contrary to the direct correlation found between LDF and other flow measurements (such as ^{133}Xe washout) in normal skin, it is questionable whether such a direct correlation exists in psoriatic skin. This is due to the fact that the depth of light penetration is difficult to define. In psoriatic patients undergoing treatment it is possible that light penetrates further into the skin as the disease resolves, since scaling, inflammation, and thickness of skin all diminish. Yet their findings indicate that any possible increase in LDF blood flow measurements induced by these changes is overwhelmed by the reduction in perfusion that accompanies the resolution of the psoriatic lesions.

Frodin et al.,[24] questioning the importance of the coal tar bath in the Ingram regimen, evaluated blood flow by LDF and transepidermal water loss by evaporimetry in addition to clinical assessments. They studied 11 patients with symmetrical lesions in the upper extremities. The patients immersed one arm in an oil emulsion bath and the other arm in a coal tar bath, followed by whole-body ultraviolet irradiation (UVB) and then by application of dithranol paste to all lesions. Clinical assessment showed a process of successive healing during the first 3 weeks. This healing process closely paralleled the normalization of transepidermal water loss studied by evaporimetry as well as the decrease of blood flow rate as measured by LDF. Using these parameters, no difference was found between the results obtained with coal tar or with the oil emulsion. They were thus able to suggest that the coal tar is unnecessary and has no advantage over oil emulsion in the Ingram regiment when erythemogenic doses of UVB are given. Like Khan et al.,[23] Frodin et al.[24] emphasize the fact that since the healing process in psoriasis influences the degree of acanthosis and scaling, different strata of blood vessels may successively contribute to the LDF signal, which could mask the purely inflammatory component of the flow. Thus, elevation in blood flow that was sometimes recorded after the first week of treatment could have resulted from thinning of the scaly layer, leading to higher recordings from deeper and larger vessels. They also note that, owing to variations in the distribution of vessels, the positioning of the probe also must be taken into account when assessing fluctuations in blood flow recorded values.

Mustakallio and Kolari[25] compared the visual estimation of the irritation caused by dithranol and 10-butyryl dithranol to the objective measurements of contact temperature and LDF. They studied the uninvolved skin of 11 psoriasis patients and found a correlation between all three methods and the dose of the irritants. LDF and the visual assessment were also significantly correlated. LDF confirmed that the early erythema responses to the irritants used were due to increased superficial blood flow. It is interesting to note that although staining of the skin by the dithranol preparation sometimes exaggerated visual estimates of erythema in this study, it did not interfere with the LDF blood flow measurements.

Broby-Johansen and Kristensen[26,27] studied the antipsoriatic effect of topical corticosteroid (clobetasol propionate) treatment with and without occlusion with a semi-permeable hydrocolloid dressing,[26] and that of the semipermeable hydrocolloid dressing alone.[27] The treatment was evaluated by comparing the clinical score (rating infiltration, erythema, and scaling on a 0 to 3 scale) to oxygen consumption and LDF,[26] as well as to temperature.[27] In both studies, untreated psoriatic plaques of the same patients served as controls. They found topical clobetasol propionate cream, when combined with occlusion, to have a pronounced, rapid effect. This treatment was found to be superior to the treatment with the steroid cream alone, which was only marginally effective. All laboratory parameters, including LDF readings, were in good agreement with the clinical score index. Yet in the other study no correlation was found between clinical and laboratory measurements.[27] Application of a semipermeable hydrocolloid dressing resulted in significant clinical improvement, but, unlike the former study, all the laboratory methods, including LDF, failed to show any significant changes.

In a more recent study, Broby-Johansen et al.[28] tried to rank the antipsoriatic effect of various topical corticosteroids applied for 1 week under a hydrocolloid dressing (one application only, with no changing) by measuring ultrasound skin thickness, LDF cutaneous blood flow, and colorimetry, in comparison to clinical assessment. The redness was the last sign to disappear in accordance with skin color measurements. With increased potency of the topical corticosteroid the parameters measured were closer to normal. The clinical scores improved, blood flow declined, skin thickness decreased, and the color normalized, approaching that of normal skin. They thus reason that a potent topical corticosteroid normalizes psoriatic skin in one week using occlusion with a hydrocolloid dressing. Consequently, they stress the value of short-term therapy with a potent corticosteroid in psoriatic patients. These authors suggest that bioengineering methods such as LDF should be employed for posttreatment assessment, since pretreatment surface variations and scaling may occasionally result in unexpected values.

VI. RECENT DEVELOPMENTS

Further improvements of the laser Doppler technique offer new possibilities and may present new findings. Progressions are illustrated by refinements such as computerization, the design of a probe holder which allows repeated measurements over the same site before and after manipulations to the skin, as was done by Braverman and Schechner,[29] through the new multichannel LDF instrument which allows simultaneous measurements of several sites. However, the most important recent development is the scanning LDF,[30-32] which records the tissue perfusion in several thousand measurement points. A map of the spatial distribution of the blood flow is obtained in a short period of time. Unlike the LDF, which continuously records the blood flow over a single point, the scanning LDF maps the blood flow distribution over a specific area. Thus, the two methods do not compete with each other, but rather are complementary. The new scanning LDF has the advantage of operating without any contact with the skin surface, so it does not disturb the local blood flow.

Speight et al.[31,32] studied its application for the objective assessment of psoriasis. Lesions of psoriasis were clearly detected by the scanning LDF as areas of increased blood flow, and the degree of this increase could be recorded as well. The mean blood flow within the psoriatic plaques was about four times greater than in uninvolved skin. Plaques which were clinically more erythematous usually had a higher mean blood flow. Within a psoriatic plaque there was greater variation than in normal skin. The presence of thick scaling decreased blood flow recorded within plaques, but values remained greater than background. They also measured the area of increased blood flow, corresponding to psoriatic plaques, as compared to their area as measured by tracing. They found a highly significant linear correlation between plaque area measured by tracing and area of increased blood flow as measured by LDF. However, the latter was significantly greater than traced area, due to a rim of increased blood flow outside the visible edge of the plaque. Likewise, using conventional LDF they found higher blood flow values beyond the visible plaque edge, as was previously described by Hull et al.[4] Thus, the scanning LDF confirmed this increase in blood flow outside the visible plaque edge and demonstrated that in some plaques it exists in all directions. The difference between area determined by tracing and area determined by scanning, due to increase in blood flow which is not clinically apparent, may be used to acquire information about the psoriasis activity. The new instrument is able to measure large areas of skin rapidly. This advanced method allows simultaneous measurement of psoriasis plaque severity in terms of mean blood flow and area of increased blood flow. Objective evaluation of therapeutic response may be obtained by serial scans.

VII. CONCLUSION

LDF is a simple, noninvasive method that has proved to be useful in monitoring the disease process and various therapeutic effects in psoriasis. This monitoring is possible since psoriatic lesions show higher LDF readings due to their tortuous and elongated blood vessels. Results obtained by LDF should not be interpreted as absolute values in psoriatic skin lesions, but rather should serve as relative values. In addition to clinical assessment, these results can provide further objective information about the disease course and thus may aid in the design and follow-up of various therapeutic modalities. Studies indicate that changes in blood flow reflect psoriatic activity. Thus, as a convenient, noninvasive technique of blood flow measurement, LDF can be used for the assessment of disease state and therapeutic modalities.

Yet it may be incorrect to compare numerically LDF measurements found by different groups of investigators, since discord between different studies may originate from the biological variation within the disease state. Using LDF, numerical comparison of data is reliable only when the same subject groups are studied at the same time.

Further developments of the laser Doppler technique may open new avenues in the study of psoriasis, as illustrated by the advanced method of scanning LDF. This method measures blood flow over a large area of the skin, allowing rapid measurement of psoriasis plaque severity in terms of mean blood flow and area of increased blood flow simultaneously. The scan image obtained reveals the distribution and intensity of the rim of increased blood flow around the psoriatic plaque. This could be used in the study of early biochemical or immunological changes in the skin, before lesions become clinically observable, and thus may help address the yet unsolved puzzle of the pathogenesis of psoriasis.

While these bioengineering studies are important for the investigation of psoriasis and other diseases, most of the results are of a more statistical value and cannot be clinically applied to an individual patient. Further improvements in LDF techniques and instrumentation and a better understanding of the involved pathophysiological mechanisms may eventually turn the laser Doppler flowmeter into a very convenient and clinically important tool for the diagnosis and follow-up of psoriasis.

REFERENCES

1. Ryan, T. J., Microcirculation in psoriasis: blood vessels, lymphatics and tissue fluid, *Pharm. Ther.*, 10, 27, 1980.
2. Brody, I., Dermal and epidermal involvement in the evolution of acute eruptive guttate psoriasis vulgaris, *J. Invest. Dermatol.*, 82, 465, 1984.
3. Telner, P., The capillary responses in psoriatic skin, *J. Invest. Dermatol.*, 36, 225, 1961.
4. Hull, S. M., Goodfield, M., Wood, E. J., and Cunliffe, W. J., Active and inactive edges of psoriatic plaques: identification by tracing and investigation by laser-Doppler flowmetry and immunocytochemical techniques, *J. Invest. Dermatol.*, 92, 782, 1989.
5. Braverman, I. M. and Yen, A., Ultrastructure of the capillary loops in the dermal papillae of psoriasis, *J. Invest. Dermatol.*, 68, 53, 1977.
6. Braverman, I. M., The role of blood vessels and lymphatics in cutaneous inflammatory processes: an overview, *Br. J. Dermatol.*, 109 (Suppl. 25), 89, 1983.
7. Ross, J. B., The psoriatic capillary, its nature and value in the identification of the unaffected psoriatic, *Br. J. Dermatol.*, 76, 511, 1964.
8. Bull, R. H., Bates, D. O., and Mortimer, P. S., Intravital video capillaroscopy for the study of the microcirculation in psoriasis, *Br. J. Dermatol.*, 126, 436, 1992.
9. Barton, S. P., Abdullah, M. S., and Marks, R., Quantification of microvascular changes in the skin in patients with psoriasis, *Br. J. Dermatol.*, 126, 569, 1992.
10. Braverman, I. M. and Sibley, J., Role of the microcirculation in the treatment and pathogenesis of psoriasis, *J. Invest. Dermatol.*, 78, 12, 1982.
11. McKenzie, A. W., Histological changes in psoriasis treated with topical fluocinolone and occlusion, *Br. J. Dermatol.*, 75, 434, 1963.
12. Gordon, M., Johnson, W. C., and Burgoon, C. F., Histopathology and histochemistry of psoriasis. II. Dynamics of lesions during treatment, *Arch. Pathol.*, 84, 443, 1967.
13. Suurmond, D., Histochemical changes in treated and untreated psoriasis, *Dermatologica*, 132, 237, 1966.
14. Mottaz, J. H., Zelickson, A. S., Thorne, E. G., and Wachs, G., Blood vessel changes in psoriatic skin, *Acta Derm. Venereol.*, 53, 195, 1973.
15. Klemp, P., Cutaneous and subcutaneous blood flow measurements in psoriasis, *Dan. Med. Bull.*, 34, 147, 1987.

16. Klemp, P. and Staberg, B., Cutaneous blood flow in psoriasis, *J. Invest. Dermatol.*, 81, 503, 1983.

17. Klemp, P. and Staberg, B., Cutaneous and subcutaneous blood flow in nonlesional skin of patients with minimal psoriatic skin manifestations, *J. Invest. Dermatol.*, 86, 584, 1986.

18. Braverman, I. M., Keh, A., and Goldminz, D., Correlation of laser Doppler wave patterns with underlying microvascular anatomy, *J. Invest. Dermatol.*, 95, 283, 1990.

19. Korotaeva, T. V., Firsov, N. N., Mach, E. S., Agababova, E. R., and Burdeiny, A. P., Microcirculatory disorders in patients with psoriatic arthritis, *Ter. Arkh.*, 63, 78, 1991.

20. Ramsay, B. and Lawrence, C. M., Measurement of involved surface area in patients with psoriasis, *Br. J. Dermatol.*, 124, 565, 1991.

21. Staberg, B. and Klemp, P., Skin blood flow in psoriasis during Goeckerman or beech tar therapy, *Acta Derm. Venereol.*, 64, 331, 1984.

22. Klemp, P. and Staberg, B., The effects of antipsoriatic treatment on cutaneous blood flow in psoriasis measured by ^{133}Xe washout method and laser Doppler velocimetry, *J. Invest. Dermatol.*, 85, 259, 1985.

23. Khan, A., Schall, L. M., Tur, E., Maibach, H. I., and Guy, R. H., Blood flow in psoriatic skin lesions: the effect of treatment, *Br. J. Dermatol.*, 117, 193, 1987.

24. Frodin, T., Skogh, M., and Molin, L., The importance of the coal tar bath in the Ingram treatment of psoriasis. Evaluation by evaporimetry and laser Doppler flowmetry, *Br. J. Dermatol.*, 118, 429, 1988.

25. Mustakallio, K. K. and Kolari, P. J., Irritation and staining by dithranol (anthralin) and related compounds, *Acta Derm. Venereol.*, 63, 513, 1983.

26. Broby-Johansen, U. and Kristensen, J. K., Antipsoriatic effect of local corticosteroids— O_2-consumption and blood flow measurements compared to clinical parameters, *Clin. Exp. Dermatol.*, 14, 137, 1989.

27. Broby-Johansen, U. and Kristensen, J. K., Antipsoriatic effect of semi occlusive treatment—O_2-consumption, blood flow and temperature measurements compared to clinical parameters, *Clin. Exp. Dermatol.*, 14, 286, 1989.

28. Broby-Johansen, U., Karlsmark, T., Petersen, L. J., and Serup, J., Ranking of the antipsoriatic effect of various topical corticosteroids applied under a hydrocolloid dressing— skin thickness, blood flow and color measurements compared to clinical assessments, *Clin. Exp. Dermatol.*, 15, 343, 1990.

29. Braverman, I. M. and Schechner, J. S., Contour mapping of the cutaneous microvasculature by computerized laser Doppler velocimetry, *J. Invest. Dermatol.*, 97, 1013, 1991.

30. Wardell, K., Naver, H. K., Nilsson, G. E., and Wallin, B. G., The cutaneous vascular axon reflex characterized by laser Doppler perfusion imaging, *J. Physiol.*, 460, 185, 1993.

31. Speight, E. L., Essex, T. J. H., and Farr, P. M., The measurement of plaques of psoriasis using a scanning laser-Doppler velocimeter, *Br. J. Dermatol.*, 127 (Suppl. 40), 34, 1992.

32. Speight, E. L., Essex, T. J. H., and Farr, P. M., The study of plaques of psoriasis using a scanning laser-Doppler velocimeter, *Br. J. Dermatol.*, 128, 519, 1993.

Chapter 9

Cutaneous Laser Doppler Flowmetry in General Medicine

Ethel Tur

CONTENTS

I. INTRODUCTION

Internal diseases and conditions like hypertension, diabetes mellitus, and pregnancy have a substantial vascular component, the evaluation of which is important for assessing the severity of the condition, its course, and sometimes its management. Among other organs, the skin microvasculature is also affected, its easy accessibility making its evaluation extremely valuable.

Laser Doppler flowmetry (LDF) is a noninvasive method that continuously follows the flow of red blood cells.[1] It operates on the Doppler principle, employing a low-power helium-neon laser emitting red light at 632.8 nm. The radiation is transmitted via an optical fiber to a transducer which is attached to the skin by double adhesive tape. The radiation is diffusely scattered and its optical frequency is shifted by the moving red blood cells. The reflected light, being coherently mixed with another portion of the light scattered from static tissues, generates a Doppler beat signal in the photodetector output current. A quantitative indication of the cutaneous blood flow can then be estimated from spectral analysis of the beat signal.

LDF, being an objective, noninvasive, real-time measurement technique, is therefore an attractive practical tool for estimating the cutaneous blood flow. Besides, LDF is relatively simple, fast, and inexpensive and can provide information which supplements the results of various other techniques.[2]

The diverse application areas of LDF include investigations of tissues other than the skin, like the buccal, nasal, or rectal mucosa, the intestine (through an endoscope), or kidney, liver, or lung (intraoperatively). This chapter deals exclusively with cutaneous LDF measurements as applied to various medical fields. Hence, in all the reviewed investigations, LDF was used to measure skin blood flow, a fact that shall not be mentioned in this chapter again. In view of the large number of studies conducted in this area, it is impossible to review each one; thus, this chapter only attempts to demonstrate the possibilities of the technique. The reader will be taken on an expedition where he will experience and inspect the fascinating views (internal conditions) using a unique type of binoculars (LDF).

After a review of some of the relevant basic considerations in experimental designs involving LDF, there will be a discussion of several conditions whose effects were examined using LDF, including smoking, environmental temperature, and pregnancy. Following that, studies of disease characteristics, attempts to identify the disease risk groups, and treatment evaluation will be outlined. The diseases covered are hypertension (including gestational hypertension), atherosclerosis, diabetes mellitus, Raynaud's phenomenon, and flushing. The chapter will be concluded with an evaluation of future prospects.

II. BASIC CONSIDERATIONS IN EXPERIMENTAL STUDIES INVOLVING LASER DOPPLER FLOWMETRY

A. APPLICATION PROCEDURES OF LASER DOPPLER FLOWMETRY BLOOD FLOW MEASUREMENTS

For monitoring internal diseases, LDF can be used to obtain both static and dynamic measurements.[3-5] Static studies, like baseline blood flow measurements, record only the steady-state blood flow, ignoring all transients. On the contrary, dynamic investigations, exemplified by reactive hyperemia,[3-5] in which the postocclusion time course of the blood flow is recorded in detail, can reveal the effects of a disease on the ability of blood vessels to respond to fast external triggers.[3-5] Other provocative methods examining the response to external triggers include the cognitive test,[6] isotonic[7] and isometric tests,[3] vasomotor reflexes,[8] intracutaneous needle stimulation,[9] and the thermal test.[3] Some of these tests are vasoconstrictive (the isometric and cognitive tests and vasomotor reflexes) and some are vasodilative (the arterial postocclusive reactive hyperemia, intracutaneous needle stimulation, the thermal test, and the isotonic test).

To optimize skin blood flow response to the different tests, vasoconstrictive stimuli should be performed on a high-blood-flow site, while vasodilative stimuli should be performed on a low-blood-flow site. In this way the magnitude of the changes induced are maximized, thereby providing a significant improvement in the quality of the data obtained. Moreover, vasoconstriction mediated by sympathetic stimulation, as in the cognitive test, should be provoked in the hands and feet, where the local blood supply is under a sympathetic vasoconstrictor control, and not in the face, which has a poor sympathetic vasoconstrictor supply.[6]

Thus, to enhance the sensitivity of the measurements, the site tested should be carefully selected. For example, the fingers, being rich with microvascular arteriovenous anastomoses, are useful for vasoconstrictive tests, while the forearms are suitable for the vasodilative tests. The forearms have several advantages as a preferred site for vasodilative tests: (1) an abundance of arterioles, capable of reactive hyperemia; (2) a local effect, rather than thermoregulatory reflexes,[10] is responsible for thermally induced vasodilatation; and (3) little inconvenience is experienced by the subjects tested.

The appropriate test also should be carefully selected in order to probe the relevant topic adequately. Before reaching a conclusion, one should bear in mind that differences in experimental settings might lead to different results. For instance, in order to study differences between young and old patients, different tests were used when addressing different questions. When aiming to study the thermoregulatory responses to cold stress, vasoconstrictor responses to inspiratory gasp, contralateral arm cold challenge, and body cooling were measured,[10] and differences were indeed found: elderly subjects had a diminished sympathetic vasoconstrictor response. In contrast, in order to evaluate the penetration of drugs through aged skin, the erythema that results from topical application of methyl nicotinate was measured;[11] no differences were found between young and old subjects, indicating that microvascular reactivity to the applied stimulus was comparable.

1. Vasoconstrictive Tests

a. Isometric Test

The laser Doppler flowmeter probe is usually placed on the left middle fingertip, and when baseline blood flow is established the subject squeezes a hard rubber ball in his right hand for 30 s. He then releases the grip and the maximum decrease in blood flow is recorded. The hemodynamic response to isometric hand grip exercise involves activation of the sympathetic nervous system, which evokes an increase in blood pressure mainly dependent on cardiac output.[3]

b. Cognitive Test

With the laser Doppler flowmeter probe on the fingertip, the subject is requested to subtract 7 sequentially from 1000 for a 2-min period. Blood flow is monitored continuously. There is a rapid fall in the blood flow to the finger at the beginning of the mental arithmetic activity and a rapid recovery at the end. The maximum decrease in blood flow is registered. This decrease is a manifestation of a sudden increase in sympathetic activity.[6]

c. Venoarteriolar Response

The venoarteriolar reflex measures the ability to decrease flow during venous stasis (normally seen in the feet on dependency) and is assumed to be dependent on an intact sympathetic nerve function.[12] The reflex occurs following increased venous pressure, which induces a constriction of the arterioles followed by a decrease in skin blood flow. An increase of venous pressure can be achieved by occlusion with a cuff (for instance, at the base of the investigated finger)[13] or by lowering the leg below heart level.[14-16] Usually, resting blood flow to the dorsum of the foot is measured with the patient resting in the supine position. Then standing flow is measured (or the flow after venous pressure increase achieved by a cuff), and the lowest reading over 5 min of standing is registered. The venoarteriolar reflex can be expressed as the percent decrease in skin blood flow on standing. The reaction is mediated by a sympathetic axon reflex, comprised of receptors in small veins and resulting in an increase in precapillary resistance.

d. Inspiratory Gasp

The subject is instructed to breathe in as deeply and quickly as possible and to hold his breath for 10 s. The percentage reduction from the resting flow is calculated. This procedure records the sympathetic vasoconstrictor reflex.[17]

2. Vasodilative Tests

a. Cutaneous Postischemic Reactive Hyperemia Test

Blood flow in the middle part of the flexor aspect of the forearms (or sometimes the proximal part of the finger) is recorded. The arm is then clamped in a pneumatic cuff and inflated to greater than 40 mmHg above systolic pressure for a period of 1 to 5 min, during which blood flow measurements are continuously recorded. The cuff is then deflated, resulting in an increase in blood flow which is recorded continuously, usually until the blood flow returns to baseline values. Any of the following parameters can be measured: (1) baseline flow; (2) peak flow above baseline flow; (3) the time required to reach the peak; (4) the ratio between the peak flow and the time required to reach it, expressing the ability of the tissue to respond to fast external triggers; (5) the time required to return to the blood flow at rest; and (6) the area under the response-time curve.[3–5]

b. Intracutaneous Needle Stimulation (Injection Trauma)

Resting blood flow is recorded; then a needle is inserted, usually in the center of the probe holder, to a depth previously set by a needle guard. The blood flow reaches a peak within 15 min of injection[9] and then gradually returns to normal over several hours, depending on the degree of trauma.

c. Thermal Test

Resting blood flow is measured in the middle part of the flexor aspect of the left forearm, while the probe is mounted through a thermostated probe holder. The temperature setting of the thermostat is adjusted to 26°C. The temperature is maintained at 26°C for 2 min before turning the setting to 28°C for the next 2 min period. This 2°C step sequence is repeated every 2 min until the temperature reaches 44°C. Blood flow is recorded at the end of each 2 min interval.[3]

d. Axon Reflex Vasodilator Response

Vasoactive substances such as substance P, capsaicin, or histamine are administered topically or intradermally.[19,20] Alternatively, acetylcholine is administered with the aid of electrophoresis.[17] The extent of the response is measured at several distances from the site of administration. The same procedure is followed for measuring the response to direct stimulation with a firm mechanical stroke with a dermograph (Lewis triple response).[17,18]

e. Isotonic Test

The subject squeezes a partially inflated blood pressure cuff with maximum effort, at which the pressure is recorded, and one third of it is calculated. The subject is then instructed to grip the cuff at this value of one third of the maximum pressure. This isotonic exercise causes vasodilatation, and an increase in skin blood flow results.[7]

B. SELECTION OF SUBJECTS

Variations in population were demonstrated by LDF with regard to sex,[21,22] age,[10] and race.[23] Obviously, this should be taken into account whenever making comparative measurements between subjects or between various groups of subjects. Assuring that subjects are matched for these variables will decrease the variance within the results.

III. NERVOUS SYSTEM

Wilkin and Trotter[6] measured skin blood flow in the finger and in the malar region during cognitive activity. They registered a decrease in digital skin blood flow, whereas no change was detected in the malar region. The absence of change in the malar region supports the view that the vasoconstriction in the finger is mediated by sympathetic stimulation, which controls the blood supply to the hands and feet, whereas the face has a poor sympathetic vasoconstrictor supply. Their study exhibits an alteration in skin blood flow due to higher mental activity unrelated to stress.

Bengtsson et al.[24] measured skin blood flow in order to evaluate the effect of sympathetic blockade produced by spinal analgesia in patients undergoing transurethral resection of the prostate. They demonstrated a reduction in skin blood flow in the shoulder and chest following spinal analgesia, whereas the skin blood flow in the lower part of the body increased. These changes persisted to a lesser extent in the postoperative period, indicating the duration of analgesia.

Willette et al.[25] evaluated the role of alpha-1 and alpha-2 adrenoceptors in mediating sympathetic responses. They measured the skin blood flow in acral regions. The alpha-2 adrenoceptor agonist which was administered intravenously was more potent than the alpha-1 adrenoceptor in increasing the cutaneous microvascular resistance and reducing the perfusion.

For further examples of the utilization of LDF for the study of the nervous system, please refer to Sections II, IV, and IX.

IV. ENVIRONMENTAL TEMPERATURE

Elam and Wallin[26] studied the effect of emotional stress following prolonged heating and cooling by measuring skin blood flow to the hands and feet. The temperature was controlled by using a water-perfused suit, achieving a skin temperature of 34°C (heating) or 22°C (cooling). In heated subjects they registered vasoconstriction, whereas vasodilatation occurred in cooled subjects (22°C). Like Wilkin and Trotter,[6] they presume that the stress-induced flow decrease in warm subjects is neurally mediated, since such hand vasoconstriction is preceded by increased skin sympathetic activity. However, the increase of flow in cold subjects is more obscure. Different neural commands at different temperatures might be a possibility. Another possibility is that arteriovenous shunts receive a vasoconstrictor sympathetic input, whereas resistant vessels receive a vasodilator input. In cold subjects with high baseline vasoconstriction, the arteriovenous shunts are closed and unable to constrict any further, so only the vasodilatation in the resistance vessels is detected. In warm subjects, thermoregulatory activity is low, the basal flow is high, and the arteriovenous shunts are open. When the emotional stress occurs, the constrictor effect on the open arteriovenous shunts will be more pronounced and will mask the vasodilatation in the resistance vessels.

Hassan et al.[27] studied the influence of central thermoregulatory mechanisms on the postural vasoconstrictor response. Following heating of the trunk with an electrical blanket, the postural fall in blood flow diminished in skin areas with relatively numerous arteriovenous shunts (e.g., plantar surface of the big toe). In contrast, areas with only a few or no arteriovenous shunts (e.g., dorsum of the foot) displayed similar postural flows before and after heating. They conclude that partial release of sympathetic vasoconstrictor tone associated with indirect heating appears to override the local postural control of cutaneous vascular tone in areas where arteriovenous anastomoses are relatively numerous. When measured at heart level, the indirect heating was accompanied by a significant increase in foot blood flow. Many experiments, utilizing both LDF and other techniques, support the view that this reflex thermoregulatory vasodilatation is mainly due to the release of sympathetic vasoconstrictor tone, induced by the elevated core temperature.[27] In a further study,[28] Hassan and Tooke showed that the normal postural fall in foot skin blood flow was preserved within the skin temperature range of 26 to 36°C, but at higher temperatures it was markedly attenuated or even abolished. They suggested that this might contribute to some of the problems of cardiovascular adaptations seen in hot environments.

Winsor et al.[29] examined the influence of room temperature on peripheral flow in healthy subjects and patients with peripheral vascular disease. They found very little effect on flow when the room temperature was increased from 24 to 30°C; therefore, this range is suitable for skin blood flow studies. In temperatures higher than 30°C the peripheral circulation increased. The relationship between skin flow and room temperature was linear in temperatures between 23 and 30°C, whereas between 30 and 35°C it was curvilinear.

V. SMOKING

Utilizing a variety of methods, several studies have demonstrated a decrease in skin blood flow immediately following smoking. These studies include Doppler ultrasound of the digits, venous plethysmography of the forearm, capillary nailfold microscopy, and LDF.[4,30-32] Although each technique provides slightly different information about skin blood flow, the various studies all suggest an abnormality in capillary blood flow and its regulation in the skin as an immediate result of cigarette smoking.

LDF, used by Tur et al.[4] to monitor the time course of cutaneous postischemic reactive hyperemia, has established that cigarette smoking has statistically significant chronic and acute effects on the cutaneous microvasculature. Lower peak flow after smoking was the significant characteristic of the acute effect, whereas the chronic effect was manifested by a longer recovery time from the reactive hyperemia in habitual smokers as compared with nonsmokers. Since reactive hyperemia serves an important role in protecting, or at least limiting, tissue damage following ischemia, the abnormal response in habitual smokers is of clinical importance. The importance of the provocative tests was illustrated in this study by the static measurements of the basic blood flow during smoking, which did not show any significant variations in four sites: forehead, postauricular, forearm, and finger. Similarly, Baab and Oberg,[33] measuring skin blood flow during and after cigarette smoking, did not find any significant changes in comparison to sham smoking an unlighted cigarette.

Likewise, Suter et al.,[34] using multilead plethysmography during smoking, did not find any significant changes in skin blood flow to the forehead or to the earlobe, while a decrease was noted in the flow to the fingers and a smaller one to the toes. However, the magnitude of the vascular response at the finger correlated positively with personality traits of neuroticism and not with the dose of nicotine. These observations, as well as differences in methodology (like the number of cigarettes smoked in each study and the subject selection), may explain the contradicting results obtained in other studies.[31,32,35,36]

Bornmyr and Svensson[35] registered a decrease in skin blood flow following smoking of two cigarettes, and the lowest values were observed within 15 min. Lecerof et al.[36] also found a decrease in skin blood flow in hypertensive habitual smokers which was maximal 15 to 30 min after smoking. Similarly, Muller et al.[31] demonstrated a decrease in fingertip blood flow during 5 min of smoking and a further decrease during the following 10-min observation period. Their subjects smoked one cigarette every hour; the decrease in blood flow due to the first up to the fourth cigarette of the day was similar, but after the eighth cigarette it was less marked.

van Adrichem et al.[37] found a decrease in the skin blood flow of the thumb in 22 smokers during smoking of one or two cigarettes. This decrease became evident during the first 2 min of smoking. The blood flow recovered only by half after 10 min of rest. In the control group of ten nonsmokers the flow did not show major variations during or following smoking, and the differences between smokers and nonsmokers were statistically significant.

Goodfield et al.[32] also found a fall in finger blood flow following the smoking of a single cigarette only in habitual smokers; nonsmokers or irregular smokers did not show such a decrease in blood flow. Following 3 days of aspirin treatment (2.4 g/day), healthy nonsmokers showed a fall in blood flow due to a single cigarette, similar to habitual smokers. Their results suggest that regular smoking sensitizes the peripheral vasculature to the vasoconstricting effects of the next cigarette and that at least part of this sensitization is mediated by the inhibition of endothelial prostacyclin synthesis.

Waeber et al.[30] have shown that cigarette smoking triggers the release of vasopressin. There was a significant correlation between the skin blood flow response to cigarette smoking and the plasma vasopressin levels after smoking. The vasopressin released may mediate some of the vasoconstriction, as evidenced by the reduction of the nicotine-induced vasoconstriction in the skin following pretreatment with a vasopressin antagonist. The vasopressin antagonist alone did not affect the skin blood flow.

Nicita-Mauro[38] studied the role of calcium ions in the pathogenesis of the vasoconstriction caused by acute smoking. Elderly habitual smokers were given the calcium channel blocker nifedipine and calcitonin, a hypocalcemizing hormone that has a vasoactive action. Both drugs prevented the LDF-measured vasoconstriction induced by cigarette smoking, and the author proposed that this was a calcium-mediated process.

These studies illuminate some of the mechanisms involved in the changes of the microvascular bed in the skin of habitual smokers and reveal both acute and chronic changes. The deleterious effects of cigarette smoking become more severe with time, although the chronic changes occur relatively early in a person's smoking history. It would be interesting to assess the correlation between these skin blood flow changes and the future development of peripheral vascular disease in a long-term study.

VI. PREGNANCY AND GESTATIONAL HYPERTENSION

The skin microvasculature may serve as an indicator of an internal state such as normal or hypertensive pregnancy. Tur et al.[3] measured a lower baseline blood flow on the fingertip in gestational hypertension than in normal pregnancy, but this measurement could not discriminate between each of these two groups and nonpregnant healthy controls. An attempt was made to identify the most discriminatory test. Among the provocative tests used (isometric test, cognitive test, reactive hyperemia test, and thermal test), the isometric test gave the most discriminative results. The control group showed a larger decrease than both the normal pregnancy group and the gestational hypertension group. Gestational hypertension showed a larger decrease than the normal pregnancy group. The cognitive test and postischemic reactive hyperemia allowed some degree of discrimination, whereas the thermal test did not show any abnormality in the pregnant groups. This study shows that the response to vasoactive stimuli, as measured by LDF, may differentiate between groups of nonpregnant subjects and subjects with normal or hypertensive pregnancy. Normal pregnancy modifies the response of the skin microvasculature to some vasoactive stimuli, whereas gestational hypertension pushes that response back toward the non-pregnancy state. However, in order to apply the method as a diagnostic tool for the individual patient, further improvements in laser Doppler techniques and instrumentation are needed.

VII. HYPERTENSION

Hypertension is associated with, or even originates from, an increased total peripheral resistance. The skin microvasculature plays a role in this peripheral resistance, hence the importance of its investigation in hypertension and the possibilities granted.

Lecerof et al.[36] studied the decrease of skin blood flow that follows smoking in 24 hypertensive habitual smokers. They measured skin blood flow after the patients had smoked two cigarettes both before any treatment and following intravenous administration of alpha-1 inhibitor doxazosin and beta-1 blocker atenolol. Skin blood flow decreased following smoking under basal conditions and also with atenolol and doxazosin, but the reduction was attenuated by doxazosin compared to atenolol. This indicates that selective alpha-1 adrenoceptors have a major effect on smoking-induced cutaneous vasoconstriction. In this respect doxazosin is preferable to atenolol for the antihypertensive treatment of patients who smoke.

Cesarone et al.[14] studied changes in the microcirculation in essential hypertension and the influence of nifedipine treatment on these changes. They measured resting blood flow to the dorsum of the foot with the patient resting in the supine position. They then measured standing flow and registered the lowest reading over 5 min of standing. The venoarteriolar reflex was expressed as the percent decrease in skin blood flow on standing. In hypertensive patients both the resting flow and the standing flow were significantly lower, and they increased after nifedipine treatment. The venoarteriolar reflex was lower in hypertensive patients, improving after nifedipine treatment, but still below normal. These findings suggest the presence of distal vasospasm. LDF could also distinguish between responders and nonresponders.[39]

Side effects of treatments may be studies by LDF. Salmasi et al.[40] compared two groups of hypertensive patients after 4 weeks of nifedipine therapy (10 mg three times daily). One group developed ankle edema as a result of treatment, whereas the other did not. Before treatment the two groups did not differ in their resting flow, supine flow, or the venoarteriolar reflex, which was normal. However, following nifedipine treatment a weaker venoarteriolar reflex was observed in the patients who developed ankle edema. This may explain the mechanism underlying the development of ankle edema in such patients.

Unlike smoking effects, where the cutaneous postischemic reactive hyperemia response disclosed differences between normotensive smokers and nonsmokers, the cutaneous postischemic reactive hyperemia response does not seem to differentiate between hypertensive patients and normotensive controls. Tur et al.[5] compared 15 patients with high blood pressure to 15 healthy controls and found no significant differences in their cutaneous postischemic reactive hyperemia response. Studying 50 diabetic patients, their subgroups with or without hypertension did not differ statistically in their response to ischemia. Similarly, Orlandi et al.[41] have compared a group of 20 patients with hypertension to a group of 20 age- and sex-matched healthy controls. Utilizing LDF, they measured skin blood flow at the peak of cutaneous postischemic reactive hyperemia and found no significant differences. Hatanaka et al.,[42] who observed a decreased fingertip blood flow in diabetic patients (utilizing LDF), have also observed normal fingertip blood flow in individuals with hypertension.

Thus, a few provocative tests were able to detect differences between hypertensive patients and controls, while other tests could not.

VIII.   ATHEROSCLEROSIS

An important determinant in the assessment of the severity of peripheral vascular disease is the adequacy of skin blood flow in the ischemic extremity. Kvernebo et al.[43] compared two groups of old and young healthy volunteers to patients with lower limb atherosclerosis and intermittent claudication and patients with lower limb atherosclerosis and critical ischemia. LDF could differentiate between the clinical groups. Elderly controls had higher flux values in the toe compared with claudicators, while claudicators had higher perfusion values than patients with critical ischemia.

Winsor et al.[29] looked for a standard set of conditions for the performance of tests to reveal peripheral vascular disease in the lower extremity. They used the ratio between toe and finger flows. The ratio narrows the range of the controls, eliminating differences in cardiac output that occur from patient to patient. In this manner they were able to distinguish between a group of healthy controls and a group of patients with peripheral vascular disease, using standard conditions of room temperature of 23°C and probe temperature of 40°C.

Belcaro and Nicolaides[15] also found the technique useful for evaluating patients with peripheral arterial disease and for differentiating among different etiologies of the disease. They showed an increase in skin blood flow with an improvement in the venoarteriolar reflex both in patients with peripheral vascular disease and in healthy controls. Patients with intermittent claudication, patients with rest pain, and those with critical foot ischemia had a significantly lower resting flow than normal. Healthy controls showed a reduction of skin blood flow on standing, which was smaller in patients with severe claudication. Patients with rest pain had higher skin blood flow values when standing, indicating a loss of vasomotor tone and an increase of flow determined by gravity. The inverse effect of an increase in blood flow with standing is associated with clinical improvement when the patients lower their limbs.

Karanfilian et al.[44] also measured skin blood flow in peripheral vascular disease, recognizing its importance in the prospect of healing of ulcers or an inevitable amputation. The baseline skin blood flow in patients with peripheral vascular disease was significantly lower than normal, and the pulse waves were attenuated or absent. In indeterminate cases the accuracy of the LDF method could be enhanced by the use of reactive hyperemia. They conclude that the method may be useful in predicting the level of successful amputation or the healing of an ischemic ulcer.

Provocative studies of this type were also conducted by Leonardo et al.[45] in patients with peripheral vascular disease to quantitatively assess arterial insufficiency of the limbs. Resting blood flow proved to be of little use, whereas the postischemic reactive hyperemia test permitted a clear distinction between healthy control subjects and patients with arterial disorders. Moreover, there was a correlation between the impairment of the LDF results and the gravity of the clinical picture, significant differences being recorded between limbs with no sign of necrosis and limbs affected by slight necrosis. In patients with pain at rest, postischemic hyperemia was absent. Following therapy with intraarterial administration of naftidrofuryl, a statistically significant improvement in several LDF-assessed parameters was achieved, and treatment of four limbs with percutaneous transluminal angioplasty resulted in a completely normal test, matching the disappearance of all symptoms caused by the peripheral occlusive arterial disease. Consequently, the authors suggested a wider use of this test in vascular diagnostic laboratories.

Kristensen et al.[46] used the postischemic reactive hyperemia test to quantitate peripheral blood flow in patients with arterial and venous ulcerations. They used the peak flow divided by the time required to reach this peak, which is a measure of the rate of ascent of the blood flow toward its peak value and which determines the time required to supply a sudden needed increase of blood flow. A reduction in this parameter was found in patients with leg ulcers, with a sensitivity comparable to the measurements of distal systolic blood pressure. They found the technique useful for detecting blood flow values that might be relevant for healing time, being related to the tissue nutrition.

These studies again demonstrate that appropriate physiological tests like hydrostatic pressure loading and, particularly, the postischemic reactive hyperemia test are more indicative of abnormalities than are static parameters like resting flow. LDF can also provide explanations for therapeutic effects, like that of CO_2 baths in occlusive arterial disease, where an improvement in skin blood flow and in the pulse waves has been observed.[47]

IX. DIABETES MELLITUS

LDF was widely used in studying the process of diabetes mellitus and the etiology of its various complications, in discriminating between different disease groups, in predicting its outcome, and following treatment. Diabetes mellitus is the most illustrative disease for the application of skin blood flow measurements, its microangiopathic changes serving as a natural target for LDF application to probe the various aspects of the disease. Damage to the microcirculation in diabetes mellitus is responsible for a great number of its grave complications. The neuropathy that further affects the microcirculation adds to the diversity and complexity of the disease, making the exposure of its yet unsolved aspects even more intriguing and rewarding. An understanding of the mechanisms responsible for microvascular complications may help in developing new treatment modalities. Furthermore, detection of early functional changes in the microcirculation might identify patients at risk at a reversible stage. The possibility of using the skin as a model for diabetic microangiopathy and its correlation to retinopathy and nephropathy are still under investigation.

A. INSULIN-DEPENDENT DIABETES MELLITUS

Rendell et al.[48] used temperature changes (with a thermal probe) for provocation and measured blood flow in insulin-dependent diabetes mellitus (IDDM) in various anatomic sites on the extremities. Healthy controls had greater blood flow values at most locations at 35°C, and this difference was even more pronounced at 44°C. Their results demonstrate that diabetic microangiopathy affects the skin microvasculature over a widespread distribution. At an elevated skin temperature of 44°C, Rendell and Bamisedun[49] found a decreased blood flow at the dorsum of the finger and toe as a function of the duration of diabetes, whereas the pulps did not show such differences. Retinopathy and nephropathy were also associated with such a decrease. It appears, therefore, that skin microangiopathy correlates with impaired retinal and renal microvasculature, and this process is apparent at sites of nutritive vasculature and not at sites of arteriovenous shunts.

Wilson et al.[50] found the response to local heating to be useful in distinguishing between healthy controls and complicated and uncomplicated IDDM patients. Measuring various values of heat-induced hyperemia, they found the time necessary to reach three times the basal flow particularly useful since it could detect microvascular deterioration before other complications were apparent. They also found cold-induced vasoconstriction helpful in identifying subjects with impaired sympathetic control. Khan et al.[10] found markedly reduced vasoconstrictor responses to contralateral arm cold challenge and body cooling and also to inspiratory gasp, presumably resulting from sympathetic neural dysfunction. Winocour et al.[51] also demonstrated altered patterns of hand skin blood flow following a cold challenge achieved by immersion of the hands in cold water. The blood flow was modified even more in patients with microvascular complications.

An impaired skin microvascular vasodilator response to both injection trauma and local thermal injury (heating the skin site to 44°C for 30 min) was demonstrated by Rayman et al.[9] in patients with IDDM. Peak response was lower in the diabetic group in widely separated skin sites, but the time course was the same. The impairment in response was not related to diabetic control. They concluded that the abnormality of the diabetic skin microvasculature response to injury might be an important factor in the development of the foot ulceration that often follows minor trauma.

In order to investigate the mechanisms underlying the impaired hyperemia to local injury, Boolell and Tooke[52] injected substance P intradermally and found a reduced peak response in IDDM patients as compared to controls, whereas the response to capsaicin was the same. Following histamine blockade with chlorpheniramine the response to capsaicin remained unaltered, while the response to substance P was reduced in both groups. Therefore, impaired skin hyperemia may represent decreased vascular reactivity to locally released substance P from peripheral nerve fibers.

Structural pathology in early diabetic neuropathy is best correlated to peroneal motor conduction velocity.[53] Transcutaneous oxygen and LDF measurements have been correlated with peroneal motor conduction velocity in IDDM patients, suggesting that hypoxia generates diabetic peripheral neuropathy and that in early neuropathy therapeutic measures to improve blood flow might arrest its progress.[53] A pilot study conducted by Rendell and Bamisedun[54] lends support to this hypothesis. Their patients with diabetic sensory neuropathy had decreased skin blood flow at 44°C which was more pronounced in the lower extremity than in the upper. Pentoxifylline treatment (400 mg three times daily) increased the skin blood flow in the lower extremity, as measured after 3 and 6 months of therapy. The authors propose a double-blind study to establish the efficacy of this treatment, using the same methods of evaluation.

Barnes et al.[55] found a higher basal resting flow in diabetic patients. The postischemic reactive hyperemia test failed to discriminate between subgroups of diabetes, and they found this test difficult to interpret. They measured skin blood flow in the forearm, and the ischemia was produced by a cuff placed above the elbow for 4 min. Similarly, Wilson et al.[50] found no differences in the postischemic reactive hyperemia response of patients with or without retinopathy and healthy controls, measuring blood flow on the dorsum of either hand or foot following a 2-min period of ischemia.

In contrast, Tooke et al.[56] did differentiate between patients with or without microvascular complications (retinopathy, nephropathy, or neuropathy) on the basis of the postischemic reactive hyperemia test following digital arterial occlusion for 1 min, and so did Walmsley and Wiles for patients with or without neuropathy (10-min occlusion, above the malleolus).[57] Such differences between different studies could originate from variations in methodology, like the site measured, the temperature, the duration of the ischemia, and choice of subjects. Houben et al.[58] could also discriminate between subgroups of IDDM on the basis of both finger skin rest flow and the postischemic reactive hyperemia test. They included patients with a short duration of IDDM and patients with a long duration with or without complications (retinopathy or incipient nephropathy). The blood flow at rest in patients with incipient nephropathy was the same as in healthy controls, unlike the remainder of the patients, who demonstrated increased blood flow. The postischemic reactive hyperemia peak flow was lower in patients with incipient nephropathy than in the rest of the patients, implying a more generalized alteration of the microcirculation. The venoarteriolar reflex was the same in all IDDM groups and the controls, suggesting that the neurogenic microvascular flow regulation was unaffected.

Parkhouse and Le Quesne[17] evaluated the nociceptive C fibers by measuring the axon reflex vasodilatation of the foot in response to electrophoresis of acetylcholine, to direct mechanical stroke with a dermograph, and to a deep inspiratory gasp. Patients with long-standing diabetes without foot complications responded with the same flare as the healthy controls. In contrast, the flare was reduced in patients with diabetes and foot complications. Impairment of the neurogenic flare correlated with the clinical diminution of pain sensation; both are components of nociceptive C fiber function. Reduction of the flare indicates an impairment of the neurogenic inflammatory response, which, along with an impairment of the protective pain sensation, may be a contributory factor in the poor and slow healing of foot lesions of diabetic patients. Since sympathetic vasoconstrictor reflexes are present in most diabetic patients with foot ulceration, the role of autonomic neuropathy in ulcer development is questioned. Similarly, Corbin et al.[59] found no clear relationship between autonomic nerve dysfunction and the degree of abnormality of blood flow as measured by the Doppler ultrasound technique.

Similar results were obtained by Walmsley and Wiles,[60] who measured the flare induced by acetylcholine iontophoresis at various current strengths. Maximum flare response was reduced in neuropathic patients, especially those with a previous history of foot ulceration, suggesting that small-fiber neuropathy affects ulcer development. The flare was also reduced in some patients with retinopathy alone and no neuropathy, suggesting an early loss of small nociceptor C fibers, preceding large-fiber neuropathy. The curve of the hyperemic response plotted against current strength did not show a rightward shift, indicating that the abnormal response was due to axonal loss rather than to dysfunction. The flare did not correlate with the cardiac autonomic function.

B. NON-INSULIN-DEPENDENT DIABETES MELLITUS

There are functional microvascular disturbances in the forearms of diabetic patients. Tur et al.[5] measured cutaneous blood flow in the forearms during the postischemic reactive hyperemia test in patients with non-insulin-dependent diabetes mellitus (NIDDM), and in nondiabetic controls. Division within the group of patients was made according to the presence or absence of proliferative retinopathy and obesity. In diabetic patients the peak flow was significantly lower than in nondiabetic controls. In diabetic patients with proliferative retinopathy the ratio between the peak blood flow following arterial occlusion and the time required to reach this peak was lower than in diabetic patients without retinopathy. Diabetic patients with a body mass index (weight/height^2) smaller than 25 kg/m^2 had a longer time to peak; in the control group, body mass index did not affect this parameter. The combination of retinopathy and body mass index lower than 25 gave the longest time to peak values. Postischemic hyperemia tests in diabetic patients thus revealed cutaneous microcirculatory changes in the forearm. Advanced retinopathy added to the functional disturbances, especially when combined with a low body mass index. This observation adds to the currently existing knowledge regarding structural microvascular disturbances.

Belcaro and co-workers[16,61] studied patients with diabetic microangiopathy and neuropathy. The type of diabetes was not stated, and probably both types of diabetes were included. Resting skin blood flow was greater in diabetic patients than in healthy controls and was even greater in patients with neuropathy. The venoarteriolar response was lower in diabetic patients than in healthy controls and was lowest in patients with neuropathy. Both the increased skin blood flow and the impaired venoarteriolar reflex are causes of edema and may contribute to the thickening of capillary basement membranes detected in diabetes. The authors advise a periodic evaluation of skin blood flow in diabetic patients.

Koltringer et al.[62] developed a hyperthermal laser Doppler flowmeter to quantify autonomic dysfunction in the skin. Their technique involves measuring the time to induction of an increase of microcirculation following hyperthermia. In a pilot study they used the technique to evaluate the autonomic function of NIDDM patients with a disease duration of 1 to 3 years, 4 to 5 years, and 5 to 6 years. All groups showed autonomic dysfunction, suggesting that it was an early complication of NIDDM, even when the disease was well controlled.

Neurovascular function tests in experimental and human diabetes were summarized by Westerman et al.[63]

C. EVALUATION OF TREATMENT

LDF was used to evaluate a variety of treatment modalities, starting with simple physical measures like elastic stockings, through old and new medications, and even a combined kidney and pancreas transplantation.

Two groups of diabetic patients with microangiopathy were followed up for 4 years by Belcaro et al.,[64] who measured the skin blood flow at rest and the venoarteriolar response. One group was instructed to wear stockings, whereas the other group was not. In the untreated group the flow at rest increased and the venoarteriolar response decreased as compared to the treated group. Thus, they concluded that elastic stockings repressed the deterioration of the microcirculation.

Flynn et al.[65] studied the effect of insulin infusion on the foot blood flow in NIDDM. The capillary blood flow, as measured by nailfold microscopy, increased, whereas the arteriovenous shunt blood flow, as measured on the pulp of the great toe by LDF, remained unchanged. These results suggest that during insulin infusion skin blood flow is redistributed, with an increase in capillary flow relative to arteriovenous shunt flow. Since blood that passes through arteriovenous shunts does not enter the capillary bed and plays no role in skin nutrition, this outcome represents an improvement in the skin nutrition.

Aman et al.[66] found that hypoglycemia, but not hyperinsulinemia, caused a regional skin vasodilation in healthy control subjects. Following a hyperinsulinemic euglycemic clamp, they induced hypoglycemia by a stepwise reduction in the intravenous glucose infusion. The increase in blood flow that was observed in healthy controls was absent in patients with NIDDM.

Belcaro et al.[67] studied the effect of a new profibrinolytic drug, defibrotide, given orally, in 117 NIDDM patients with microangiopathy. A group of 53 patients was treated with diet and oral antidiabetic medications only, while a group of 64 patients received defibrotide in addition. The microcirculation was evaluated by the venoarteriolar response and by the rise in skin blood flow following local heating from 36 to 40° C. Both are decreased in diabetic patients, and indeed this decrease was observed in both groups at the beginning of the study. Following 6 months of treatment, patients in the new drug treatment group improved their microcirculatory parameters (venoarteriolar reflex) in association with an improvement in signs and symptoms, whereas no such improvements occurred in the other group. Several patients were followed up for 18 months, and the deterioration of the microcirculatory parameters was slower in the group treated with defibrotide. Thus, LDF was successfully applied for the objective noninvasive evaluation of a new drug treatment.

Jorneskog et al.[13] studied the effect of combined kidney and pancreas transplantation in nine patients with severe late diabetic complications. In a previous study[68] they demonstrated that the skin microvascular reactivity was better in a group of diabetic patients who had undergone such a transplantation as compared to a group of diabetic patients with a similar disease severity waiting for transplantation. They then tried to assess whether the transplantation could reverse or halt diabetic complications.[13] They measured skin blood flow at rest, postischemic (1-min occlusion) reactive hyperemia, and venoarteriolar reflex at 2 and 38 months following transplantation. The blood flow at rest was higher 38 months following transplantation as compared to 2 months, but the delayed time to peak during hyperemia was even more impaired at 38 months. This delayed time to peak response following a short occlusion is probably due to a disturbed function of the smooth muscle cells of the vessel wall, and the results may indicate a progression of structural changes in spite of an improved metabolism. The venoarteriolar reflex was also impaired at both time points, probably due to neuropathy, since this reflex depends on an intact sympathetic nerve function. However, although not statistically significant, they observed an improvement in four out of five patients with the most impaired reflex, which might indicate that diabetic neuropathy could be improved after transplantation.

These and other studies indicate that LDF can be useful in the investigation of some pathophysiological mechanisms of diabetes and as an adjunct in the determination of the severity of the disease and the efficacy of its control. The skin is, therefore, very valuable as a model for microangiopathy.

X. RAYNAUD'S PHENOMENON

The intermittent blanching that occurs in Raynaud's phenomenon is believed to result from an active microvascular vasoconstriction and emptying. Therefore, LDF is evidently useful for the investigation of the pathophysiological mechanisms underlying Raynaud's phenomenon and for evaluating treatments.

Shawket et al.[69] used the LDF technique to clarify the etiology of Raynaud's phenomenon. Three vasodilators were administered intravenously, and the response of calcitonin-gene-related peptide was compared with that of endothelium-dependent adenosine triphosphate and endothelium-independent prostacyclin. The first vasodilator induced an increase in blood flow in the hands of patients but not in healthy controls, which might reflect a deficiency of endogenous calcitonin-gene-related peptide release in Raynaud's phenomenon.

The role of the histaminergic and peptidergic axes in primary Raynaud's phenomenon was studied by Bunker et al.[19,20] They measured the digital blood flow response to intradermal injections of saline, histamine, histamine-releasing agent (compound 48/80); substance P, and calcitonin-gene-related peptide. They found no evidence of local deficiency in histamine release or in the response to histamine,[20] even at low temperatures,[19] and the patients reacted normally to the neuropeptides substance P[19,20] and calcitonin-gene-related peptide,[20] providing a rationale for treating Raynaud's phenomenon with vasoactive peptides.

Increased and prolonged vasoconstrictor response to cold was measured in patients who had received combination chemotherapy for testicular cancer.[70] This is a characteristic response in Raynaud's phenomenon of the vasospastic type. A correlation with neuropathy as an etiological factor could not be established. The authors proposed that a prospective study before and after chemotherapy might give additional information about the development of Raynaud's phenomenon in these patients.

Suichies et al.[71] measured digital skin blood flow of both hands during local heating of only one hand. Patients with Raynaud's phenomenon showed a decreased digital blood flow during stepwise cooling in both hands, but the reaction in the cooled hand was more pronounced and more consistent.

Walmsley and Goodfield[22] found that patients with Raynaud's phenomenon had an abnormal vascular response to temperature change. They studied the hyperemic response to local skin warming, and the patients showed vasodilatation at lower skin temperatures than normal, independent of central sympathetic control. A knowledge of skin temperature is therefore important for the interpretation of blood flow studies in Raynaud's phenomenon.

Similar differences were observed by Allen et al.[72] when, after cooling the finger, a group with Raynaud's had lower mean blood flow values than controls. However, this method could not distinguish between individual subjects. Therefore, they used provocative testing (occlusion) under different degrees of finger and body cooling. They then detected an increase in the number of fingers of patients exhibiting vasospasm as the severity of cooling increased.

Occupational exposure to vibrations causes Raynaud's phenomenon. For prevention and treatment of vibration syndrome an objective test was developed by Kurozawa et al.,[73] who combined finger cooling with LDF. They found significant differences among four groups: subjects without vibration exposure, subjects with exposure but with no signs of white fingers, subjects with few attacks, and subjects with frequent attacks.

Kristensen et al.[74] measured digital blood flow reactivity to cooling and rewarming in patients with Raynaud's phenomenon suffering from scleroderma. They found that during cooling the reduction in blood flow was the same as for healthy controls, but the patients had a longer rewarming period.

Goodfield et al.[75] also studied Raynaud's phenomenon in systemic sclerosis. They used the postischemic reactive hyperemia test following arterial occlusion for 2 min. When the resting blood flow was very low the hyperemic response was absent. After warming the hand in warm water the hyperemic response was restored and its magnitude corrected, but its time course was longer, with a delay to achieving maximum flow as compared to controls. This may be related to changes in the vessels themselves or to connective tissue sclerosis limiting the rate of response.

Engelhart and Seibold[76] studied groups of patients with primary Raynaud's phenomenon, systemic sclerosis, and undifferentiated connective tissue disease as compared to a control group. In addition to local and central temperature changes, they used provocative tests which evoke sympathetic tone, like the isometric test. They found considerable differences in both the level of vessels involved and the relative importance of local finger temperature. Finger temperature was the principal determinant of arterial flow in systemic sclerosis, and arterial flow was the principal determinant of microvascular perfusion. The patients with systemic sclerosis were unable to maintain nutritive flow when triggered with cold or a reflex. Thus, LDF proved valuable in discriminating between various etiologies of Raynaud's phenomenon and also in investigating the underlying pathophysiological mechanisms.

In a study of the acute effect of smoking in Raynaud's phenomenon,[32] cigarette smoking caused a decrease in blood flow only in smokers, but the same pattern was demonstrated in subjects with or without Raynaud's phenomenon.

Wollersheim et al.[77] found that LDF measurements of cutaneous postocclusive reactive hyperemia could determine the degree of obstructive vascular disease in groups of patients with Raynaud's phenomenon, but could not discriminate among individuals in the subgroups. Another provocative test, the cognitive test, was used by Martinez et al.,[78] who detected two subgroups within the patients with Raynaud's phenomenon. The first subgroup showed a reduction in blood flow similar to healthy controls, whereas the second showed a paradoxical increase, suggesting an organic etiology.

LDF was used to evaluate various treatment modalities in Raynaud's phenomenon. Engelhart[79] studied ketanserin, an antagonist of the serotonin-2 (5-HT-2) receptor, in nine patients with generalized scleroderma. He found no improvement, measuring finger systolic pressure and LDF after cooling and rewarming of the finger. Thus, he concluded that ketanserin in the doses used (20 mg three times a day in the first week and 40 mg three times a day for 4 weeks) was not effective in the treatment of Raynaud's phenomenon in generalized scleroderma. When given to patients with primary Raynaud's phenomenon,[80] ketanserin normalized digital blood flow as measured by LDF. Pretreatment with alpha-adrenoceptor antagonists did not abolish this effect, suggesting that, in contrast to the effects on the systemic circulation, the mechanism underlying digital vasodilatation after ketanserin administration does not involve alpha-adrenoceptor antagonism.

Bunker et al.[81] applied the vasodilator hexyl nicotinate at various sites on the upper limb and demonstrated an increase in blood flow both in patients with Raynaud's phenomenon (13 with primary Raynaud's disease and 12 with systemic sclerosis and Raynaud's phenomenon) and in normal subjects. They even demonstrated an increase in flow rate by increasing the drug concentration.

Rustin et al.[82] studied the effect of nifedipine on perniosis. On a finger or a toe adjacent to a diseased one they measured the baseline blood flow and followed it after administration of nifedipine. An increase in blood flow could be demonstrated after intake of the drug.

XI. FLUSHING AND FACIAL PALLOR

Flushing, a transient reddening of the face and other body sites, is caused by vasodilatation, which may be provoked by many pharmacological and physiological reactions. Flushing may be associated with alcohol and conditions such as menopause, carcinoid, mastocytosis, or drugs (e.g., nicotinic acid).[83] Flushing occurs after orchidectomy for carcinoma of the prostate, and Frodin et al.[84] have found that LDF measurements correspond closely to the intensity of the attacks experienced by the patients and also to sweating as measured by evaporimetry.

Wilkin[85] showed that the LDF technique is appropriate for quantitative assessment of alcohol-provoked flushing. Comparing LDF to the change in malar thermal circulation index, he found a linear correlation between the two methods. Moreover, the LDF method was more specific and more sensitive.

These findings led the way to other quantitative studies of facial flushing and its treatment. Comparing the therapeutic effects of isotretinoin and oxytetracycline in rosacea, Irvine et al.[86] measured the cutaneous blood flow over the cheeks. They found no reduction in the oxytetracycline-treated group, while the group treated with isotretinoin showed a statistically significant reduction both when measurements were taken at room temperature and at 34°C.

Others[83] have used cutaneous blood flow measurements for monitoring various therapeutic modalities for facial flushing.

Systemic administration of nicotinic acids produces a generalized cutaneous erythema partially mediated by prostaglandin biosynthesis. To examine the mechanism further, local cutaneous vasodilatation was studied following topical application of methyl nicotinate.[87] Pretreatment with prostaglandin inhibitors (indomethacin, ibuprofen, and aspirin) significantly suppressed the erythemal response, while doxepin had no effect on this response.

Wiles et al.[88] infused high levels of arginine vasopressin, comparable to those attained during physical stress, and produced a marked facial pallor in healthy men. The pallor was objectively verified by LDF measurements, which were consistent with a fall in nutritional blood flow to the skin. In contrast, blood flow to the finger rose, indicating an increased blood flow through arteriovenous shunts. Thus, LDF assisted in determining that arginine vasopressin has a selective vasoactive effect in the skin.

XII. OTHER DISEASES

Examples for the application of LDF in the evaluation of treatment are given in the Sections IX, X, and XI. Other diseases, like leprosy, have distinct skin lesions which may be followed up by LDF. Agusni et al.[89] measured blood flow in leprosy lesions during treatment and related it to the clinical appearance and histopathology of the lesions. They found a clear relationship between the granuloma fraction and the blood flow elevation. The amount of hyperemia, as measured by LDF, was found to be useful in monitoring the early changes of reversal reaction during chemotherapy.

The use of LDF to investigate microvascular physiology in patients with sickle-cell disease is illustrated in the study of Rodgers et al.[90] They found large local oscillations in skin blood flow to the arm, occurring simultaneously at sites separated by 1 cm, suggesting a synchronization of rhythmic flow in large domains of microvessels. The periodic flow demonstrated in their study may be a compensatory mechanism to offset the deleterious altered rheology of erythrocytes in sickle-cell disease.

LDF has a potential for use in oncology, since it can be employed to monitor vascularization in tumors and adjacent skin. For instance, it may serve as an additional tool in the diagnosis of pigmentary skin lesions. Melanomas showed higher laser Doppler blood flow readings than basal cell carcinomas, and both showed higher readings than benign lesions.[91]

XIII. CONCLUSION AND FUTURE PROSPECTS

LDF has proved to be useful in monitoring disease processes when a disease alters the skin blood flow. However, results obtained by LDF should not be interpreted as absolute values, but rather should serve as relative estimates. Another disadvantage of the LDF is that results obtained by different instruments or by the same instrument in different subjects cannot be compared. Furthermore, biological variations within the disease state can also produce discord between different studies. There is a need, therefore, for technical improvements, calibration standards, better probe designs, and more reproducible measuring procedures before LDF becomes a useful clinical tool.

An important recent development is the scanning LDF,[18] which, during a relatively short period of time, sequentially and remotely scans the tissue perfusion in several thousand measurement points. A map of the spatial distribution of the blood flow is thus obtained. Unlike regular LDF, which continuously records the blood flow over a single point, the scanning LDF maps the blood flow distribution over a specific area. Thus, the two methods do not compete with each other, but rather are complementary. The new scanning LDF has the advantage of operating without any contact with the skin surface, and therefore it does not disturb the local blood flow. However, while this technique has not yet been fully explored in the study of internal diseases, it is safe to assume that its utilization to further study the involved pathophysiological mechanisms, in conjunction with other modalities, will allow for further adaptations and enhance its efficiency.

Other improvements in laser Doppler instrumentation and techniques offer new possibilities and may lead to new findings. Progress is illustrated by the introduction of advanced computerization and the development of new probes, such as multisite probes allowing simultaneous measurements at several sites and multisubject probes allowing simultaneous measurements of several subjects. These developments are accompanied by new multichannel LDF instruments capable of simultaneous collection of data from many independent probes. Finally, a new probe holder design[92] permits repeated measurements over the same site before and after manipulations to the skin.

All these new technical innovations may finally improve the accuracy and repeatability of laser Doppler flowmetry to a level which will justify its use as a clinical tool. In view of the many existing investigational applications of LDF in general medicine, its future clinical value is more than promising.

REFERENCES

1. Tur, E., Cutaneous blood flow: laser Doppler velocimetry, *Int. J. Dermatol.*, 30, 471, 1991.
2. Swain, I. D. and Grant, L. J., Methods of measuring skin blood flow, *Phys. Med. Biol.*, 34, 151, 1989.
3. Tur, E., Tamir, A., and Guy, R. H., Cutaneous blood flow in gestational hypertension and normal pregnancy, *J. Invest. Dermatol.*, 99, 310, 1992.
4. Tur, E., Yosipovitch, G., and Oren-Vulfs, S., Chronic and acute effects of cigarette smoking on skin blood flow, *Angiology*, 43, 328, 1992.
5. Tur, E., Yosipovitch, G., and Bar-On, Y., Skin reactive hyperemia in diabetic patients, *Diabetes Care*, 14, 958, 1991.
6. Wilkin, J. K. and Trotter, K., Cognitive activity and cutaneous blood flow, *Arch. Dermatol.*, 123, 1503, 1987.
7. Weinstein, L., Janjan, N., Droegemueller, W., and Katz, M. A., Forearm plethysmography in normotensive and hypertensive pregnant women, *South. Med. J.*, 74, 1230, 1981.
8. Low, P. A., Neumann, C., Dyck, P. J., Fealey, R. D., and Tuck, R. R., Evaluation of skin vasomotor reflexes by using laser-Doppler velocimetry, *Mayo Clin. Proc.*, 58, 583, 1983.
9. Rayman, G., Williams, S. A., Spencer, P. D., Smaje, L. H., Wise, P. H., and Tooke, J. E., Impaired microvascular hyperaemic response to minor skin trauma in type I diabetes, *Br. Med. J.*, 292, 1295, 1986.
10. Khan, F., Spence, V. A., and Belch, J. J. F., Cutaneous vascular responses and thermoregulation in relation to age, *Clin. Sci.*, 82, 521, 1992.
11. Roskos, K. V., Bircher, A. J., Maibach, H. I., and Guy, R. H., Pharmacodynamic measurements of methyl nicotinate percutaneous absorption: the effect of aging on microcirculation, *Br. J. Dermatol.*, 122, 165, 1990.
12. Shami, S. K. and Chittenden, S. J., Microangiopathy in diabetes mellitus. II. Features, complications and investigation, *Diabetes Res.*, 17, 157, 1991.
13. Jorneskog, G., Tyden, G., Bolinder, J., and Fagrell, B., Skin microvascular reactivity in fingers of diabetic patients after combined kidney and pancreas transplantation, *Diabetologia*, 34, S135, 1991.
14. Cesarone, M. R., Laurora, G., and Belcaro, G. V., Microcirculation in systemic hypertension, *Angiology*, 43, 899, 1992.
15. Belcaro, G. and Nicolaides, A. N., Microvascular evaluation of the effects of nifedipine in vascular patients by laser-Doppler flowmetry, *Angiology*, 8, 693, 1989.
16. Belcaro, G. and Nicolaides, A. N., The venoarteriolar response in diabetics, *Angiology*, 42, 827, 1991.

17. Parkhouse, N. and Le Quesne, P. M., Impaired neurogenic vascular response in patients with diabetes and neuropathic foot lesions, *N. Engl. J. Med.,* 318, 1306, 1988.

18. Wardell, K., Naver, H. K., Nilsson, G. E., and Wallin, B. G., The cutaneous vascular axon reflex characterized by laser Doppler perfusion imaging, *J. Physiol.,* 460, 185, 1993.

19. Bunker, C. B., Foreman, J. C., and Dowd, P. M., Vascular responses to histamine at low temperatures in normal digital skin and Raynaud's phenomenon, *Agents Actions,* 33, 197, 1991.

20. Bunker, C. B., Foreman, J. C., and Dowd, P. M., Digital cutaneous vascular responses to histamine and neuropeptides in Raynaud's phenomenon, *J. Invest. Dermatol.,* 96, 314, 1991.

21. Maurel, A., Hamon, P., Macquin-Mavier, I., and Largue, G., Cutaneous microvascular flow studied with laser-Doppler, *Presse Med.,* 20, 1205, 1991.

22. Walmsley, D. and Goodfield, M. J. D., Evidence for an abnormal peripherally mediated vascular response to temperature in Raynaud's phenomenon, *Br. J. Rheumatol.,* 29, 181, 1990.

23. Gean, C. J., Tur, E., Maibach, H. I., and Guy, R. H., Cutaneous responses to topical methyl nicotinate in black, Oriental and Caucasian subjects, *Arch. Dermatol. Res.,* 281, 95, 1989.

24. Bengtsson, M., Nilsson, G. E., and Lofstrom, J. B., The effect of spinal analgesia on skin blood flow evaluated by laser Doppler flowmetry, *Acta Anaesthesiol. Scand.,* 27, 206, 1983.

25. Willette, R. N., Hieble, J. P., and Sauermelch, C. F., The role of alpha adrenoceptor subtypes in sympathetic control of the acral-cutaneous microcirculation, *J. Pharmacol. Exp. Ther.,* 256, 599, 1991.

26. Elam, M. and Wallin, B. G., Skin blood flow responses to mental stress in man depend on body temperature, *Acta Physiol. Scand.,* 129, 429, 1987.

27. Hassan, A. A. K., Rayman, G., and Tooke, J. E., Effect of indirect heating on the postural control of skin blood flow in the human foot, *Clin. Sci.,* 70, 577, 1986.

28. Hassan, A. A. K. and Tooke, J. E., Effect of changes in local skin temperature on postural vasoconstriction in man, *Clin. Sci.,* 74, 201, 1988.

29. Winsor, T., Haumschild, D. J., Winsor, D., and Mikail, A., Influence of local and environmental temperatures on cutaneous circulation with use of laser Doppler flowmetry, *Angiology,* 40, 421, 1989.

30. Waeber, B., Schaller, M. D., Nussberger, J., Bussien, J. P., Hofbauer, K. G., and Brunner, H. R., Skin blood flow reduction induced by cigarette smoking. Role of vasopressin, *Am. J. Physiol.,* 247, 895, 1984.

31. Muller, P. H., Imhof, P. R., Mauli, D., and Milovanovic, D., Human pharmacological investigations of a transdermal nicotine system, *Methods Find. Exp. Clin. Pharmacol.,* 11, 197, 1989.

32. Goodfield, M. J. D., Hume, A., and Rowell, N. R., The acute effects of cigarette smoking on cutaneous blood flow in smoking and non-smoking subjects with and without Raynaud's phenomenon, *Br. J. Rheumatol.,* 29, 89, 1990.

33. Baab, D. A. and Oberg, P. A., The effect of cigarette smoking on gingival blood flow in humans, *J. Clin. Periodontol.,* 14, 418, 1987.

34. Suter, T., Buzzi, R., and Battig, K., Cardiovascular effects of smoking cigarettes with different nicotine deliveries: a study using multilead-plethysmography, *Psychopharmacologia,* 80, 106, 1983.

35. Bornmyr, S. and Svensson, H., Thermography and laser-Doppler flowmetry for monitoring changes in finger skin blood flow upon cigarette smoking, *Clin. Physiol.,* 11, 135, 1991.

36. Lecerof, H., Bornmyr, S., Lilja, B., De Pedis, G., and Hulthen, U. L., Acute effects of doxazosin and atenolol on smoking-induced peripheral vasoconstriction in hypertensive habitual smokers, *J. Hypertension,* 8, S29, 1990.

37. van Adrichem, L. N. A., Hovius, S. E. R., van Strik, R., and van der Meulen, J. C., Acute effects of cigarette smoking on microcirculation of the thumb, *Br. J. Plast. Surg.,* 45, 9, 1992.

38. Nicita-Mauro, V., Smoking, calcium, calcium antagonists, and aging, *Exp. Gerontol.,* 25, 393, 1990.

39. Cesarone, M. R., Laurora, G., De Sanctis, M. T., Incadella, L., Marelli, C., and Belcaro, G. V., Skin blood flow and venoarteriolar response in essential hypertension, *Minerva Cardioangiol.,* 40, 115, 1992.

40. Salmasi, A. M., Belcaro, G., and Nicolaides, A. N., Impaired venoarteriolar reflex as a possible cause for nifedipine-induced ankle oedema, *Int. J. Cardiol.,* 30, 303, 1991.

41. Orlandi, C., Rossi, M., and Finardi, G., Evaluation of the dilator capacity of skin blood vessels of hypertensive patients by laser Doppler flowmetry, *Microvasc. Res.,* 35, 21, 1988.

42. Hatanaka, H., Matsumoto, S., Ishikawa, K., Kawasaki, T., Kubota, S., Takagi, K., Tanke, G., Yoshimura, Y., Oimomi, M., and Baba, S., Fundamental studies on the measurement of skin blood flow by a Periflux laser Doppler flowmeter and its clinical application, *Rinsho Byori,* 32, 1025, 1984.

43. Kvernebo, K., Slagsvold, C. E., Stranden, E., Kroese, A., and Larsen, S., Laser Doppler flowmetry in evaluation of lower limb resting skin circulation. A study in healthy controls and atherosclerotic patients, *Scand. J. Clin. Lab. Invest.,* 48, 621, 1988.

44. Karanfilian, R. G., Lynch, T. G., Lee, B. C., Long, J. B., and Hobson, R. W., The assessment of skin blood flow in peripheral vascular disease by laser Doppler velocimetry, *Am. Surg.,* 50, 641, 1984.

45. Leonardo, G., Arpaia, M. R., and Del Guercio, R., A new method for the quantitative assessment of arterial insufficiency of the limbs: cutaneous postischemic hyperemia test by laser Doppler, *Angiology,* 38, 378, 1987.

46. Kristensen, J. K., Karlsmark, T., Bisgaard, H., and Sondergaard, J., New parameters for evaluation of blood flow in patients with leg ulcers, *Acta Derm. Venereol.,* 66, 62, 1986.

47. Hartmen, B., Drews, B., Burnus, C., and Bassenge, E., Increase in skin blood circulation and transcutaneous oxygen partial pressure of the top of the foot in lower leg immersion in water containing carbon dioxide in patients with arterial occlusive disease, *Vasa,* 20, 382, 1991.

48. Rendell, M., Bergman, T., O'Donnell, G., Drobny, E., Borgos, J., and Bonner, F., Microvascular blood flow, volume and velocity measured by laser Doppler techniques in IDDM, *Diabetes,* 38, 819, 1989.

49. Rendell, M. and Bamisedun, O., Diabetic cutaneous microangiopathy, *Am. J. Med.,* 93, 611, 1992.

50. Wilson, S. B., Jennings, P. E., and Belch, J. J. F., Detection of microvascular impairment in type I diabetes by laser Doppler flowmetry, *Clin. Physiol.,* 12, 195, 1992.

51. Winocour, P. H., Mitchell, W. S., Gush, R. J., Taylor, L. J., and Baker, R. D., Altered hand skin blood flow in type 1 (insulin-dependent) diabetes mellitus, *Diabetic Med.,* 5, 861, 1988.

52. Boolell, M. and Tooke, J. E., The skin hyperemic response to local injection of substance P and capsaicin in diabetes mellitus, *Diabetic Med.,* 7, 898, 1990.

53. Young, M. J., Veves, A., Walker, M. G., and Boulton, A. J. M., Correlations between nerve function and tissue oxygenation in diabetic patients: further clues to the aetiology of diabetic neuropathy, *Diabetologia,* 35, 1146, 1992.

54. Rendell, M. and Bamisedun, O., Skin blood flow and current perception in pentoxifylline-treated diabetic neuropathy, *Angiology,* 43, 843, 1992.

55. Barnes, M. D., Peppiatt, T. N., and Mani, R., Glimpses into measurements of the microcirculation in skin, *J. Biomed. Eng.,* 13, 185, 1991.

56. Tooke, J. E., Ostergren, J., Lins, P. E., and Fagrell, B., Skin microvascular blood flow control in long duration diabetics with and without complications, *Diabetes Res.*, 5, 189, 1987.
57. Walmsley, D. and Wiles, P. G., Myogenic microvascular responses are impaired in long-duration type 1 diabetes, *Diabetic Med.*, 7, 222, 1990.
58. Houben, A. J., Schaper, N. C., Slaaf, D. W., Tangelder, G. J., and Nieuwenhuijzen Kruseman, A. C., Skin blood flux in insulin-dependent diabetic subjects in relation to retinopathy or incipient nephropathy, *Eur. J. Clin. Invest.*, 22, 67, 1992.
59. Corbin, D. O. C., Young, R. J., Morrison, D. C., Hoskins, P., McDicken, W. N., Housley, E., and Clarke, B. F., Blood flow in the foot, polyneuropathy and foot ulceration in diabetes mellitus, *Diabetologia*, 30, 468, 1987.
60. Walmsley, D. and Wiles, P. G., Early loss of neurogenic inflammation in the human diabetic foot, *Clin. Sci.*, 80, 605, 1991.
61. Belcaro, G., Nicolaides, A. N., Volteas, A. N., and Leon, M., Skin flow the venoarteriolar response and capillary filtration in diabetics. A 3-year follow-up, *Angiology*, 43, 490, 1992.
62. Koltringer, P., Langsteger, W., Lind, P., Klima, G., Wakonig, P., Eber, O., and Reisecker, F., Autonomic neuropathies in skin and its incidence in non-insulin-dependent diabetes mellitus, *Horm. Metab. Res.*, 26, S87, 1992.
63. Westerman, R. A., Low, A. M., Widdop, R. E., Neild, T. O., and Delaney, C. A., Non-invasive tests of neurovascular function in human and experimental diabetes mellitus, in *Endothelial Cell Function in Diabetic Microangiopathy. Problems in Methodology and Clinical Aspects*, Vol. 9, Molinatti, G. M., Bar, R. S., Belfiore, F., and Porta, M., Eds., S. Karger, Basel, 1990, 127.
64. Belcaro, G., Laurora, G., Cesarone, M. R., and Pomante, P., Elastic stockings in diabetic microangiopathy. Long-term clinical and microcirculatory evaluation, *Vasa*, 21, 193, 1992.
65. Flynn, M. D., Boolell, M., Tooke, J. E., and Watkins, P. J., The effect of insulin infusion on capillary blood flow in the diabetic neuropathic foot, *Diabetic Med.*, 9, 630, 1992.
66. Aman, J., Berne, C., Ewald, U., and Tuvemo, T., Cutaneous blood flow during a hypoglycemic clamp in insulin-dependent diabetic patients and healthy subjects, *Clin. Sci.*, 82, 615, 1992.
67. Belcaro, G., Marelli, C., Pomante, P., Laurora, G., Cesarone, M. R., Ricci, A., Girardello, R., and Barsotti, A., Fibrinolytic enhancement in diabetic microangiopathy with defibrotide, *Angiology*, 43, 793, 1992.
68. Jorneskog, G., Tyden, G., Bolinder, J., and Fagrell, B., Does combined kidney and pancreas transplantation reverse functional diabetic microangiopathy?, *Transplant Int.*, 3, 167, 1990.
69. Shawket, S., Dickerson, C., Hazleman, B., and Brown, M. J., Selective suprasensitivity to calcitonin-gene-related peptide in the hands in Raynaud's phenomenon, *Lancet*, 2, 1354, 1989.
70. Heier, M. S., Nilsen, T., Graver, V., Aass, N., and Fossa, S. D., Raynaud's phenomenon after combination chemotherapy of testicular cancer, measured by laser Doppler flowmetry. A pilot study, *Br. J. Cancer*, 63, 550, 1991.
71. Suichies, H. E., Aarnoudse, J. G., Wouda, A. A., Jentink, H. W., de Mul, F. F. M., and Greve, J., Digital blood flow in cooled and contralateral finger in patients with Raynaud's phenomenon. Comparative measurements between photoelectrical plethysmography and laser Doppler flowmetry, *Angiology*, 43, 134, 1992.
72. Allen, J. A., Devlin, M. A., McGrann, S., and Doherty, C. C., An objective test for the diagnosis and grading of vasospasm in patients with Raynaud's syndrome, *Clin. Sci.*, 82, 529, 1992.
73. Kurozawa, Y., Nasu, Y., and Oshiro, H., Finger systolic blood pressure measurements after finger cooling using the laser-Doppler method for assessing vibration-induced white finger, *J. Occup. Med.*, 34, 683, 1992.

74. Kristensen, J. K., Englhart, M., and Nielsen, T., Laser Doppler measurement of digital blood flow regulation in normals and in patients with Raynaud's phenomenon, *Acta Derm. Venereol.*, 63, 43, 1983.
75. Goodfield, M., Hume, A., and Rowell, N., Reactive hyperemic responses in systemic sclerosis patients and healthy controls, *J. Invest. Dermatol.*, 93, 368, 1989.
76. Engelhart, M. and Seibold, J. R., The effect of local temperature versus sympathetic tone on digital perfusion in Raynaud's phenomenon, *Angiology*, 41, 715, 1990.
77. Wollersheim, H., Reyenga, J., and Thien, T., Postocclusive reactive hyperemia of fingertips, monitored by laser Doppler velocimetry in the diagnosis of Raynaud's phenomenon, *Microvasc. Res.*, 38, 286, 1989.
78. Martinez, R. M., Saponaro, A., Dragagna, G., Santoro, L., Leopardi, N., Russo, R., and Tassone, G., Cutaneous circulation in Raynaud's phenomenon during emotional stress. A morphological and functional study using capillaroscopy and laser Doppler, *Int. Angiol.*, 11, 316, 1992.
79. Engelhart, M., Ketanserin in the treatment of Raynaud's phenomenon associated with generalized scleroderma, *Br. J. Dermatol.*, 119, 751, 1988.
80. Brouwer, R. M. L., Wenting, G. J., and Schalekamp, M. A. D. H., Acute effects and mechanism of action of ketanserin in patients with primary Raynaud's phenomenon, *J. Cardiovasc. Pharmacol.*, 15, 868, 1990.
81. Bunker, C. B., Lanigan, S., Rustin, M. H. A., and Dowd, P. M., The effects of topically applied hexyl nicotinate lotion on the cutaneous blood flow in patients with Raynaud's phenomenon, *Br. J. Dermatol.*, 119, 771, 1988.
82. Rustin, M. H. A., Newton, J. A., Smith, N. P., and Dowd, P. M., The treatment of chilblains with nifedipine: the results of a pilot study, a double-blind placebo-controlled randomized study and a long-term open trial, *Br. J. Dermatol.*, 120, 267, 1989.
83. Tur, E., Ryatt, K. S., and Maibach, H. I., Idiopathic recalcitrant facial flushing syndrome, *Dermatologica*, 181, 5, 1990.
84. Frodin, T., Alund, G., and Varenhorst, E., Measurement of skin blood flow and water evaporation as a means of objectively assessing hot flushes after orchidectomy in patients with prostate cancer, *Prostate*, 7, 203, 1985.
85. Wilkin, J. K., Quantitative assessment of alcohol-provoked flushing, *Arch. Dermatol.*, 122, 63, 1986.
86. Irvine, C., Kumar, P., and Marks, R., Isotretinoin in the treatment of rosacea and rhinophyma, in *Acne and Related Disorders, Proceedings of an International Symposium*, Marks, R. and Plewig, G., Eds., Martin Dunitz, London, 1989, 301.
87. Wilkin, J. K., Fortner, G., Reinhardt, L. A., Flowers, O. V., Kilpatrick, S. J., and Streeter, W. C., Prostaglandins and nicotinate-provoked increase in cutaneous blood flow, *Clin. Pharmacol. Ther.*, 38, 273, 1985.
88. Wiles, P. G., Grant, P. J., and Davies, J. A., The differential effect of arginine vasopressin on skin blood flow in man, *Clin. Sci.*, 71, 633, 1986.
89. Agusni, I., Beck, J. S., Potts, R. C., Cree, I. A., and Ilias, M. I., Blood flow velocity in cutaneous lesions of leprosy, *Int. J. Lepr.*, 56, 394, 1988.
90. Rodgers, G. P., Schechter, A. N., Noguchi, C. T., Klein, H. G., Nienhuis, A. W., and Bonner R. F., Periodic microcirculatory flow in patients with sickle-cell disease, *N. Engl. J. Med.*, 311, 1534, 1984.
91. Tur, E. and Brenner, S., Cutaneous blood flow measurements for the detection of malignancy in pigmented skin lesions, *Dermatology*, 184, 8, 1992.
92. Braverman, I. M. and Schechner, J. S., Contour mapping of cutaneous microvasculature by computerized laser Doppler velocimetry, *J. Invest. Dermatol.*, 97, 1013, 1991.

Chapter 10

Assessment of Skin Blood Flow in Vascular Diseases

Pierre G. Agache and Anne-Sophie Dupond

CONTENTS

I. INTRODUCTION

The last decade has been marked by important progress in skin blood flow assessment techniques, both by improvement of already established instruments such as those used in photoplethysmography, capillaroscopy, thermography, and skin photometry and by the emergence of new techniques such as transcutaneous oxygen pressure, thermal clearance, and laser Doppler flowmetry. Previous techniques such as epicutaneous xenon clearance, although of recognized value, have been virtually abandoned due to difficulty of use or association with hazards for both patients and operators. Of such an impressive panel only a few techniques have emerged to date in the current practical management of the so-called vascular diseases. These are skin color assessment, photoplethysmography, thermography, laser Doppler flowmetry, and transcutaneous oxygen pressure measurement.

Skin color assessment, or chromametry has been increasingly used as a way to measure erythema (see Part 3 of this book) thanks to a series of commercially available devices, the reliability of which has been demonstrated. However, there is no report of their use in vascular diseases, probably because erythema, or blanching, is an overall phenomenon which involves only subepidermal vessels, which are not known to be involved in other conditions with the exception of inflammation, telangiectasias, and port wine stains. The same holds true with capillaromicroscopy, but to a lesser extent since this technique allows a direct observation of capillary flow and vessel shape, which is of interest, but is more difficult to perform on sites other than the fingers. With videocapillaromicroscopy and image analysis salient progress has been made recently. This has allowed some investigations to be made in peripheral arterial occlusive disease, as will be seen later.

Thermography uses either cholesterol liquid crystals applied onto the skin or an infrared camera. The first is less precise and has the disadvantage of occluding the skin surface, thus altering the skin temperature. The latter has recently benefited from continuous recording over time of a single point or a line with two- or three-dimensional representation,[1] allowing the researcher to follow up a spell of vasoconstriction or vasodilation. Unfortunately, it is still an expensive device which needs a temperature-regulated room and a strict thermal environment to give reliable data.[2] It is a powerful and irreplaceable technique in vascular

0-8493-8371-4/95/$0.00+$.50
© 1995 by CRC Press, Inc.

155

disease exploration, since it is the only way to get a mapping of skin perfusion with warm core blood in a wide surface area and to follow up thermal changes on a large scale.

Photoplethysmography utilizes the absorption of red light by blood in the subdermal plexus. While the amount of light entering the skin is constant, the backscattered light is assessed through a photovoltaic cell and accordingly follows in intensity the variation in amount of blood in the same skin volume. This technique can be used in two ways. The oldest method was to record the amplitude and shape of systolic-diastolic variation in skin blood content. This had been replaced by laser Doppler flowmetry, although the assessed physical phenomenon is not the same. A more recent use involves damping down the pulse variation of backscattered light and recording the baseline signal, which is related to the amount of blood per unit skin volume, in arbitrary units since the light diffusion volume is unknown. This is what is today called photoplethysmography. This technique is used by either dermatologists or angiologists to investigate the congestion of a skin area—for example, the cheek in rosacea in supine and sitting positions or the goiter area in primary varicose or perforating vein insufficiency (see Chapter 11). Up to now there have not been other uses of this technique in vascular diseases.

Thermal clearance is an old technique which has been revived recently but is no longer commercially available. This explains the paucity of studies using such a device. It is a record of the heat needed for maintaining a small skin surface area at a constant temperature.[1] The more heat needed, the higher the skin blood flow. This technique is a measurement of heat washout and consequently refers mostly to return vessels, as do other clearance techniques. Accordingly, the signal preferably should be linked to the amount of blood perfusing throughout the skin rather than passing through deeper arteriovenous shunts, as opposed to techniques mostly based on arterial inflow such as laser Doppler flowmetry.

Today with the exception of photoplethysmography, which is used almost exclusively in venous insufficiency, only two techniques are commonly used in vascular diseases management—namely, laser Doppler flowmetry (LDF) and the assessment of transcutaneous oxygen pressure ($tcPO_2$).

The principle of LDF has been presented in Chapter 3 of this book. It is mostly a measurement of blood flow in the lower dermis since the bulk of blood is located in the subdermal plexus and around sweat glands and hair bulbs, which are both deeply located. By contrast, $tcPO_2$ relies on the subepidermal plexus blood flow. This technique is a measurement of the oxygen partial pressure (as compared to atmospheric pressure) just above the skin surface and, consequently, of the oxygen outflow.[2] At normal skin temperature $tcPO_2$ is very low—barely measurable—and variable, owing to the permanently changing skin blood flow due to vasomotor changes. When the skin is heated up to 44° C, an intense vasodilation is obtained which results in an increased stable flow also allowing a much greater amount of oxygen to be freed from oxyhemoglobin. Accordingly, $tcPO_2$ becomes stable and measurable. Its value depends on three parameters: the arterial oxygen partial pressure (paO_2), the subepidermal blood flow, and the absorption coefficient of oxygen for superficial skin layers. The latter is a function of the structure and thickness of this part of the skin. Accordingly, $tcPO_2$ measurement has three main purposes: (1) assessment of paO_2, mostly in premature infants, in neonates, and in adults under general anesthesia; (2) assessment of blood perfusion pressure within the subepidermal plexus; and (3) assessment of the oxidative metabolism of skin. Whereas the first two uses are common and widely documented, the last one is only at its start. Only the second purpose is relevant to vascular diseases. Since venous insufficiency is dealt with in Chapter 11 of this book, only arterial occlusive disease will be considered here.

II. ARTERIAL INSUFFICIENCY

Peripheral arterial occlusive disease, or, more simply, arterial insufficiency (AI), is a very common disease which results in reduced blood flow in lower limb tissue such as muscle and skin, with gangrene as its most severe outcome. This stage and the one immediately preceding it have been called leg critical ischemia since 1981 and are defined, from the recent European Consensus Conference[3] as either persistent pain in the extremities at rest or gangrene, together with an ankle systolic pressure below 50 torr (or below 30 torr in diabetic patients).*

The mechanism of skin blood flow reduction is mostly attributed to a decrease in perfusion pressure, i.e., in the arteriovenous pressure gradient, but also to some extent is due to other local phenomena such as a decrease in vasomotion and impaired distribution of blood within the skin. All these events can be assessed by two methods: $tcPO_2$ and LDF.

A. TRANSCUTANEOUS OXYGEN PRESSURE IN ARTERIAL INSUFFICIENCY

The relationship between $tcPO_2$ and skin perfusion pressure (and skin blood flow as a consequence) was made clear through the experiments of Wyss et al.,[4] Matsen et al.,[5] and Eickhoff et al.[6] on foot $tcPO_2$ at 44° C (Figure 1). They demonstrated that when the skin perfusion pressure was high, $tcPO_2$ increased slowly and linearly with perfusion pressure. Accordingly, under normal conditions $tcPO_2$ is not a good indicator of skin blood flow. By contrast, when perfusion pressure was low the relationship was curvilinear and steep and could be used as a sensitive index of skin blood flow. In this range, $tcPO_2$ values were found to be stable and reliable in the same subject. For example, from daily measurements over 3 weeks current coefficients of variation are below 10%.[7,8]

The most commonly used parameter in AI is the absolute $tcPO_2$ value, assessed on the dorsal foot close to the toes. For some special purposes it is also taken at various levels of the ischemic lower limb. The site of measurement has some importance since the range in healthy people lies around 70 on the foot and calf, 74 on the thigh and 80 on the chest, with lower values in older subjects according to the equation $tcPO_2$ (torr) $= 88.6 - 0.33$ age (years) (SD of slope and intercept are 4% and 19%, respectively).[8] Also, $tcPO_2$ is higher in females than in males by about 11% and is lower in smokers.[9]

As mentioned earlier, whatever the local perfusion pressure, $tcPO_2$ is strongly influenced by the cardiorespiratory function, which is not infrequently impaired in atherosclerotic people. To obviate such a source of error Hauser and Shoemaker[10] suggested the use of the limb-to-chest $tcPO_2$ ratio, which they named the regional perfusion index (RPI), rather than limb $tcPO_2$ alone. Over absolute values this index also has the advantage of being insensitive to aging. Its coefficient of variation was in the same range as basal $tcPO_2$ values.

In AI the dorsal foot resting $tcPO_2$ value is reduced in parallel to the degree of ischemia and follows Leriche and Fontaine's staging into four groups of increasing severity.[11,12] However, while a reduction has a diagnostic value, many patients in stage 1 (asymptomatic) or even in stage 2 (claudicant) have a reclining-position $tcPO_2$ in the normal range. Accordingly, several sensitizing procedures have been proposed.

Reactive hyperemia following 3-min total occlusion has been found to be of interest not so much in the rebound amplitude as in the length of time to regain half of the pretest figure, which increases with the degree of ischemia. Unfortunately, the range for normal people differs widely with studies, from 38 ± 7 s[13] to 87 ± 27 s.[14]

Other authors assessed $tcPO_2$ during and following exercise, either by treadmill,[10,15] regular stamping,[16] or dorsiplantar flexion of the foot when lying supine.[17] During exercise $tcPO_2$ at the thigh and calf of healthy limbs remains stable, while foot $tcPO_2$ falls significantly but rapidly returns to baseline after termination. Patients with stage 1 or 2 AI experience similar trends during exercise and recovery, but with lower figures. Some authors have concluded

---

*1 torr = 1 mmHg = 133.4 Pa. The term was coined in honor of Torricelli, the well-known Italian scientist (17th century) who first measured atmospheric pressure using a tube filled with mercury.

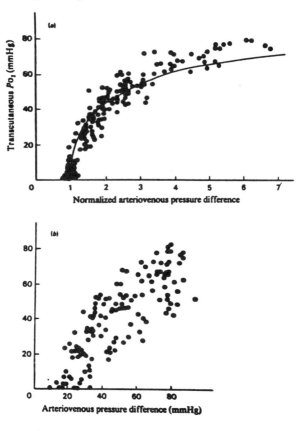

Figure 1   (a) tcPO$_2$ data from 15 subjects vs. arteriovenous pressure difference normalized for the value of the tcPO$_2$ zero intercept for each subject. The line shows predicted venous PO$_2$. (b) Un-norm alisea data shown for comparison. (From Wyss, C. R., Matsen, F. A., King, R. V., Simmons, C. V., and Burgess, E. M., *Clin. Sci.,* 60, 449, 1981. With permission.)

that tcPO$_2$ following exercise has a 100% sensitivity and specificity in detecting the presence of AI,[10,18] but this opinion has not been shared by others, perhaps owing to different types of exercises used.

Posture tests are a means of assessing the influence of either increasing hydrostatic pressure or reducing the arteriovenous gradient in the tested area. By passing from a supine to a standing position, as a result of elevated hydrostatic pressure within veins, tcPO$_2$ increases because skin heating (44° C) impairs the venoarteriolar vasoconstrictive reflex. The extent of increase is a good indicator of the possibility of improving skin perfusion and has a good prognostic value.[10,19]

By contrast, the arteriovenous perfusion pressure can be lowered by raising the foot of the recumbent subject, thus reducing tcPO$_2$. This was found as well in stage 1 AI patients, but with a reduction greater than in healthy controls. This maneuver allowed the researcher to discriminate between group 1 patients and normal subjects.[19]

Finally, Hauser and Shoemaker stressed the usefulness of RPI linked with exercise for the same purpose; a 10% decrease of thigh RPI and a 15% decrease of calf RPI (not of foot

RPI) following a treadmill exercise were significantly associated with the asymptomatic, so-called healthy leg of claudicants, as compared with normal controls.[10] These findings were confirmed by Modesti et al.[19]

The stage of critical ischemia can be defined as a state where tcPO$_2$ falls below 10 torr.[4] These low values and even a zero tcPO$_2$ do not mean absence of skin blood flow; some positive perfusion pressure is still maintained at this stage, but all oxygen delivered by blood flow is supposed to be taken up by epidermis metabolism. When recumbent tcPO$_2$ is found to be low it is worth switching to a sitting or upright position. The critical ischemia is confirmed if tcPO$_2$ does not rise above 10 torr when sitting. Other authors suggest that oxygen breathing be performed at this stage: the absence of a rise in tcPO$_2$ would compete with ankle pressure lower than 45 torr (as measured with a Doppler probe) for predicting amputation. For a period averaging 7 months Bongard and Kraehenbuehl [20] followed up 26 patients with grade 4 AI and in whom vascular surgery had failed or not been possible. No amputation was performed in the group having showed an increase in tcPO$_2$ of at least 10 torr after 6 min of oxygen breathing; conversely, all patients in whom the pedal tcPO$_2$ had increased less than 10 torr after oxygen breathing had to undergo amputation. Since several amputated patients had a resting pedal tcPO$_2$ above 10 torr, the authors conclude that an increase in the pedal tcPO$_2$ during oxygen inhalation is more predictive of amputation than resting tcPO$_2$ alone and even more than an ankle blood pressure lower than 45 torr as assessed by a Doppler probe.

Where amputation is needed, some authors have used tcPO$_2$ assessment at various sites on the leg to predict the lowest level of amputation permitting stump wound healing. Franzeck et al.[14] and Wyss Et al.[4] suggested 20 torr, whereas other authors proposed 40 torr[21,22] or even 50 torr.[23] Conversely, the critical tcPO$_2$ value below which no healing would be obtained was set at 40 torr,[23] 36 torr,[21] 20 torr,[4] or 5 torr.[14] While admitting Franzeck's et al. prognostic values, Depairon et al.[24] reported satisfactory healing in four patients whose tcPO$_2$ was below 10 torr at the amputation level. Finally, Kram et al.[25] have suggested an interesting approach to solve the problem: First, tcPO$_2$ measurements are made at both anterior and posterior calf, the lower value of which is taken; if it is less than 20 torr, a measurement of brachial tcPO$_2$ is done and the calf-to-brachial ratio is computed. The stump will heal if amputation is made at the level where the ratio is higher than 0.2.

Whatever the AI grade, tcPO$_2$ assessment can help follow up a patient and evaluate the effect of treatment.[26] As an example, Borzykowski and Krahenbuhl[12] found dorsal foot mean tcPO$_2$ values of 28.7 ± 22.3 and 41.1 ± 13.7, respectively, before and immediately after a 3-h intravenous infusion of 600 ng naftidrofuryl in 15 patients with AI whose three did not respond. In the same way continuous tcPO$_2$ monitoring during vascular surgery was used by Neidhardt et al;[27] when the only unobstructed distal blood vessel was clamped, tcPO$_2$ fell to zero almost immediately, whereas it only steadily decreased if some other blood supply existed. When the clamp was removed, a correct limb perfusion was checked through a significant rise in tcPO$_2$ relative to the preoperative figure. Its persistence is also of major importance; in a presented case a sudden fall indicated a postoperative thrombosis.

tcPO$_2$ monitoring is also currently used by radiologists during lumbar or thoracic sympatholysis with the probe fixed to the involved skin nervous territory. A sudden rise in tcPO$_2$ is observed only when the sympatholytic product is injected at the right place.

B. LASER DOPPLER FLOWMETRY IN ARTERIAL INSUFFICIENCY

In contrast to the great number of literature data on tcPO$_2$ in AI, fewer studies have been published on the use of LDF. The probe is usually placed on either the first toe ball or the dorsal foot close to the first toe web. The resting blood flow in the supine position at the casual skin temperature (i.e., below 35° C) is lower in AI as compared to a normal limb. However, there is no clear-cut relationship with the degree of ischemia, and several authors have indicated that it is of little use. Hence, sensitizing procedures are needed to obtain reliable results.

The simplest one is a postural test including elevation of the foot while the patient is lying supine and/or tilting to a sitting or upright position. In healthy subjects when the probe is not heated such maneuvers cause a decreased LDF signal and are not only related to arteriovenous differential pressure.[28] This is due to the arteriovenular vasoconstrictive sympathetic reflex on one hand and other poorly known reflexes on the other hand, the latter preventing a clear-cut decrease in blood flow when the foot is raised. However, in severe AI the venoarteriolar reflex can be impaired and the LDF signal decreased on dependency diminished, suppressed, or even replaced by an increased signal.[29,30] This was investigated by Ubbink et al.[31] in 76 patients and 12 asymptomatic subjects by using LDF at an unheated skin temperature and at 36° C and tcPO$_2$ at 37° C and 44° C. Both skin blood flow and oxygenation diminished with decreasing ankle-to-brachial systolic blood pressure ratio (ankle index). Upon changing from the supine to the sitting position skin perfusion and oxygenation were reduced in healthy controls but enhanced in patients with an ankle index less the 30%, indicating disturbed posturally induced vasoconstriction. By increasing the local skin temperature, skin perfusion was enhanced in healthy controls but unchanged in ischemic limbs because the microvessels were already maximally dilated.

A comparative study of the deep thermoregulatory skin blood flow and the more superficial nutritive flow was performed in 21 patients with AI and 11 age-matched controls using LDF and videocapillaromicroscopy, respectively, on the big toe.[32] In recumbency both types of flows were reduced in diseased legs. After switching to the upright position the LDF signal increased ischemic legs, whereas it decreased in control subjects as well as in contralateral legs of patients; however, the capillary flow was reduced in the three groups, although to a lesser extent in the ischemic limbs. These data suggest a suppression of the postural reflex at the thermoregulatory level, but not at the nutritive level, and confirm that LDF mostly assesses deeper skin blood flow together with that of arteriovenous shunts.

Raising the temperature of the probe up to 40° C is a way to relate linearly the LDF signal to the blood perfusion pressure, as demonstrated in normal subjects by Enkema et al.[33] and Matsen et al.,[28] thereby suppressing the above-mentioned reflexes. Winsor et al.[29] studied the effect of probe temperature over the range 10 to 46° C on the skin blood flow in 22 subjects with AI and 22 healthy controls when the probe was fixed on the hallux ball. Whereas a significant difference was observed at all temperatures, a clear separation between the two groups with no overlap was observed only at 40° C and 45° C. Accordingly, the authors suggested employing LDF with the probe at 40° C in patients with AI. The room temperature had only a minor effect when it was below 30° C.

As shown by Leonardo et al.,[35] heating the probe up to 44° C does not at all impair the capacity of the hyperemic response to a 3-min proximal occlusion. A postischemic hyperemia test was used by these authors in 138 lower limbs of patients affected with AI and 71 limbs of healthy controls. Following 15 min of rest in the supine position, a 3-min total occlusion was performed on the upper part of the thigh. The LDF signal was recorded on the pulp of the hallux. All the flow parameters were found to be statistically different between normals and AI patients, but the one that made the clearest distinction between the two groups as a whole was the time needed for recovering the baseline resting flow value: 9.2 ± 0.7 s and 67.0 ± 5.6 s in control and AI limbs, respectively. On the other hand, the time lapse to reach the peak flow was the best way to distinguish normal limbs from those with early-stage AI. Also four limbs with AI recovered a normal postocclusive hyperemia LDF pattern following transluminal angioplasty.

Vasodilating drugs are widely used in the treatment of AI even though their effectiveness has been questioned. Leonardo et al.[36] studied the effect in AI limbs of an intra-arterial bolus of 100 mg naftidrofuryl on hallux pulp skin blood flow and hallux subcutaneous blood flow using LDF and strain-gauge plethysmography, respectively. Immediately after injection a steep rise in LDF signal was observed which proved to be inversely correlated to baseline values, and there was no change in subcutaneous flow. Then a sudden fall occurred with a strong reduction in sphygmic amplitude while the subcutaneous flow remained unchanged, but with a less intense reduction in sphygmic amplitude. A few seconds later the LDF signal recovered and set up to slightly elevated values as compared to baseline. The fall in the LDF signal was accompanied by a burning sensation in the toes and occurred about 10 sec after injection. It was interpreted by the authors as a transient blood steal by muscle because a strong rise in calf volume was found by strain-gauge plethysmography of this area, peaking 1 min after injection.

The effect of oxygen inhalation (40% FIO_2 for 6 min) while assessing the LDF signal and capillary flow on the dorsal foot close to the first toe web was investigated by Bongard et al.[37] in 17 legs of 11 patients with critical ischemia and in 13 legs of healthy subjects. In normal-leg subjects, oxygen inhalation, as expected, induced a significant decrease of LDF signal and blood cell velocity, as assessed by videocapillaromicroscopy, and a $tcPO_2$ increase. Ischemic legs behave differently whether oxygen inhalation induces a rise in $tcPO_2$ or not. While the first group reacted qualitatively, as did normal legs, the second group showed an increase in blood flow displayed by both LDF and capillaromicroscopy. The authors conclude that breathing oxygen is a valuable maneuver to diagnose severe critical ischemia when followed by $tcPO_2$ and either LDF or capillaromicroscopy and that it may be of therapeutic benefit. In this study the LDF was used at the casual skin temperature, but after subtracting the residual signal (called "biological zero")[38] obtained when arterial occlusion had been made proximally at the base of the tested hallux.

Another situation often encountered by dermatologists is having to decide whether or not an elderly patient with a leg ulcer caused by venous insufficiency can be prescribed compressive bandaging when an arterial insufficiency has been found in the same limb. Usually an evaluation of the tolerance of the external pressure cannot be based on clinical grounds only because the patient does not walk enough to disclose an intermittent claudication, and if a congestive erythema would arise in dependency it would be either veiled by the bandage or inapparent due to the lipodermatosclerotic changes in the skin. In that case a LDF measurement at the big toe, which is always outside the bandage, may be helpful because in pure lipodermasclerotic lesions of venous origin the flow is increased as compared to the normal limb; consequently, and decrease in LDF signal amplitude would suggest ischemia. This problem has been addressed by Matsen et al.[39] in normal volunteers by raising the foot above heart level when supine, thereby reducing the perfusion pressure, as if IA were present. They showed that application of external pressure on the level or elevated limb would decrease the perfusion pressure by the value of applied pressure. However, application of an external pressure had no influence on perfusion pressure provided it remained lower than hydrostatic pressure. One can deduce that keeping a compressive bandaging on when in bed may easily induce critical ischemia in the case of AI, since according to Laplace's law the external pressure would increase when the limb volume decreased, whereas there is no risk in either standing or sitting if the bandage pressure is kept lower than 50 torr (68 cm H_2O). This was verified by Boccalon et al.[40] in six patients following 2 H 30 medium-pressure elastic bandaging. No change in either LDF signal on the hallux ball or ankle blood pressure was found.

One of the most interesting phenomena evidenced by LDF in normal subjects is a wave pattern of skin blood flow different from and superimposed on the pulsed variations, called vasomotion.[41] The frequency was found to vary according to skin site; on dorsal foot it ranged between 4 and 10 min^{-1}. It is not a consistent finding, and there is no agreement on its prevalence (ca. 80% according to Bollinger et al.[42]) (Figure 2). Vasomotion disappears during local heating[43] or local or systemic anesthesia, it can be induced or enhanced after a systemic thermal challenge and in the postischemic phase following proximal occlusion

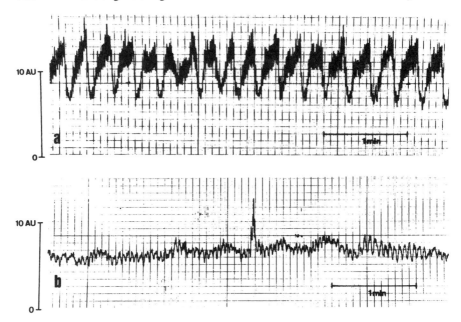

Figure 2   (a) Vasomotion in the forefoot of a normal subject. (b) Typical high-frequency waves recorded in the forefoot of a patient with rest pain due to peripheral arterial occlusive disease. (Modified from Bollinger, A., Hoffmann, U., and Franzeck, U. K., *Blood Vessels*, 28, 21, 1991. With permission.)

mostly if of long (more than 4 min) duration, or seemingly any local increase in blood flow.[44] It remains unchanged during oxygen breathing and proximal nerve blockade. From these data it seems that vasomotion appears after a peak in erythrocyte flux and disappears before reaching resting flux level;[45] consequently it would be a local response of the myogenic vessel wall to an increase in intravascular pressure. This would be in accordance with Poiseuille's law, from which it can be demonstrated that the resistance of a vessel with a constant diameter is greater than that of a vessel of the same average diameter in which the diameter changes sinusoidally Thus vasomotion, or flowmotion, might be a local response, "that compensates for reduced levels of microcirculatory blood flow."[42]

A similar wave pattern has been demonstrated at the base of the first and second toe in AI patients.[46] Both the frequency histogram and the Fourier amplitude spectrum of the LDF signal have documented a large-amplitude, low-frequency component (less than 10 min⁻¹), although these are altered or suppressed in severe limb ischemia.[37,46] In such patients and occasionally in healthy volunteers a high-frequency, low-amplitude component was also found (18.6 ± 4.5 min⁻¹) which was synchronous with respiratory movements, disappeared during apnea, and followed the frequency of controlled fast or slow respiration, thus suggesting that variations in vena cava pressure are transmitted to dilated skin venules and arterioles through the extremely low arteriovenous pressure gradient.[47] Accordingly, these high-frequency waves would include a poor prognosis. When low-frequency waves were suppressed, videocapillaromicroscopy could still observe vasomotion (ca. 4 min⁻¹), indicating a dissociation between nutritive superficial flow and deeper bulk skin flow in AI.[48,49] In Bongard et al.'s experiment,[37] breathing oxygen allowed the LDF vasomotion pattern to reappear in all AI subjects in whom it was absent, presumably because the local vascular tone is oxygen dependent. By contrast, capillaromicroscopy vasomotion was not influenced by oxygen inhalation.

In conclusion, LDF appears to be of value either to detect AI, to assess the severity of the disease, or to evaluate the efficacy of treatments. However, the resting LDF signal at casual skin temperature is rarely of value, and sensitizing procedures such as posture maneuvers, heating the probe to 44° C, reactive hyperemia, or breathing oxygen should be performed to achieve reliable data recording.

Finally, it should be borne in mind that it is a measurement of the bulk blood flow, including the deeper dermis blood flow, which is involved with thermoregulation, and this is much greater than the capillary nutritive skin blood flow. The latter only seems to be involved in trophic alteration induced by reduction in perfusion pressure, and it is better assessed by tcPO$_2$.

C.   ARTERIAL INSUFFICIENCY AND DIABETICS

Microcirculatory disturbances associated with AI in diabetic patients are more complex since diabetes per se is able to induce vascular changes. These are of two main types: first, a tendency for blood to shift from superficial into deep vessels through an abnormal opening of arteriovenous shunts; and second, a weakening of the venoarteriolar vasoconstrictive reflex associated with neuropathy. Accordingly, any skin blood flow alteration possibly related to AI should be interpreted in accord with this altered reactivity.

In diabetics without apparent AI, resting flow was found to be reduced in both lower and upper extremities as compared to normal subjects when assessed by LDF.[50] Also, tcPO$_2$[51] and blood volume[50] were lower and the capillary loops were reduced in number.[52] Upon warm challenge the increase was less intense than in nondiabetic subjects.[52,53] During and following cold immersion tests digital arteries often failed to clamp down as in normal subjects,[52,53] but the recovery time was longer.[54,55] The hyperemic response after total proximal occlusion was decreased and there was a prolonged recovery time.[51,56] Altogether these data suggest a permanent shunting of blood through deeper arteriovenous vessels. Since shunting is a way to prevent heat loss, Wiles et al.[57] wanted to see whether hypoglycemia would facilitate heat loss. In eight type 1 diabetic patients, using LDF on the interdigital skin web, they found the skin blood flow to be increased during hypoglycemic spells, while the whole finger blood flow, as measured by venous occlusion plethysmography, changed in the opposite way. Altogether these data suggest an influence of blood glucose level on arteriovenous anastomoses closure which might constitute a worsening factor in AI.

When diabetic neuropathy has taken place, the main effect is a weakening or suppression of the vasoconstrictive venoarteriolar reflex. This can be easily demonstrated by LDF measurement at skin temperature, on reclining followed by standing or sitting position, or before and during proximal venous occlusion. In a study by Rayman et al.[58] the big toe pad skin blood flow was assessed using LDF in diabetic patients with or without neuropathy and compared with that of age-matched healthy controls. When the foot was passively lowered 50 cm below the heart level there was a fall in skin blood flow to 18.1% of baseline level in controls, 28.9% in diabetic patients without neuropathy, and 53.5% in patients with neuropathy. These differences were significant. According to the authors these results also may account for the edema seen in some patients with diabetic neuropathy. Nevertheless, not all patients with neuropathy have impaired venoarteriolar reflex.

III.   RAYNAUD'S PHENOMENON AND DISEASE

Because this condition is related to a hypersensitivity to cold, thermography assessments have had and still retain great value. Presently it stands as a unique tool in skin angiology since it provides a mapping of skin perfusion by core blood which is warmer than skin tissue by at least 2° C and is able to assess quantitatively the gradient of temperature fall along the extremities and to follow up temperature in a given skin area.

Zitnan et al.[59] discovered Raynaud's phenomenon (RP) in 37 out of 144 patients suffering from systemic lupus erythematosus by using thermography assessment of blood flow following a hand cooling test. Partsch[60] underlined the usefulness of this technique to diagnose the vasospasm characteristic of RP and assess the sensitivity to drugs. O'Reilly et al.,[61] using thermography, demonstrated a significantly lower casual skin temperature in fingers of subjects with RP or Raynaud's disease (RD) than in healthy controls. Also, following a hand cold bath the temperature was lower and the time required to recover was longer. They concluded that thermography was a very valuable technique to diagnose RP or RD and assess the results of treatment. The effect of isoxsupine (20 mg daily) on RP vs. placebo was successfully evaluated by Wesseling and Wouda.[62] Another type of treatment, superficial electrostimulation, was tested by Kruk et al.[63] using thermography and a hand cooling test (7 to 9° C) in 29 patients with RP, 21 patients with Buerger's disease, and 10 healthy controls. The treatment was impressively effective; no more vasospastic attacks occurred in 72% of patients with RP. A chemically stable prostacyclin derivative was given by Yamaoka et al.[64] to nine patients with RP plus connective tissue diseases at a dosage of 0.1 to 0.2 mg/kg/min intravenously twice a day for 2 to 4 weeks. Using thermography assessment following a hand cooling test a significantly higher finger temperature was found after treatment, while vasospastic attacks were reduced in frequency and severity in seven out of nine patients; the improvement persisted 6 months later. With the same technique and in four similar patients Katoh et al.[65] observed a beneficial effect of intravenously injected prostaglandin E_1. However, Kyle et al.[66] using a prostacyclin analogue in 13 patients with severe RS, could not find warmer skin using thermographic assessment, although vasospastic attacks decreased in frequency.

Recently, LDF was also used in assessing RP severity or sensitivity to treatment. The authors addressed the following questions:

1. Is there a peculiar LDF pattern in this condition?
2. Can the prognosis be assessed through LDF?
3. Does LDF help follow up the course of the disease under treatment?

To answer the first question, Kristensen et al.[67] assessed skin blood flow during a cold-water immersion test in nine patients and ten controls. The site of measurement was the pad of the second or middle finger. Indirect and direct cooling were performed by first dipping one hand into water at 0° C for 10 s, then the other hand for the same duration, while blood flow was assessed in one hand only. A decreased flow was observed to the same extent in patients and in controls, but the return to normal was delayed in the former: 5.4 ± 1.0 min as compared to 2.9 ± 1.1 min. Also, 5 min after the bath the flow recovery was 33% in patients and 85% in controls. Boccalon et al.[68] attempted to differentiate patients with Raynaud's disease (28 cases) and Raynaud's syndrome (19 cases) from healthy subjects (14 cases) by using LDF measurements obtained under progressive cooling. Hands were placed in a box where the temperature was shifted from 25° C to 5° C in 15 min. They observed a decreased mean ratio of fluxes at 25° C vs. 5° C in patients relative to controls: 1.06 RP and 1.10 in RD, as opposed to 1.34 in controls. Also, the mean flux at baseline temperature was 1.29 ± 0.95 V, 2.94 ± 1.49 V, and 3.77 ± 0.88 V in each group, respectively. The conclusion was that LDF could distinguish between the three groups on a statistical basis. With this in mind, Shawket et al.[69] compared the effect on LDF signal of intravenously injected calcitonin-gene-related peptide, the endothelium-dependent adenosine triphosphate, and the endothelium-independent prostacyclin. Only the first product was able to induce vasodilation in RD patients but not in healthy controls. If confirmed, these data might have both diagnostic and therapeutic significance in that condition.

The second question was addressed by Englehart and Kristensen[70] in 1987 using both local and systemic cooling tests. The 48 patients were classified in three groups: a low-severity group where simultaneous local and systemic cooling were necessary to stop the

LDF signal, a middle-severity group where local cooling alone sufficed to stop blood flow, and a high-severity group where systemic cooling alone brought the LDF signal to zero. The proportions of patients with progressive scleroderma were 30%, 50%, and 100% in the three groups, respectively, lending support to the prognostic value of LDF assessment in a cold test.

Goodfield et al.[71] used a warming test in 15 patients with Raynaud's syndrome as compared with 15 healthy controls. The hand was dipped in a hot bath and measurement was taken on the pad of the second finger. At 28° C the blood flow increase was greater in controls than in patients, but this difference disappeared at 35° C. It was concluded that vasodilation could be achieved by warming in such patients, the intensity of which could have prognostic value.

Evaluation of treatment efficacy through LDF measurement was investigated by Goodfield and Rowell[72] in 12 patients with RP and progressive scleroderma. The treatment consisted of dipping the hand for 5 min in hot water every 4 h in daytime every other week for 6 weeks. Vasospastic attacks were significantly reduced in number during warming periods, with a concomitant increase in blood flow. Bunker et al.[73] evaluated the efficacy of topically applied 0.1% hexyl nicotinate as assessed by LDF in 25 patients with Raynaud's syndrome and 5 healthy controls. Drug application and LDF measurement were made on volar forearm and fingers. An increased blood flow was observed in all patients when 1% hexyl nicotinate had been used. The effect of nifedipine on LDF signal (10 mg twice a day for 2 weeks) was estimated in 38 patients by Cesarone et al.[74] The blood flow was assessed on the index pad following cooling at 20° C for 1 min. After treatment, full recovery following cooling took 14.4 ± 4.3 min as compared to 23.1 ± 4.3 min before treatment. In a similar manner the effect of ketanserin three times daily (vs. placebo), 60 mg for 1 week and then 120 mg for 4 weeks, was studied by Engelhart[75] in nine patients with RP and progressive scleroderma. At the end of the treatment period a nonsignificant difference was observed between ketanserin- and placebo-treated groups. Captopril, an angiotensin-converting enzyme inhibitor, given in 15 patients with RP at a dosage of 25 mg three times a day for 6 weeks induced a significant rise in skin blood flow in the extremities, but without modification of either severity or frequency of vasospastic attacks.[76] More recently, prostaglandin E_1 administered intravenously in 17 patients suffering from scleroderma with RP was found to increase skin blood flow significantly after 3 weeks of treatment as assessed by LDF.[77]

From these studies it is evident that LDF has proved to be a valuable tool in assessing skin blood flow impairment in the extremities in RP and RD from a diagnostic, prognostic, and therapeutic follow-up point of view, mostly when used in conjunction with cooling or warming tests.

IV. MISCELLANEOUS DISEASES

Blood flow assessment is rarely used or has not been a subject of investigation in acrosyndromes such as perniosis, acrodynia, and erythermalgia. Long ago the authors of this chapter found an increased subcutis blood flow in a case of unilateral foot erythermalgia[78] using epicutaneous xenon clearance—35 vs. 16 ml min^{-1}/100 ml tissue during a spell as compared to baseline—while in contralateral limb the dermal blood flow was reduced: 13 vs. 24 ml min^{-1}/100 ml tissue, respectively. In a case of hand acrodynia responding to dihydroergotamine the increased dermal blood flow was reduced following treatment—18 vs. 32 ml min^{-1}/100 ml tissue—while the subcutis blood flow increased: 30 vs. 17 ml min^{-1}/100 ml tissue. These observations lent support to a selective increase in subcutis blood flow in erythermalgia and in dermal blood flow in acrodynia. Unfortunately, other cases have not been observed since then and no literature data have been found concerning these diseases.

Using LDF the response of cutaneous blood flow to autonomic stimuli was evaluated in eight patients with sympathetic dystrophy and eight healthy control subjects by Bej and

Schwartzman.[79] Affected limbs were found to have significantly increased blood flow during the Valsalva maneuver and cold pressure test, while blood flow decreased in normal controls. No significant difference had been found in limb temperature or baseline blood flow. Moreover, vasomotion was observed by LDF in controls, but not in patients. These results support a central abnormality of the sympathetic nervous system in reflex sympathetic dystrophy.

Concerning chilblains, Rustin et al.,[80] in a double-blink vs. placebo study of nifedipine (20 to 60 mg per day for 6 weeks) in ten patients, found a 180% increase in finger blood flow using LDF. A dosage higher than 20 mg daily was necessary to elicit a response in five patients. In a presented case vasomotion amplitude was increased fivefold and wave frequency decreased (2.4 to 1.0 min^{-1}) following administration.

An interesting investigation on vasomotion as assessed by LDF was performed by Rodgers et al.[81] in six patients with sickle-cell disease. The probe was placed on volar forearm. The resting flow was normal as compared with that of six healthy control. All six patients demonstrated a vasomotion of 8.5 to 10 min^{-1} frequency with a 20 to 82% fractional amplitude range. Only one of 60 sites measured in the six normal subjects had comparable oscillations in flow. Moreover, a similar blood flow pattern was observed in two patients whose hematocrit values were higher, although their hemoglobin S concentrations were lower. In contrast, lowering the percentage of sickle erythrocytes to less than 40% by exchange transfusions led to a reduction of vasomotion. These data show the value of LDF in following the functioning of terminal arterioles in response to a rise in rheological impedance.

The same authors found in two thalassemic subjects a resting LDF signal ranging within normal values, but with an increased fractional (systolic-diastolic) amplitude, suggesting a decreased peripheral vascular resistance.

REFERENCES

1. Dittmar, A., Skin thermal conductivity. A reliable index of skin blood flow and skin hydration, in *Cutaneous Investigation in Health and Disease,* Leveque, J. L. Ed., Marcel Dekker, New York, 1989, 323.

2. Agache, P. G., Agache, A., and Lucas, A., Transcutaneous oxygen pressure measurement: usefulness and limitations, in *Vascular Medicine,* Boccalon, H., Ed., Excepta Medica, Amsterdam, 1993, 527.

3. Dormandy, J. A., Second European Consensus document on critical leg ischaemia, *Angiology,* 84 (4) (Suppl.), 1991.

4. Wyss, C. R., Matsen, F. A., King, R. V., Simmons, C.V., and Burgess, E. M., Dependence of transcutaneous oxygen tension on local arteriovenous pressure gradient in normal subjects, *Clin. Sci,* 60, 499, 1981.

5. Matsen, F. A., Wyss, C. R., and Robertson, C. L., The relationship of transcutaneous PO_2 and laser Doppler measurements in a human model of arterial insufficiency, *Sug. Gynecol. Obstet.* 159, 418, 1984.

6. Eickhoff, J. H., Ishimara, S., and Jacobsen, E., Effect of arterial and venous pressure on transcutaneous oxygen tension, *Scand. J. Clin. Lab. Invest.,* 40, 755, 1980.

7. Coleman, L. S., Dowd, G. S. E., and Bentley, G., Reproducibility of tcPO$_2$ measurement in normal volunteers, *Clin Phys. Physiol. Meas.,* 7, 259, 1986.

8. Agache, P. G., Lucas, A., and Agache, A., Influence of age on tcPO$_2$, in *Aging Skin: Properties and Functional Changes,* Leveque, J. L. and Agache, P., Eds., Marcel Dekker, New York, 1992, 125.

9. Lucas A., Agache, P., Risold, J. C., and Cuenot, C., Variation de la tcPO$_2$ en fonction de la consommation de tabac chez le sujet sain, *J. Mal. Vasc.,* 14, 363, 1989.

10. Hauser, C. J. and Shoemaker, W. C., Use of transcutaneous PO$_2$ regional perfusion index to quantify tissue perfusion in peripheral vascular disease, *Ann. Surg.,* 197, 337, 1983.

11. Dowd, G. S. E., Linge, K., and Bentley, G., Transcutaneous PO_2 measurement in skin ischaemia, *Lancet*, 2, 48, 1982.

12. Borzykowski, J. and Krahenbuhl, B., Mesure transcutanée de la PO_2 du pied pour suivre l'évolution des artériopathies occlusives des membres inférieurs, *Vasa*, 10, 1, 1981.

13. Lusiani, L., Visona, A., Nicolin, P., Papesso, B., and Pagnan, A., $tcPO_2$ measurement as a diagnostic tool in patients with peripheral vascular disease, *Angiology*, 39, 873, 1988.

14. Franzeck, U. K., Talka, P., and Bernstein, E. F., Transcutaneous PO_2 measurements in health and peripheral occlusive disease, *Surgery*, 91, 156, 1982.

15. Gardner, A. W., Skinner, J. S., Cantwell, B. W., Smith, L. K., and Diethrich, E. B., Relationship between foot $tcPO_2$ and ankle systolic blood pressure at rest and following exercise, *Angiology*, 42, 481, 1991.

16. Oghi, S., Ito, K., Hara, H., and Mori, T., Continuous measurement of $tcPO_2$ on stress test in claudicants and normals, *Angiology*, 37, 27, 1986.

17. Oghi, S., Ito, K., and Mori, T., Quantitative evaluation of skin circulation in ischemic legs by transcutaneous measurement of oxygen tension, *Angiology*, 32, 833, 1981.

18. Byrne, P., Provan, J. C., Ameli, P. H., and Jones, D. P., The use of $tcPO_2$ tension measurements in the diagnosis of peripheral vascular insufficiency, *Ann. Surg.*, 200, 159, 1984.

19. Modesti, P. A., Boddi, M., Pogesi, L., Gensini, G. F., and Serneri, G. G. N., Transcutaneous oxymetry in evaluation of the initial peripheral artery disease in diabetics, *Angiology*, 38, 457, 1987.

20. Bongard, O. and Kraehenbuehl, B., Predicting amputation in severe ischaemia. The value of $tcPO_2$ measurement, *J. Bone Joint Surg.*, 70B, 465, 1988.

21. Burgess, E. M. and Matsen, F. A., Determining amputation levels in peripheral vascular disease, *J. Bone Joint Surg.*, 63A, 1493, 1981.

22. Dowd, G. S. E., Linge, K., and Bentley, G., Measurement of transcutaneous oxygen pressure in normal and ischemic skin, *J. Bone Joint Surg.*, 65B, 79, 1983.

23. White, R. A., Noland, L., Harley, D., Long J., Klein, S., Tremper, K., Nelson, R., Tabrisky, N., and Schoemaker, W., Non-invasive evaluation of peripheral vascular disease using transcutaneous oxygen tension, *Am. J. Surg.*, 144, 68, 1982.

24. Depairon, M., Kraehenbuehl, B., and Vaucher, J., Détermination du niveau d'amputation par la mesure transcutanée de la PO_2 et la pression artérielle systolique distale, *J. Mal. Vasc.*, 11, 229, 1986.

25. Kram, H. B., Appel, P. L., and Shoemaker, W. C., Multisensor transcutaneous oximetric mapping to predict below-knee amputation would healing: use of critical PO_2, *J. Vasc. Surg.* 9, 796, 1989.

26. Hauss, J., Spiegel, H. U., Schonleben, K., and Bunte, H., Monitoring of tissue PO_2 for invasive diagnostics in angiology and vascular surgery, *Angiology*, 38, 13, 1987.

27. Neidhardt, A., Costes, Y., Sav, P., Roullier, M., Christophe, J. L., and Bachour, K., Apport en chirurgie vasculaire du monitorage per-opératoire de la circulation périphérique par la pression cutanée d'oxygéne $tcPO_2$, *Conv. Med. (Paris)*, 6, 9, 1986.

28. Matsen, F. A., Wyss, C. R., Robertson, C. L., and Öberg, P. A., The relationship of $tcPO_2$ and laser Doppler measurements in a human model of local arterial insufficiency, *Surg. Gynecol. Obstet.*, 159, 418, 1984.

29. Winsor, T., Haumschild, D. J., Winsor, W., Wang, Y., and Luong, T. N., Clinical application of laser Doppler flowmetry for measurement of cutaneous circulation in health and disease, *Angiology*, 38, 727, 1987.

30. Belcaro, G., Vasdekis, S., Rulo, A., and Nicolaides, A. N., Evaluation of skin blood flow and venoarteriolar response in patients with diabetes and peripheral vascular disease by laser Doppler flowmetry, *Angiology*, 40, 953, 1989.

31. Ubbink, D. T., Jacobs, M. J., Tangelder, G. J., Slaaf, D. W., and Reneman, R. S., Posturally induced microvascular constriction in patients with different stages of leg ischaemia: effect of local skin heating, *Clin. Sci.*, 81, 43, 1991.

32. Ubbink, D. T., Jacobs, M. J., Slaaf, D. W., Tangelder, G. J., and Reneman, R. S., Microvascular reactivity differences between the two legs of patients with unilateral lower limb ischaemia, *Eur. J. Vasc. Surg.*, 6, 269, 1992.

33. Enkema, L., Holloway, G. A., Piraino, D. W., Harry, D., Zick, G. L., and Kenny, M. A., Laser Doppler velocimetry vs. heater power as indicators of skin perfusion during transcutaneous O_2 monitoring, *Clin. Chem.*, 27, 391, 1981.

34. Winsor, T., Haumschild, D. J., Winsor, D., and Mikail, A., Influence of local and environmental temperatures on cutaneous circulation with use of laser Doppler flowmetry, *Angiology*, 40, 421, 1989.

35. Leonardo, G., Arpaia, M. R., and Del Guercio, R., A new method for the quantitative assessment of arterial insufficiency of the limbs: cutaneous postischemic hyperemia test by laser Doppler, *Angiology*, 38, 378, 1987.

36. Leonardo, G., Arpaia, M. R., and Del Guercio, R., Evaluation of the effect of vaso-active drugs on cutaneous microcirculation by laser Doppler velocimetry, *Angiology*, 37, 12, 1986.

37. Bongard, O., Bounameaux, H., and Fagrell, B., Effects of oxygen inhalation on skin microcirculation in patients with peripheral arterial occlusive disease, *Circulation*, 86, 878, 1992.

38. Bongard, O. and Fagrell, B., Variations in laser Doppler flux and flow motion patterns in the dorsal skin of the human foot, *Microvasc. Res.*, 39, 212, 1990.

39. Matsen, F. A., Wyss, C. R., Krugmire, R. B., Simmons, C. W., and King, R. V., The effects of limb elevation and dependency on local arteriovenous gradients in normal human limbs with particular reference to limbs with increase tissue pressure, *Clin. Orthop. Rel. Res.*, 150, 187, 1980.

40. Boccalon, H., Binon, J. P., Ginestet, M. C., and Puel, P., Effets hémodynamiques artériels et microcirculatoires immédiats de la contention élastique chez l'artéritique variqueux, *Phlebologie*, 41, 837, 1988.

41. Salerud, E. G., Tenland, T., Nillson, G. E., Öberg, P. A., Rhythmical variations in human skin blood flow, *Int., J. Microcirc. Clin. Exp.*, 2, 91, 1983.

42. Schmidt, J. A., Borgstrom, P., Allegra, C., and Intaglietta, M., Vasomotion during conditions of reduced tissue perfusion, in *Vascular Medicine*, Boccalon, H., Ed., Excerpta Medica, Amsterdam, 1993, 541.

43. Kastrup, J., Buelow, J., and Lassen, N. A., Vasomotion in human skin before and after local heating recorded with laser Doppler flowmetry, *Int., J. Microcirc. Clin. Exp.*, 8, 205, 1989.

44. Wilkin, J. K., Periodic cutaneous blood flow during postocclusive reactive hyperemia, *Am. J. Physiol.*, 250, H756, 1986.

45. Wilkin, J. K., Poiseuille, periodicity and perfusion: rhythmic oscillatory vasomotion in the skin, *J. Invest. Dermatol.*, 93, 1135, 1989.

46. Seifert, H., Jaeger, K., and Bollinger, A., Analysis of flow motion by the laser Doppler technique in patients with peripheral arterial occlusive disease, *Int. J. Microcirc. Clin. Exp.* 7, 223, 1988.

47. Bollinger, A., Hoffmann, U., and Franzeck, U. K., Evaluation of flux motion in man by the laser Doppler technique, *Blood Vessels*, 28, 21, 1991.

48. Fagrell, B., Microcirculatory methods for evaluating effect of vasoactive drugs in clinical practice, *Acta Pharmacol. Toxicol.*, 59, (Suppl. 6), 103, 1986.

49. Bongard, O. and Fagrell, B., Discrepancies between total and nutritional skin microcirculation in patients with peripheral arterial occlusive disease, *Vasa*, 19, 105, 1990.

50. Rendell, M., Bergman, T., O'Donnell, G., Drobny, E., Borgos, J., and Bonner, R. F., Microvascular blood flow, volume and velocity measured by laser Doppler technique in insulin-dependent diabetes mellitus, *Diabetes*, 38, 819, 1989.

51. Railton, R., Newman, P., Hislop, J., and Harrower, A. D., Reduced tcPO$_2$ and impaired vascular response in type 1 (insulin-dependent) diabetes, *Diabetologica,* 25, 340, 1983.

52. Mitchell, W. W., Winocour, P. H., Gush, R. J., Taylor, L. J., Baker, L. D., Anderson, D. C., and Jayson, M. I., Skin blood flow and limited joint mobility in insulin-dependent diabetes mellitus, *Br. J. Rheumatol.,* 28, 195, 1989.

53. Winocour, P. H. Mitchell, W. S., Gush, R. J., Taylor, L. J., and Baker, R. D., Altered handskin blood flow in type 1 (insulin-dependent) diabetes mellitus, *Diabetic Med.,* 5, 861, 1988.

54. Hatanaka, H., Matsumoto, S., Kitamura, Y., Maeda, Y., Hata, F., Oimomi, M., and Baba, S., Skin blood flow in diabetic patients during cold loading, *Kope J. Med. Sci.,* 35, 131, 1989.

55. Haver, J. L., Boland, O. M., Ewing, D. J., and Clarke, B. F., Hand skin blood flow in diabetic patients with autonomic neuropathy and microangiopathy, *Diabet. Care,* 14, 897, 1991.

56. Tooke, J. E., Ostergren, J., Lins, P. E., and Fagrell, B., Skin microvascular blood flow control in long duration diabetics with and without complications, *Diabetes Res.,* 5, 189, 1987.

57. Wiles, P. G., Grant, P. J., Stickland, M. H., Dean, H. G., Wales, J. K., and Davies, J. A., Regional variation in skin blood flow response to hypoglycemia in type 1 (insulin-dependent) diabetic patients without complications, *Diabetologica,* 31, 98, 1988.

58. Rayman, G., Hassan, A., and Tooke, J. E., Blood flow in the skin of the foot related to posture in diabetes mellitus, *Br. Med. J.,* 292, 87, 1986.

59. Zitnan, D., Tauchmannova, H., et al., Raynaud's phenomenon and systemic lupus erythematosus, *Bratisl. Lek. Listy,* 75, 529, 1981.

60. Partsch, H., Complementary examinations in the more frequent vascular diseases, *Hautarzt,* 36, 203, 1985.

61. O'Reilly, D., Taylor, L., El-Hadidy, K., and Jayson, M. I. V., Measurement of cold challenge responses in primary Raynaud's phenomenon and Raynaud's phenomenon associated with systemic sclerosis, *Ann. Rheum. Dis.,* 51, 1193, 1992.

62. Wesseling, H., and Wouda, A. A., Sublingual and oral isoxsupine in patients with Raynaud's phenomenon, *Eur. J. Pharamacol.,* 20, 329, 1981.

63. Kruk, M., Szczypiorski, P., Borkowski, M., and Szamowska, P., Thermographic assessment of superficial electrostimulation of blood microcirculation in the skin in peripheral vascular diseases, *Pol. Tyg. Lek.,* 40, 485, 1985.

64. Yamaoka, K., Miyasaka, N., Sato, K., et al., Therapeutic effects of Cs-570, a chemically stable prostacyclin derivative, on Raynaud's phenomenon and skin ulcers in patients with collagen vascular diseases, *Int. J. Immuother.,* 3, 271, 1987.

65. Katoh, K., Kawai, T., Narita, M., et al., Use of prostaglandin E$_1$ (lipo PGE$_1$) to treat Raynaud's phenomenon associated with connective tissue disease: thermographic and subjective assessment, *J. Pharm. Pharmacol.,* 44, 442, 1992.

66. Kyle, M. V., Belcher, G., and Hazelman, B. L., Placebo controlled study showing therapeutic of Iloprost in the treatment of Raynaud's phenomenon, *J. Rheumatol.,* 19, 1403, 1992.

67. Kristensen, J. K., Englehart, M., and Nielsen, T., Laser Doppler measurement of digital blood flow regulation in normals and in patients with Raynaud's phenomenon, *Acta Derm. Venereol.,* 63, 43, 1983.

68. Boccalon, H., Ginestret-Venerandi, M. C., and Puel, P., Phénoméne de Raynaud. Doppler au laser. Caisson isotherme. Exploration de sujets normaux et pathologigues, *J. Mal. Vasc.,* 10, 11, 1985.

69. Shawket, S., Dickerson, C., Hazleman, B., et al., Selective suprasensitivity to calcitonin gene related peptide in the hands in Raynaud's phenomenon, *Lancet,* 2, 1354, 1989.

70. Englehart, M. and Kristensen, J. K., Colour changes during "Raynaud's phenomenon" and finger blood supply during direct and indirect cooling procedures, *Clin. Exp. Dermatol.,* 12, 339, 1987.

71. Goodfield, M. J. D., Hume, A., and Rowell, N. R., The effect of simple warming procedures on finger blood flow in systemic sclerosis, *Br., J. Dermatol.,* 118, 661, 1988.

72. Goodfield, M. J. D. and Rowell, N. R., Hand warming as a treatment for Raynaud's phenomenon in systemic sclerosis, *Br. J. Dermatol.,* 119, 643, 1988.

73. Bunker, L. B., Lanigan, S., Rustin, M. H. A. et al., The effects of topically applied hexyl nicotinate lotion on the cutaneous blood flow in patients with Raynaud's phenomenon, *Br. J. Dermatol.,* 119, 771, 1988.

74. Cesarone, M. R., Laurora, G., Smith, S. R. G., et al., Laser Doppler flowmetry in the assessment of mild Raynaud's phenomenon and its treatment, *Panminera Med.,* 32, 151, 1990.

75. Engelhart, M., Ketanserin in the treatment of Raynaud's phenomenon associated with generalized scleroderma, *Br. J. Dermatol.,* 119, 751, 1988.

76. Rustin, M. H. A., Almond, N. E., Beacham, J. A., et al., The effects of captopril on cutaneous blood flow in patients with primary Raynaud's phenomenon, *Br. J. Dermatol.,* 117, 751, 1987.

77. Elsmann, H. J., Rabe, E., Schuler-Pyrtek, P., and Bauer, R., Laser Doppler flowmetry in prostaglandin E_1 therapy of scleroderma, *Z. Hautkr.,* 66, 533, 1991.

78. Risold, J. C., Laurent, R., Debrand, J., and Agache, P., Applications cliniques des méthodes d'exploration vasculaire: érythermalgie, maladie de Buerger, acrodynie, *Ann. Med. Nancy,* 19, 197, 1980.

79. Bej, M. D. and Schwartzman, R. J., Abnormalities of cutaneous blood flow regulation in patients with reflex sympathetic dystrophy as measured by laser Doppler fluxmetry, *Arch. Neurol.,* 48, 912, 1991.

80. Rustin, M. H. A., Newton, J. A., Smith, N. P., and Dowd, P. N., The treatment of chilblains with nifedipine: the results of a pilot study, a double-blind placebo-controlled randomized study and a long-term open trial, *Br. J. Dermatol.,* 120, 267, 1989.

81. Rodgers, G. P., Schechter, A. N., Nogushi, C. T., Klein, H. G., Nienhuis, A. W., and Bonner, R. T., Periodic microcirculatory flow in patients with sickle-cell disease, *N. Engl. J. Med.,* 311, 1534, 1984.

Chapter 11

Light Reflection Rheography (Photoplethysmography) and Venous Insufficiency

Volker Wienert

CONTENTS

ABSTRACT: In the last few years, light reflection rheography (LRR) has established itself as a screening method for clarifying the venous hemodynamics in the leg, and nowadays it can be considered as being equal to phlebodynamometry (sanguinary, dynamic venous pressure measurement). The measuring principle of LRR is based on the detection and evaluation of radiation fractions which are emitted from numerous selective infrared light sources into the skin and then reflected from the venous plexus to a photodetector. With this measuring technique the blood-filled plexus appears dark and the emptied plexus light. Thus, the emptying of the venous plexus during the exercises performed by the patients and the refilling of the plexus during the rest period can be recorded as a curve. In this chapter the handling of the instrument is described and application errors are indicated. Indication and contraindication for examination are demonstrated. The technical differences between LRR and other photoplethysmographs (PPG) are dealt with in detail.

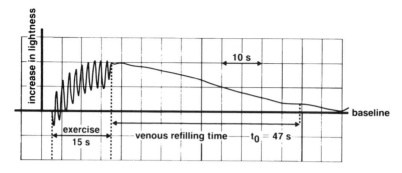

Figure 1   LRR trace of a healthy person.

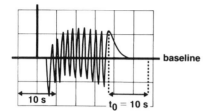

Figure 2   LRR trace of a patient with venous insufficiency.

I. THE INSTRUMENT

A. MEASURING PRINCIPLE

The basic part of LRR is the measuring head, which consists of a built-in, centrally positioned photodetector, three infrared light sources with focusing lenses attached to the photodetector at equal distances, and a temperature sensor. Via a cable the measuring head is connected to an electronic measuring and evaluation unit which consists of a digital display and a recorder. The infrared beams are irradiated into the skin and reflected from the venous plexus to the photodetector. The blood-filled plexus appears dark and the blood-depleted plexus light. Standing patients show a high blood content in the cutaneous venous plexus which results in a low degree of reflection. After applying the muscle joint pump there is a profound drop in the venous pressure of the veins in the leg and there is an emptying of the venous plexus. The skin becomes lighter and the degree of reflection increases. After the exercise program has ended, the venous plexus in normal subjects gradually refills via the arterial inflow; as a rule, this takes longer than 25 s. If there is a pronounced venous insufficiency, the reflux in the microcirculation is greater. Therefore, the quicker the venous plexus refills once the muscle activity has ended, the shorter the venous refilling time t_0 (Figures 1 and 2). To begin with, LRR only enables the general quantification of the venous hemodynamics in the leg; it gives information as to whether the venous outflow is good, bad, or of limited value. Only on the basis of selective occlusion tests (tourniquet tests) can one differentiate whether, for example, there is damage in the superficial, deep, or communicating venous system of the leg.

B. HARDWARE

The instrument consists of the measuring head (sensor), the battery-operated recording and evaluating unit, and the loading unit.

1. The Sensor

Three gallium-arsenide light-emitting diodes (LED) are used as radiation sources with a maximum emission wavelength of 940 nm (Figure 3). The reflected light fractions, not visible

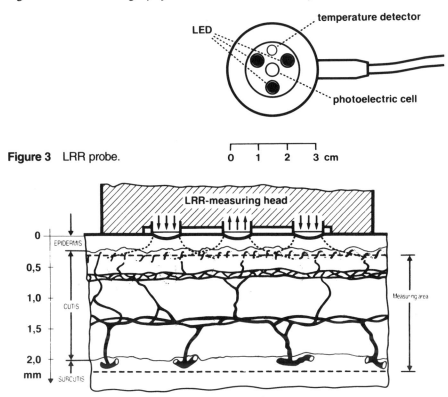

Figure 3   LRR probe.

Figure 4   Vascularization of the skin in the lower limb.

to the naked eye, can be detected via a silicon photo-transistor. The application of several light sources enables a homogeneous screening of the skin. Since the light diodes are equipped with lenses, the measuring light falls perpendicularly onto the surface of the skin, thereby preventing a reflection at the skin. Only at a depth of ca. 0.3 mm can the vessels be optically scanned (Figure 4). The distance of the three IR light sources to each other and to the centrally positioned detector is optimized in such a way that reflections can be detected at a skin depth of 0.3 to 2.0 mm. Thus, the various quasi-horizontal venous plexuses definitely can be reached by the radiation. The gallium-arsenide light-emitting diodes are therefore selected because a particularly favorable "measuring window" for the optical scanning of the skin can be obtained at a wavelength of 940 nm: ca. 85% of the incident radiation successfully reaches the superficial blood vessels through the epidermis. The additional integrated temperature sensor measures the actual skin temperature in the measuring area, which is recorded on the digital display in degrees centigrade. The sensor has a diameter of 30 mm and weighs about 10 g. It is attached to the skin using a double-sided adhesive foil band.

2.   The Light Reflection Rheography Instrument

The LRR instrument consists of a measuring unit, an evaluation unit, a digital display panel, and an analog recorder. A modulated signal triggers the sensor of the measuring instrument. The optical filter assembled before the photodetector enables an external light-independent measurement. A compensating circuit automatically calibrates the brightness contrast of the skin of each patient (so-called resting degree of reflection) so that each person has the same starting condition for an examination. The pen of the x-t recorder (paper width: 140 mm;

maximum recorder width: 120 mm) always has the same starting position before the examination. The actual LRR signal changes are indicated by means of a digital voltmeter and are registered on the recorder. In order to standardize the exercise program, a timing generator is built into the instrument. It optically and acoustically supports the patients when they perform ten maximal dorsal flexions within a period of 15 s.

C. OTHER PHOTOPLETHYSMOGRAPHS

The measuring principle of the LRR is not different from other photoplethysmographs. However, from the technical point of view the system does show definite optimizations.[1,7,8] The technical differences of the system include the following: the transmitter and the receiver in the LRR head are constructed to optimize radiation, and three emitters are assembled next to each other so that a homogeneous planar light field can be perpendicularly irradiated deep into the skin up to the dermal venous plexus. Hence, a reflection at the surface of the skin cannot occur. The centrally positioned detector receives the reflection signal. Analog or digital photoplethysmographs have only a receiver next to the transmitter. Not only is light thereby spot-irradiated, but undesired reflections at the skin surface are also recorded which could possibly result in false measuring signals. The reproducibility of the method can, therefore, be queried. In contrast to conventional PPG instruments, LRR is not influenced by external light; light in the examination room or solar radiation has no influence on the measurement results. Only LRR shows a long-term constancy. The applied measuring head delivers a straight line to the recorder, the so-called baseline; this means that there is a constant signal until the patient begins his exercise program.

The venous hemodynamics of the lower extremities depends on the skin temperature. Lowering the skin temperature, which is normally 28 to 32° C, to 20° C results in a 37% prolongation of the refilling time in a healthy subject and even 50% in a venous patient.[15] The actual skin temperature which is shown on the digital display is continuously detected by means of the temperature sensor integrated in the LRR sensor. This temperature measuring is not possible with conventional PPG instruments. The curve in LRR is recorded analogically. The digital recording used in the digital PPG may contain sources of errors: the use of a microchip with a certain algorithm and an appropriate filter technique results in "embellished" curves which are often false. This can be clarified by the following example: within seconds, the microcomputer detects whether the recorded curve has reached a plateau of 4 s. When this occurs, the recording process is stopped and the refilling time t_0 is defined. Any experienced doctor who has carried out many LRR measurements is well aware of the fact that one cannot always expect textbook-like curves. A deflection of the baseline can be observed quite often. Using LRR one would first of all continue with the recording and wait for the curve to return to the baseline. The digital technique, on the contrary, would register a totally incorrect refilling time (too short) due to the automatic switch-off.

II. HANDLING OF THE INSTRUMENT

A. TAKING MEASUREMENTS

Shoes, stockings, and long trousers are removed by the patient for the examination (Figure 5). Patients are asked to sit on a chair and lean back to ensure maximal body relaxation. Their feet are placed on a mat in such a way that a 110° angle is formed between the lower and upper leg. The LRR instrument is put into operation by turning on the main switch; the green light diode indicates that the instrument is functioning. The "sensitivity" switch must be set on "normal" and the adjacent switch on "calibrate". The protecting cap of the felt-tip pen is removed and the pen is placed in the appropriate holder. The exercise program is explained to the patient and then practiced with him: the patient performs a maximum of ten uninterrupted dorsiflections in the ankle joint in time to a metronome. The tip of the foot must be raised to its maximum with the heel resting on the floor. The LRR measuring head

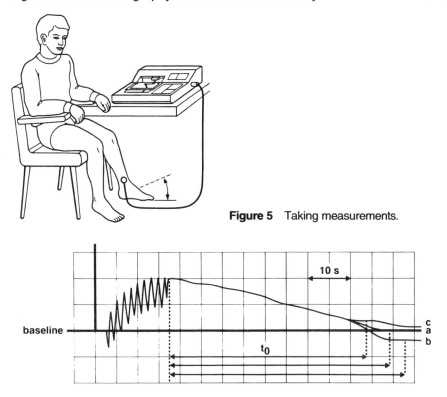

Figure 5 Taking measurements.

Figure 6 Determining the venous refilling time (a) in the baseline, (b) under the baseline, and (c) above the baseline.

is removed from the instrument and a foil ring is attached to it. The second protecting foil is also removed. The sensor is attached to the medial lower leg approximately 10 cm above the malleolus of clinically healthy skin. Should this not be possible, a more proximal position must be chosen. At first the measuring head is attached to the lower right leg with the cable hanging down toward the floor. The digital display (millivolts) of the instrument must reach a stable starting value. By turning the upper right knob to "° C" the temperature sensor is switched on; the degrees appear on the display and the control value is set at 28 to 32° C. Should the temperature be too low, the legs of the patient must be gradually warmed up by placing the person near a heater, for example. When the desired temperature has been reached, the turning knob is set to "R_0". The recorder is put into operation by turning the switch to "manual" and the adjacent switch to "measuring". At first the baseline is recorded. The patient is instructed to start the exercise program upon hearing the first peep tone. The "start" key is pressed and recording of the LRR curve commences. The refilling time starts once the exercise program is ended. It is necessary to wait until the curve is displayed as a constant horizonal line over 4 s (5 s = one square). It can return to the level of the baseline, but can also proceed above or below it (Figure 6). The recording is ended by switching to "automatic". Should a refilling time "less than 25 s" be recorded (= pathological), a new curve must be recorded after tourniquets have been simultaneously applied to the upper and lower leg. The hemodynamics of the left leg is similarly measured. When the procedure is ended, the switch is turned to "fast feed" and the recording paper is torn off. The result of the test is indicated in the form of a printed curve. The measuring head (without foil) is again attached to the instrument. By turning the main switch to "0" and the adjacent one to "calibrate,"

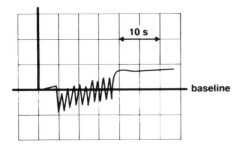

Figure 7 The influence of eczema on the LRR trace.

the instrument is switched off. Should the instrument not be used that day, the felt-tip pen is removed from the holder and covered with the protective cap. The following information is written on the curves: surname, first name, date of birth, examination date, right leg, and left leg. The application of tourniquets is denoted above the curve with ''++''. The refilling time (t_o) is measured (5 s = one square) and recorded below the curve.

Checklist

Instrument: Green light on—instrument ready for operation
             Recorder paper fed in
             Felt-tip pen ready for use
             Skin temperature 28 to 32° C
             Switch sensitivity ''normal''
             Adjacent switch ''calibrate''

Equipment: Adhesive rings
            Tourniquets

B. APPLICATION ERRORS

In order to obtain comparable measuring results, the sensor always should be positioned approximately 10 cm above the inner malleolus. If it was additionally fixed in a proximal position, this would result in a shortening of the refilling time. If there is too much hair on the lower leg or ointment rubbed in, the adhesive ring will not stick. The measuring head would partly lose contact with the skin so that a radiation transmitter and also the photodetector might not be optimally positioned on the skin. This would result in an incorrect recording of the curve. Therefore, excessive hair and ointment residues should be removed. Hyperemized skin (e.g., eczema, erysipelas) is unsuitable as a measuring location because no emptying of the venous plexus will occur when the muscle pump is used (Figure 7). This applies to the presence of a varicose ulcer as well as an insufficient perforating vein since there is no orthological flow rate present (Figure 8). The sensor is not to be attached to the skin with adhesive plaster because any pressure on the skin area would change the flow rate in the microcirculation as well as the measuring results.

The test should not be performed on patients under the age of 20, or rather those not having reached complete physical maturity. In a group of 13-year-old healthy boys and girls, it could be demonstrated that 50% of the children had refilling times of less than 25 s. This finding can be pathologically assessed in accordance with defined international norms.[11] The patient must be able to move his ankle freely. The required movement cannot be performed if there is, for example, an ankylosis or a partial ankylosis of the ankle joint. A LRR measurement would thereby lose its comparability and therewith its indication. Although rare,

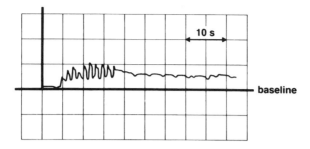

Figure 8 The influence of probe positioning on the LRR trace (here: insufficient perforator).

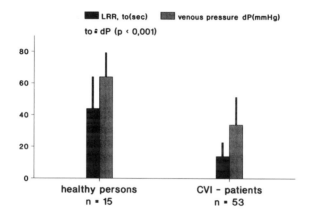

Figure 9 LRR and direct measurements of venous pressure. (CVI = Chronic Venous Insufficiency, P = Pressure.)

there may be patients who are not mentally able to perform the required program movements. More frequently, however, patients whose lower legs are too short are unable to form an angle of 110° with the upper leg when they place their foot on the mat. A book placed under the floor mat can be helpful.

C. MEASUREMENT INTERPRETATION

A diagnosis can be made only in conjunction with the clinical findings. The only relevant and most important parameter is the venous refilling time (t_o) in seconds; the objective, qualitative, and quantitative venous function can thereby be documented. One can generally state that the bigger the t_o, the better the venous hemodynamics of the leg. LRR can absolutely compete with (golden standard) phlebodynamometry. An extensive phlebographically controlled group of healthy subjects and patients with venous insufficiency was studied to compare the function parameter to (LRR) and the decrease of venous pressure (dP[mmHg]). It could be demonstrated[3,10] that, in all groups, measuring t_o shows a higher specificity and sensitivity than measuring the drop in venous pressure (Figure 9). The reproducibilities of both methods were compared. The peripheral venous hemodynamics was simultaneously tested with LRR and with sanginous venous pressure measurement. This was done 40 times on different days and on one and the same test persons.[2] The results of both methods partially showed a distinct scattering range which was nonetheless comparable.

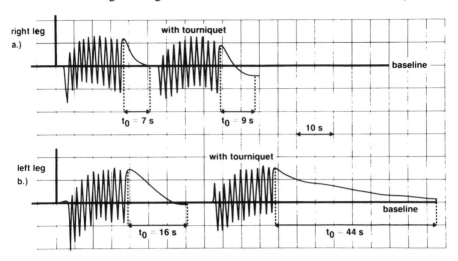

Figure 10 LRR trace (without and with tourniquet) of a patient with long saphenous incompetence on the (a) right and (b) left leg.

The venous refilling time t_o is nowadays standardized and internationally applicable:

Grade I: t_o = 25 to 20 s (low flow impairment)
Grade II: t_o = 20 to 10 s (average flow impairment)
Grade III: t_o = below 10 s (high flow impairment)

The venous refilling time is determined only after the final value of the curve is reached; i.e., the curve has to form a plateau over a period of at least 4 s which either corresponds with the baseline or which is above or below the baseline.

D. PITFALLS AND TROUBLESHOOTING

Since errors most likely occur in patients, the following prerequisites must be fulfilled:

- The patient must be over the age of 20.
- The patient must be mentally able to carry out the required exercise program.
- The patient must be able to move his ankle joint freely.
- A book is to be placed under the foot if the lower leg is too short.
- An orthological skin temperature (28 to 32° C) must be recorded.
- The sensor must be placed on healthy skin.
- The sensor application site must have no hair or ointment residues.

III. APPLICATIONS OF THE INSTRUMENT

A. PREOPERATIVE AND POSTOPERATIVE DIAGNOSIS

When there is evidence of a stem varicosis of the vena saphena magna an operation is advisable, particularly if the deep leg venous system is completely in function.[14] A decision can be made with the aid of LRR. In Figure 10, for example, the LRR curve of the right leg of a patient with stage IV stem varicosis of the vena saphena magna shows a refilling time of 7 s, which quantifies and qualifies the hemodynamics of the leg as generally bad. Even after selective occlusion (tourniquet test) of the vena saphena magna, the refilling time does not improve very much; i.e., it remains low. One must start with the principle that there is

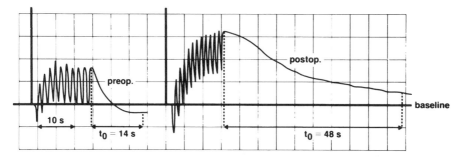

Figure 11 LRR trace (preoperative and postoperative) of a patient with long saphenous incompetence.

an impairment in the deep venous system: either a deep or a muscle venous insufficiency or a postthrombotic syndrome should there be no other perforating insufficiency. In this case, an operation is not advisable. A stem varicosis of the vena saphena magna is also found in the left leg of the same patient. At first one can conclude from the curve that there is a considerable restriction of the venous outflow. However, an orthological flow is detected after occlusion of the vena saphena magna; i.e., the deep leg venous system functions and an operation is indicated. It must be noted, however, that in patients with stage I to II stem varicosis of the vena saphena magna stage it is not always possible to differentiate between healthy subjects and patients with venous impairment merely on the basis of the venous refilling time. The reason for this is that these patients sometimes exhibit orthological refilling times.

The result of the improved hemodynamics also can be documented after an operation. In Figure 11, for example, this female patient suffered from a stem varicosis of the left vena saphena magna. The preoperative refilling time was 14 s, and after the operation it was 48 s.

B. DIFFERENTIAL DIAGNOSIS OF LEG DROPSY
Only venous edema, irrespective of whether it is induced by extrafascial or intrafascial venous insufficiency, shows pathological refilling times. This is not the case with lymphatic edema and lipedema or with cardiac, renal, or hepatic edema.

C. DIFFERENTIAL DIAGNOSIS OF LEG PAIN
Orthopedically and neurologically induced pain does not produce a pathological refilling time.

D. DIFFERENTIAL DIAGNOSIS OF ULCUS CRURIS
Only venous ulcus cruris results in pathological refilling times; ulcus arteriosum, microangiopathicum, infectiosum, hemopoeticum, and neoplasticum do not. Whereas ulcus cruris postthromboticum shows no improvement even in a tourniquet test, the occlusion test does bring about a prolongation of the refilling time in ulcus cruris varicosum provided that there is not an additional insufficiency in the deep venous system.

E. DIAGNOSIS DURING PREGNANCY
An aggrevated venous flow in the legs of many women could be detected in the last trimester since it is manifested in a shortened refilling time. Compression therapy should be introduced in pathological refilling times.[4]

F. THERAPY CONTROL

An improved venous function could be ascertained after thrombectomy or fibrinolysis and after treatment with a venous drug.

G. QUANTIFICATION OF THE POSTTHROMBOTIC SYNDROME

Depending on the refilling time, postthrombotic syndrome (PTS) can be differentiated into decompensated (less than 15 s) and compensated forms (more than 15 s). Yet a differential diagnosis of thrombosis, postthrombosis, or deep and muscle venous insufficiency is not possible. This can only be diagnosed by means of duplex sonography or phlebography.

H. DIAGNOSIS OF LEG PHLEBOTHROMBOSIS

LRR does not show pathological refilling time in a thrombosis of the pelvic vein or the femoral vein. In hemodynamically effective calf vein thrombosis or in multilayer thrombosis, LRR can be taken into consideration if prefindings with orthological refilling times are available. If there are no prefindings, LRR is only suitable to exclude the presence of a thrombosis. However, some authors are of a different opinion. With the aid of LRR they are able to "positively diagnose" deep leg phlebothrombosis.[5,6,12]

The advantages of LRR are manyfold. It is noninvasive, painless, without risk, highly accepted, rapid, and easy to handle. Few staff are required, and there is good reproducibility and high sensitivity, specifity, and repeatability. Test recording is rapid, and diagnosis and evaluation can be done quickly. LRR can often replace preoperative phlebography. A prerequisite for an optimal application is for the doctor to have experience in phlebology in general; he must be able to use the instrument and interpret the findings. Although LRR is not complicated to use, the test results depend on an exact execution and a number of specific parameters (e.g., skin temperature, ankle joint mobility). In order to be able to interpret LRR curves correctly one must know and appropriately evaluate these factors in combination with clinical examinations.

REFERENCES

1. Abramowitz, H. B., Queral, L. A., Flinn, W. R., Nora, P. F., Jr., Bergan, L. K., Peterson, L. K., and Yao, J. S. T., The use of photoplethysmography in the assessment of venous insufficiency: a comparison to venous pressure measurements, *Surgery,* 86, 434, 1979.
2. Blazek, V., Wienert, V., and May, R., Gesicherte Erkenntnisse und neue Anwendungsmöglichkeiten der Lichtreflexionsrheographie, *Phlebol. Proktol.,* 14, 74, 1984.
3. Drews, S., Gebhardt, V., Straub, H., Völkel-Nuhn, R., and Barmeyer, J., Wertigkeit der Lichtreflexionsrheographie und Venendruckmessung in der venösen Funktionsdiagnostik, in Kriessmann, A., Eds., *Fortschritte in der Angiologie,* Huber, Bern, 1988, 79.
4. Karl, C., Wienert, V., Blazek, V., and Linden, D., Die venöse Hämodynamik der unteren Extremitäten in der Gravidität und Wochenbett, *Swiss Med.,* 5, 25, 1983.
5. Kuhlmann, Th. P., Sistrom, C. L., and Chance, J. F., Light reflection rheography as a noninvasive screening test for deep venous thrombosis, *Ann. Emerg. Med.,* 50, 513, 1992.
6. Mitrani, A. A., Gonzales, M. L., O'Connell, M. T., Guerra, J., Harward, R. B., and Gardner, L. B., Detection of clinically suspected deep vein thrombosis using light reflection rheography, *Am. J. Surg.,* 161, 646, 1991.
7. Nicolaides, A. N. and Miles, C., Photophlethysmography in the assessment of venous insufficiency, *J. Vasc. Surg.,* 5, 405, 1987.
8. Norris, C. S., Beyrau, A., and Barnes, W., Quantitative photophlethysmography in chronic venous insufficiency: a new method of noninvasive estimation of ambulatory venous pressure, *Surgery,* 94, 758, 1983.

9. Schmeller, W. and Schadewinkel, M., Die Beinvenenhämodynamik in Abhängigkeit vom Bewegungsausmaß im oberen Sprunggelenk, *Phlebol. Proktol.*, 16, 33, 1987.

10. Shepard, A. D., Mackey, W. C., and O'Donnell T. F., Jr., Correlation of venous pressure measurements with LRR, in *Die Licht-Reflexions-Rheographie,* May, R. and Stemmer, R., Eds., Perimed, Erlangen, Germany, 1984, 15.

11. Sick, H. and Wienert, V., Licht-Reflexions-Rheographie—Beinvenendiagnostik, Normwerte der 10- bis 30-jährigen, *Phlebol. Proktol.*, 14, 78, 1985.

12. Thomas, P. R. S., Buttler, C. M., Bowman, J., Grieve, N. W. T., Bennett, C. E., Taylor, R. S., and Thomas, M. H., Light reflection rheography: an effective non-invasive technique for screening patients with suspected deep venous thrombosis, *Br. J. Surg.*, 78, 207, 1991.

13. Wienert, V. and Blazek, V., Eine neue Methode zur unblutigen dynamischen Venendruckmessung, *Hautarzt,* 33, 498, 1982.

14. Wienert, V., Blazek, V., and Mayer, O., Nicht-invasive hämodynamische Untersuchungen vor und nach Babcock-Stripping, *Phlebol. Proktol.*, 15, 72, 1986.

15. Wienert, V. and Rütten, M., Der Einfluß unterschiedlicher Temperaturen auf die venöse Hämodynamik der unteren Extremität, *Phlebol. Proktol.*, 13, 25, 1984.

Chapter 12

Port Wine Stains and Treatment

Sean W. Lanigan and John A. Cotterill

CONTENTS

I. INTRODUCTION

The nevus flammeus, or port wine stain, is a congenital abnormality affecting between 0.3 and 1%[1,2] of the population and is usually present from birth. Most lesions occur on the face, and the sexes are affected equally.[3] The color of the nevus varies from pink to a deep purple, the change to purple often occurring with increasing age.[4] They present a devastating cosmetic disability to the patient which persists.[5]

On the face, the lesion is most commonly sited in the distribution of the trigeminal nerve[6] (usually of the first two branches) and is generally, but not always, unilateral, being sharply marginated at the midline. After the face, the upper trunk is most commonly affected; any area of the body may be involved, although the genitals are rarely affected.[7] In contrast to capillary "strawberry" hemangiomata, which will usually regress spontaneously,[8] port wine stains usually do not involute, their growth being commensurate with the patient's. Port wine stains, particularly in older patients, may develop solitary or multiple hemangiomatous tumors, eventually producing a cobblestone pattern[9,10] and increasing the cosmetic disability. The histopathological changes seen within a port wine stain are progressive, with minimal abnormalities at birth. However, progressive ectatic dilatation and enlargement of thin-walled, mature dermal blood vessels develop after birth. Initially, these changes are situated in the most superficial dermis, which always remains the most markedly affected, but eventually the ectasia involves deeper dermal blood vessels with the development of papular and nodular lesions. The color of port wine stains correlates with the histological changes of vascular ectasia and erythrocyte content in the superficial dermis, and port wine stains have been divided into a juvenile type, which is pink with small relatively erythrocyte-free vessels, and the adult, mature type, which is purple in color with large, blood-filled ectatic vessels.[11]

A histological study[14] of port wine stains using antibodies to S100 protein to demonstrate cutaneous nerves showed a significant reduction in perivascular nerves when compared to normal skin. It has been hypothesized that a lack of neural modulation of vascular flow may be important in the pathogenesis of port wine stains.[15]

The development of laser therapy as a successful treatment for port wine stains[16-18] has led to renewed interest in these vascular nevi, particularly with regard to developing objective methods of assessment both for increased understanding of their pathogenesis and for quantifying changes induced by treatment, including identifying those nevi which will respond optimally to a particular laser.

II. OBJECTIVE ASSESSMENTS
OF PORT WINE STAINS

Much of the practice of treating port wine stains relies entirely on the subjective methods of assessment of clinical observation and color photography. With advances in the treatment of these nevi it has become increasingly important to develop more objective methods of assessment to assist in predicting the outcome of therapy, quantification of results, and understanding their pathophysiology.

Most objective assessments of treated port wine stains have been by histological changes;[19,20] however, in a treatment designed to improve cosmetic appearances, noninvasive methods of assessment are desirable. It has been demonstrated that the color of port wine stains correlates with erythrocyte filling of ectatic blood vessels[4] and that the color of this nevus influences its response to laser therapy.[11,18] Therefore, objective assessments based on the color of port wine stains may be of value in predicting response to therapy.

Noninvasive methods of assessing port wine stains have been based on the abnormal vasculature of the nevi and have included thermography,[21] transcutaneous microscopy,[22] tristimulus colorimetry,[23,24] and laser Doppler flowmetry.[25] Laser Doppler flowmetry is a noninvasive method of measuring cutaneous blood flow.[26] Light from a helium-neon laser is conducted by an optical fiber to the skin surface and penetrates the skin to a depth of 1 to 1.5 mm over a surface area of 1 mm.[2] Backscattered light is collected by 2 fibers leading to separate photodetectors. Some light will be Doppler shifted as it reflects back from vascular beds. The unshifted light is extracted by subtraction of one signal from the other. The machine gives an output in millivolts which is linearly proportional to red blood cell flux (number of red blood cells × velocity). The machine provides a reproducible and objective measurement of blood flow in the area.[27,28] These measurements are comparable with those obtained using radioactive xenon washout and photopulse plethysmography.[29]

By sequential biopsy studies in normal subjects it has been demonstrated that the wave patterns of the cutaneous microcirculation produced by laser Doppler velocimetry can be correlated with the types of microvessels beneath the tissue probe.[30] By using two wavelengths of laser light it is possible to distinguish between blood flow within different areas of the dermis, providing information concerning capillary and large microvascular blood flows.[31]

Since the majority of the ectatic vessels within a port wine stain reside within the superficial dermis, information provided by red blood cell flux as measured by laser Doppler velocimetry may be of value in assessing these nevi. This chapter will consider the use of the laser Doppler velocimeter in assessing the vascular dynamics of port wine stains and in evaluating the changes in laser-treated port wine stains.

III. STUDIES ON THE VASCULAR DYNAMICS
OF PORT WINE STAINS

A number of studies have been published where laser Doppler velocimetry has been utilized to study reactivity of the microvasculature within port wine stains.[32–35] Comparisons have been made between the port wine stain and normal contralateral skin. All investigators used a Periflux®* laser Doppler velocimeter. Other evaluations included skin temperature,[33–35]

---

*Registered trademark of Perimed Corporation, Stockholm, Sweden.

Table 1   **Baseline measurements in port wine stains in relation to color**

	Pink-Red (n = 2)	Red (n = 2)	Red-Purple (n = 3)	Purple (n = 4)
Reflectance (absolute units)	294.3 ± 12	338.0 ± 6	292.1 ± 14	291.7 ± 16
Laser flowmetry (mV)	41.0 ± 5	52.5 ± 15	45.2 ± 7	48.2 ± 3

Note: Values are means ± SEM.
From Lanigan, S. W. and Cotterill, J. A., *Br. J. Dermatol.*, 118, 805, 1988. With permission.

histology,[33] and reflectance spectrophotometry.[34,35] All investigators found that baseline measurements in untreated port wine stains revealed significant elevations of red blood cell flux in the majority compared to normal contralateral skin. Skin surface temperature was similarly elevated within the port wine stain.

Apfelberg et al.[32] studied the responses of five patients to local heating of their port wine stains by heating the laser Doppler probe to 40° C. Three patients demonstrated increased flux associated with presumed vasodilatation. The changes in flux after application of ice to the skin were investigated in eight patients. Results were variable and unpredictable. A reactive hyperemia following discontinuation of hypothermia was seen in normal skin and in five port wine stains. Injection of epinephrine with Xylocaine® in five port wine stains produced a profound decrease in blood flow/perfusion.

Bongard et al.[33] studied 12 patients with facial port wine stains. After stable baseline measurements were obtained, occlusion of the arterial supply was performed by pressing the laser Doppler probe against the skin until the flux reached its lowest stable value. Reactive hyperemia after release of pressure was recorded. There was significant ($p < 0.05$) impairment of post-occlusive reactive hyperemia within untreated port wine stains compared to normal skin. Respiration had no effect on rhythmic activity recorded, and there was no relationship between laser Doppler flux and histological changes.

Lanigan and Cotterill in two studies[34,35] compared laser Doppler flowmetry and reflectance spectrophotometry in the assessment of dynamic changes within a port wine stain. Reflectance spectrophotometry is based on the theory that the logarithm of the inverse reflectance (LIR) of the skin approximates the sum of the absorbances of the skin's constituent pigments. The wavelengths of light used to illuminate the skin (566 and 640 nm) provide information on both oxy- and deoxyhemoglobin combined as a total blood content of the skin area assessed, and a reproducible objective measurement of this contribution to skin color can be made.[36,37]

Assessments were performed on 11 patients with facial port wine stains.[34] Baseline assessments revealed significant ($p < 0.05$) elevation of both reflectance spectrophotometry and laser Doppler flowmetry readings in port wine stains compared to normal contralateral skin. Reflectance and laser Doppler measurements did not show any consistent relationship with the color of the port wine stain (Table 1), although the two patients with red port wine stains had the highest recordings with both instruments. Cooling the skin by the application of ice produced a fall in laser Doppler velocimetry of similar magnitude and duration in port wine stains, normal contralateral skin, and normal controls. Changes in skin surface temperature paralleled changes in red blood cell flux. There were no changes in reflectance spectrophotometry in any group.

After local heating of the skin with a glass container at 43° C an increase in laser Doppler velocimetry was observed. The changes were of similar magnitude and duration in the three groups studied. A rise in temperature paralleled the changes in red blood cell flux. No changes in reflectance spectrophotometry were observed.

A subsequent study[35] investigated the vascular reactivity of port wine stains to the percutaneous application of vasoactive substances and compared this with normal contralateral skin. Again, laser Doppler velocimetry and reflectance spectrophotometry were the main

Table 2   **Vasodilator group measurements**

Group	Mean[a]
Reflectance spectrophotometry ($n = 14$)[b]	
1. Normal skin	126.5 (91.2–161.8)
2. Normal skin + vasodilator	158.2 (117.7–198.7)
3. Port wine stain	203.7 (167.7–239.7)
4. Port wine stain + vasodilator	220.3 (198.2–252.4)
Laser Doppler flowmetry ($n = 11$)[c]	
1. Normal skin	22.2 (14.1–30.3)
2. Normal skin + vasodilator	66.3 (40.1–92.5)
3. Port wine stain	39.3 (29.5–49.1)
4. Port wine stain + vasodilator	56.7 (43.0–70.4)

[a]95% confidence limits; [b]Significant differences ($p < 0.05$): 1 vs. 3, 1 vs. 4. Values are in absolute units; [c]Significant differences ($p < 0.05$): 1 vs. 2, 1 vs. 3, 1 vs. 4. Values are in millivolts.
From Lanigan, S. W. and Cotterill, J. A., *Br. J. Dermatol.*, 122, 618, 1990. With permission.

Table 3   **Vasoconstrictor group measurements**

Group	Mean[a]
Reflectance spectrophotometry ($n = 16$)[b]	
1. Normal skin	117.5 (79.3–155.7)
2. Normal skin + vasoconstrictor	97.3 (57.2–137.4)
3. Port wine stain	213.8 (158.2–269.4)
4. Port wine stain + vasoconstrictor	194.5 (156.3–232.7)
Laser Doppler flowmetry ($n = 6$)[c]	
1. Normal skin	25.3 (10.3–40.3)
2. Normal skin + vasoconstrictor	23.2 (13.8–32.6)
3. Port wine stain	61.6 (11.9–111.3)
4. Port wine stain + vasoconstrictor	61.5 (16.7–106.3)

[a]95% confidence limits; [b]Significant differences ($p < 0.05$): 1 vs. 3, 1 vs. 4, 2 vs. 3, 2 vs. 4. Values are in absolute units; [c]Analysis of variance: $F = 3.27$ ($p = 0.042$); no significant differences. Values are in millivolts.
From Lanigan, S. W. and Cotterill, J. A., *Br. J. Dermatol.*, 122, 619, 1990. With permission.

methods of objective assessment. The effects of a topical rubrifacient containing methyl nicotinate and capsaicin were assessed in 14 patients. There was a significant increase in red blood cell flux in normal skin after application of the rubrifacient; levels exceeded those of contralateral port wine stain skin, where the elevation from baseline was not significant (Table 2). Changes in reflectance spectrophotometry revealed a tendency for a smaller increase in hemoglobin content in port wine stain skin after application of vasodilator compared with normal skin ($p = 0.052$).

The vasoconstrictor effect of an adhesive tape impregnated with flurandrenolone (4 μ g/ml[2]) was assessed in 16 patients, 6 of whom were subjected to laser Doppler velocimetry. The corticosteroid tape produced a reduction in reflectance spectrophotometry and laser Doppler flowmetry in both normal and port wine stain skin which did not achieve statistical significance (Table 3).

The percentage reduction in hemoglobin content as measured by reflectance spectrophotometry was significantly greater in normal skin compared to port wine stains ($p < 0.05$). The percentage change in blood flow measured by the laser Doppler was significantly greater in normal skin compared with contralateral port wine stains ($p < 0.05$).

A. SUMMARY

It is difficult to compare the four studies discussed since methodology varies, and conclusions based on small groups of patients should be made with caution. However, within-patient controls were used and laser Doppler velocimetry could demonstrate, as expected, an increased red cell flux within port wine stain nevi. Reactive hyperemia and responses to percutaneous vasoactive substances are impaired within port wine stains, which is consistent with the hypothesis of an impaired neural regulation of port wine stains.[14] There was no consistent relationship between laser Doppler velocimetry and port wine stain color.

IV. STUDIES ON THE CHANGES IN BLOOD FLOW WITHIN PORT WINE STAINS AFTER LASER TREATMENT

Successful treatment of a port wine stain results in lightening of the affected skin to approximate that of normal nonlesional skin. Histological studies of laser-treated port wine stains reveal damage to the ectatic dermal vessels, which become narrowed with reduced numbers of erythrocytes within them.[11,19] These changes may be either confined to the blood vessels or part of more wide-ranging thermal effects within lesional skin, depending on the laser used.[11,19,38] Laser Doppler velocimetry has been utilized to study the changes in red blood cell flux within laser-treated port wine stains. Since it has been demonstrated that red cell flux is elevated within untreated port wine stains compared to nonlesional skin,[32-35] it may be possible to quantify changes to provide an objective measurement of outcome.

Johnston and colleagues[25] evaluated 43 patients with facial port wine stains using laser Doppler velocimetry and reflectance spectrophotometry. Multiple point measurements were made. There were marked differences in laser Doppler determinations within the same area, and the authors concluded that spectrophotometry, but not laser Doppler velocimetry, was useful in predicting outcomes after argon laser therapy.

Apfelberg et al.[32] analyzed changes in laser Doppler flowmetry in argon laser-treated port wine stains. The changes following laser therapy were similar to those following injection of epinephrine. There was no correlation between clinical assessment of fading of the port wine stain post-laser therapy and objective measurements of blood flow/perfusion.

Sheehan-Dare et al.[39] utilized laser Doppler velocimetry and reflectance spectrophotometry to assess objectively the changes in port wine stains treated by an argon laser at different pulse durations. There were significant ($p < 0.005$) reductions in both measurements after treatment. There were no differences in objective measurement changes between the two pulse durations investigated.

These studies and those investigating the dynamic changes within port wine stains expose the difficulties of measuring changes in laser Doppler velocimetry as a point source within a port wine stain when there is such great variation within the vascular architecture of a port wine stain, even in the same area.[40] The development of a laser Doppler scanning device[41] that creates an image of tissue perfusion may help to overcome this limitation. Braverman and Schechner[42] evaluated a probe holder to allow a laser Doppler probe to move in increments over a field to produce a contour map of the cutaneous microvasculature which was analyzed by a computer. This allowed more precision and reliability in the use of the laser Doppler velocimeter, but creation of the contour maps was time-consuming.

Troilius et al.[43] studied 13 patients with port wine stains who were treated with an argon laser. Skin blood perfusion and temperature were mapped with laser Doppler imaging and thermography. Light from the laser Doppler imager was scanned over the patient's port wine stain using an optical mirror system. The scanning system and perfusion sampling at each site were controlled by an AT-IBM-compatible personal computer. A color-coded image showing spatial distribution of tissue perfusion was generated on the monitor. A metallic reflector was positioned around the boundaries of the port wine stain to facilitate identification of the demarcation between normal and port wine stain skin. In 4 of the 13 patients there was a significantly ($p < 0.01$) higher blood flow recorded in the port wine stain compared to normal skin. In two of these patients skin temperature was also elevated in the port wine stain using a real-time thermal imaging system (Thermovision 870®*). Immediately after argon laser treatment there was pronounced hyperemia in the border between port wine stain and normal skin. Perfusion was also elevated in nine patients in the center of the port wine stain. After 3½ months, reactive hyperemia in the border between port wine stain and normal skin had disappeared. In five patients a blanched port wine stain and higher perfusion were observed within the treated area in comparison with the surrounding skin. The authors conclude that neither laser Doppler imaging techniques nor thermography can unambiguously predict clinical results after argon laser treatment of port wine stains.

Finally, because laser treatment of port wine stains may be painful, a variety of anesthetic techniques have been used to reduce the pain of laser treatment. One study investigated the effects of anesthetic agents on cutaneous perfusion in patients undergoing laser treatment for port wine stains.[44]

Kennard and Whitaker[44] performed a prospective double-blind, placebo-controlled evaluation of the iontophoresis of 4% lidocaine HCl and 4% lidocaine HCl with epinephrine (1:50,000) for local anesthesia during pulsed dye laser treatment of port wine stains. Changes in cutaneous perfusion after iontophoresis were measured using a Periflux® laser Doppler velocimetry flowmeter. Iontophoresis of plain lidocaine HCl did not increase perfusion of port wine stains above baseline, although this could occur in normal skin.[45] However, iontophoresis of lidocaine HCl with epinephrine produced a significant decrease in skin perfusion in port wine stains. The authors found no consistent effect of iontophoresis on the outcome of pulsed dye laser treatment of port wine stains even when clinical blanching was present.

A. SUMMARY

It is disappointing that laser Doppler velocimetry currently does not provide a useful objective measurement of treatment outcome after laser treatment of port wine stains.[46] Although changes can be detectable between treated and untreated lesional skin, there is poor correlation with observed or measured color changes. The reasons for the disappointing results have not been entirely elucidated to date. The complex three-dimensional vascular bed of a port wine stain, the relationship between flow and color, the alterations after laser treatment, both in treated and adjacent skin, and the healing process all undoubtedly contribute.

More selective vascular damage from newer lasers and scanning laser Doppler velocimetry together may provide more information in the future concerning tissue perfusion and laser therapy. Currently, laser Doppler velocimetry remains a research tool in investigating port wine stains and their treatment. Future technology involving, for instance, pulsed photothermal radiometry may be adapted to characterize port wine stains and the effects of laser treatment.[46]

REFERENCES

1. Jacobs, A. H. and Walton, R. G., The incidence of birthmarks in the neonate, *Paediatrics*, 58, 218, 1976.
2. Pratt, A. G., Birthmarks in infants, *Arch. Dermatol. Syphilol.*, 67, 302, 1953.
3. Mulliken, J. B., Capillary (port-wine) and other telangiectatic stains, in *Vascular Birthmarks, Hemangiomas and Malformations,* Mulliken, J. B., and Young, A. E., Eds., W. B. Saunders, Philadelphia, 1988, chap. 10.
4. Barsky, S. H., Rosen, S., Geer, D. E., and Noe, J. M., The nature and evolution of port wine stains: a computer-assisted study, *J. Invest. Dermatol.*, 74, 154, 1980.

---

*Registered trademark of Agema Infrared Systems AB, Stockholm, Sweden.

5. Lanigan, S. W. and Cotterill, J. A., Psychological disabilities amongst patients with port wine stains, *Br. J. Dermatol.,* 121, 209, 1989.

6. Enjolras, O., Riche, M. C., and Merland, J. J., Facial port wine stains and Sturge-Weber syndrome, *Pediatrics,* 76, 48, 1985.

7. Moroz, B., The course of hemangiomas in children, in *Vascular Birthmarks, Pathogenesis and Management,* Ryan, T. J. and Cherry, G. W., Eds., Oxford Medical, Oxford, 1987, chap. 2.

8. Finn, M. C., Glowacki, J., and Mulliken, J. B., Congenital vascular lesions: clinical application of a new classification, *J. Pediatr. Surg.,* 18, 894, 1983.

9. Finley, J. L., Noe, J. M., Arndt, K. A., and Rosen, S., Port wine stains morphologic variations and developmental lesions, *Arch. Dermatol.,* 120. 1453, 1984.

10. Lanigan, S. W. and Cotterill, J. A., Pyogenic granulomas, port wine stains and laser therapy, *Lasers Med. Sci.,* 3, 7, 1988.

11. Noe, J. M., Barsky, S. H., Geer, D. E., and Rosen, S., Port wine stains and the response to argon laser therapy: successful treatment and the predictive role of color, age, and biopsy, *Plast. Reconstr. Surg.,* 65, 130, 1980.

12. Mulliken, J. B. and Glowacki, J., Hemangiomas and vascular malformations in infants and children: a classification based on endothelial characteristics, *Plast. Reconstr. Surg.,* 69, 412, 1982.

13. Ohmori, S. and Huang, C. K., Recent progress in the treatment of port wine staining by argon laser: some observations on the prognostic value of relative spectro-reflectance (RSR) and the histological classification of the lesions, *Br. J. Plast. Surg.,* 34, 249, 1981.

14. Smoller, B. R. and Rosen, S., Port-wine stains. A disease of altered neural modulation of blood vessels?, *Arch. Dermatol.,* 122, 177, 1986.

15. Rosen, S. and Smoller, B. R., Port wine stains: a new hypothesis, *J. Am. Acad. Dermatol.,* 17, 164, 1987.

16. Cosman, B., Experience in the argon laser therapy of port-wine stains, *Plast. Reconstr. Surg.,* 85, 119, 1980.

17. Morelli, J. G. and Parrish, J. A., Lasers in dermatology: a selective historical review, *Photodermatology,* 2, 303, 1985.

18. Tan, O. T., Sherwood, K., and Gilchrest, B. A. Treatment of children with port wine stains using the flashlamp-pulsed tunable dye laser, *N. Engl. J. Med.,* 320, 416, 1989.

19. Tan, O. T., Carney, M., Margolis, R., Eki, Y., Boll, J., Anderson, R. R., and Parrish, J. A., Histologic responses of port-wine stains treated by argon, carbon dioxide and tunable dye lasers, *Arch. Dermatol.,* 122, 1016, 1986.

20. Solomon, H., Goldman, L., Henderson, B., Richfield, D., and Franzen, M., Histopathology of the laser treatment of port-wine lesions, *J. Invest. Dermatol.,* 50, 141, 1968.

21. Patrice, T., Dreno, B., Weber, J., Le Bodic, L., and Barriere, H., Thermography as a predictive tool for laser treatment of port-wine stains, *Plast. Reconstr. Surg.,* 76, 554, 1985.

22. Shakespeare, P. G. and Carruth, J. A. S., Investigating the structure of the port-wine stain by transcutaneous microscopy, *Lasers Med. Sci.,* 1, 107, 1986.

23. Neumann, R. A., Knobler, R. M., and Lindmaier, A. P., Photoelectric quantitative evaluation of argon laser treatment of port wine stains, *Br. J. Dermatol.,* 124, 181, 1991.

24. Queille-Roussel, C., Poncet, M., and Schaefer, H., Quantification of skin-color changes induced by topical corticosteroid preparations using the Minolta Chroma Meter, *Br. J. Dermatol.,* 124, 264, 1991.

25. Johnston, M., Heuther, S., and Dixon, J. A., Spectrophotometric and laser-Doppler evaluation of normal skin and port-wine stain treated with argon laser, *Lasers Surg. Med.,* 3, 149, 1983.

26. Nilsson, G. E., Tenland, T., and Oberg, A. P., Evaluation of a laser-Doppler flowmeter for measurements of tissue blood flow, *IEEE Trans. Biomed. Eng.,* 27, 597, 1980.

27. Drouard, V., Wilson, D. R., Maibach, H. I., and Guy, R. H., Quantitative assessment of UV-induced changes in microcirculatory flow by laser Doppler velocimetry, *J. Invest. Dermatol.*, 83, 188, 1984.
28. Biggaard, H. and Kristensen, J. K., Quantitation of microcirculatory blood flow changes in human cutaneous tissue induced by inflammatory mediators, *J. Invest. Dermatol.*, 83, 184, 1984.
29. Johnson, J. M., Taylor, W. F., Shephard, A. P., and Park, M. K., Laser-Doppler measurement of skin blood flow: comparison with plethysmography, *J. Appl. Physiol.*, 56, 798, 1984.
30. Braverman, I. M., Keh, A., and Goldminz, D., Correlation of laser Doppler wave patterns with underlying microvascular anatomy, *J. Invest. Dermatol.*, 95, 283, 1990.
31. Gush, R. J. and King, T. A., Discrimination of capillary and arterio-venular bloodflow in skin by laser Doppler flowmetry, *Med. Biol. Eng. Comput.*, 29, 387, 1991.
32. Apfelberg, D. B., Smith, T., and White, J., Preliminary study of the vascular dynamics of port-wine hemangioma with therapeutic implications for argon laser treatment, *Plast. Reconstr. Surg.*, 83, 820, 1989.
33. Bongard, O., Malm, M., and Fagrell, B., Reactivity of the microcirculation in port wine stain hemangiomas evaluated by Doppler fluxmetry, *Scand. J. Plast. Reconstr. Hand Surg.*, 24, 157, 1990.
34. Lanigan, S. W. and Cotterill, J. A., Objective assessments of port wine stains: response to temperature change, *Br. J. Dermatol.*, 118, 803, 1988.
35. Lanigan, S. W. and Cotterill, J. A., Reduced vasoactive responses in port wine stains, *Br. J. Dermatol.*, 122, 615, 1990.
36. Dawson, J. B., Barker, D. J., Ellis, D. J., Grassam, E., Cotterill, J. A., Fisher, G. W., and Feather, J. W., A theoretical and experimental study of light absorption and scattering by *in vivo* skin, *Phys. Med. Biol.*, 25, 695, 1980.
37. Feather, J. W., Ryatt, K. S., Dawson, J. B., Cotterill, J. A., Barker, D. J., and Ellis, D. J., Reflectance spectrophotometric quantification of skin color changes induced by topical corticosteroid preparations, *Br. J. Dermatol.*, 106, 437, 1982.
38. Lanigan, S. W. and Cotterill, J. A., The treatment of port wine stains with the carbon dioxide laser, *Br. J. Dermatol.*, 123, 229, 1990.
39. Sheehan-Dare, R. A., Lanigan, S. W., and Cotterill, J. A., Argon laser treatment of port wine stains: comparison of the effects of 0.2s and 1s pulse duration, *Lasers Med. Sci.*, 5, 271, 1990.
40. Niechajev, I. A. and Clodins, L., Histology of port wine stain, *Eur. J. Plast. Surg.*, 13, 79, 1990.
41. Nilsson, G. E., Jakobsson, A., and Wardell, K., Imaging of tissue blood flow by coherent light scattering, in *Proc. IEEE 11th Annu. EMBS Conf.*, Seattle, 1989, pp. 391–392.
42. Braverman, I. M. and Schechner, J. S., Contour mapping of the cutaneous microvasculature by computerized laser Doppler velocimetry, *J. Invest. Dermatol.*, 97, 1013, 1991.
43. Troilius, A., Wardell, K., Bornmyr, S., Nilsson, G. E., and Ljunggren, B., Evaluation of port wine stain perfusion by laser Doppler imaging and thermography before and after argon laser treatment, *Acta Derm. Venereol.*, 72, 6, 1992.
44. Kennard, C. D. and Whitaker, D. C., Iontophoresis of lidocaine for anaesthesia during pulsed dye laser treatment of port wine stains, *J. Dermatol. Surg. Oncol.*, 18, 287, 1992.
45. Bezzant, J. L., Stephen, R. L., Petelenz, T. J., and Jacobsen, S. C., Painless cauterization of spider veins with the use of iontophoretic local anaesthesia, *J. Am. Acad. Dermatol.*, 19, 869, 1988.
46. Stuart Nelson, I., Milmer, T. E., Jacques, S. L., Tran, N. H., and Berns, M. W., Use of pulsed photothermal radiometry to characterize port wine stains, *Lasers Surg. Med.*, 5 (Suppl.), 59, 1993.

Chapter 13

Ultraviolet Light Responses

Véronique Epstein-Drouard

CONTENTS

I. SKIN AND ULTRAVIOLET RAYS

A. UV SUN RAYS

When emitted, the sun's spectrum is continuous from 100 nm to 500 m. Its filtration through the earth's atmosphere results more particularly in

- Absorption by the ozone layer of the wavelengths below 290 nm, which are not compatible with life (ultraviolet [UV] C)
- Absorption by the earth's atmosphere of part of the infrared rays.

So, the sunlight we receive on the ground is composed of a continuous spectrum from 290 nm to 3000 nm.[1] Depending on the authors,[1-4] this spectrum has been arbitrarily classified into the following components:

- UVB: from 280 or 290 nm to 315 or 320 nm
- UVA: from 315 or 320 nm to 390, 400 or 420 nm
- Visible light: from 390, 400, or 420 nm to 750, 780, or 800 nm
- Infrared: from 750, 780, or 800 nm to 3000 nm

Its quality varies with altitude, latitude, season, time of day, hydrometry, and pollution. Fog or clouds interfere with its diffusion. The nature of the ground may indirectly increase its effect by reflection, particularly if there is snow and sand.

UV sun rays may be reproduced experimentally using sophisticated experimental designs which allow variation in UV irradiation doses (amount and wavelength).

B. SKIN RESPONSE TO UV RAYS

Each spectral region corresponds to different cutaneous responses, the shorter wavelengths being the most destructive.

1. Ultraviolet (UV) C

UVC rays are the least penetrating: less than 10% reach keratinocytes; most of them stop at the stratum corneum level. Their cutaneous effects, apart from erythema and thickening of the epidermis, are essentially cancers.

2. Ultraviolet (UV) B

The penetration of UVB rays depends on their wavelength: at 290 nm UVB rays reach only the horny layer, while at 320 nm they reach the epidermis at the stratum malpighi level.[2]

UVB is responsible for the most important skin reactions to sunlight; it is known as sunburn radiation. Acute exposure leads to a more or less intense erythema which attains its maximum in 6 to 24 h. A 20 to 50 mJ/cm^2 dose of UVB energy is sufficient to cause erythema.[5] Chronic exposure may lead to an increase in epidermis thickness, wrinkling, aging, cell damages, elastosis, and skin cancers. However, UVB is still necessary because it allows vitamin D synthesis and melanin production. The development of a long-lasting tan follows erythema and appears about 3 days after exposure.

3. Ultraviolet (UV) A

UVA rays are the only ones to penetrate in the dermis considerably. Immediate, direct, but temporary tanning of the skin is one of the most visible results of UVA exposure. Erythema may be produced, but it requires doses 300 to 1000 times greater than needed with UVB.[6] UVA is well known to produce photoactivation and peroxidation leading either to skin aging, photosensitization, or skin cancer.[1,3,7]

II. UV-INDUCED ERYTHEMA ASSESSMENT

A. MINIMAL ERYTHEMA DOSE

The most common method of evaluating skin photosensitivity is the determination of the minimal dose of radiation required to elicit a minimum perceptible erythema, the so-called minimal erythema dose (MED). MED is currently defined as the threshold erythemal dose, measured in seconds of exposure to a particular source, sun or lamp, required to produce an erythema which develops after 6 h and is still just visible after 24 h. However, even this definition is subject to variation depending on the author.

The sensitivity and reproducibility of the MED is a function of a number of variables, among which are the subject, the ultraviolet radiation increments used by the investigator, and the erythema assessment method.

The erythema threshold of the subject depends on the thickness of the stratum corneum, the melanin concentration, and the hormonal and nervous condition of the subject.[8]

The larger the UV increments used by the investigators the less reliable is the MED. The recommendation of a geometric series based on the ratio of $\sqrt{2}$ was made by Schulze[26] in the 1950s. Arithmetic progression is still discussed and a wide range of permutations are used.[9]

The estimation of the minimum perceptible erythema is still currently done visually.

B. VISUAL ASSESSMENT

The most common method of evaluating skin photosensitivity is visual assessment. This approach is widely used not only to determine the MED, but also to grade the approximate intensity of redness. Although a well-educated and experienced observer will consistently grade degrees of erythema, inter-grader evaluations may be subject to large variations.[10]

Recognition of the limited analytical value of the visually derived data has led to various work on other quantification techniques, such as reflectance spectroscopy, skin thermometry, and laser Doppler velocimetry.[11–14]

C. VASCULAR CHANGES TO UV RAYS

Two main theories have been postulated to explain the mechanism of UV erythema. The first was developed in 1899 by Finsen, who suggested a direct dilating effect of UV rays (UVR) on dermal blood vessels.[15] The second, described by Lewis in 1927, suggested that UVR acts by releasing a vasoactive material which diffuses to the dermal vessels to produce dilatation.[16] Further investigations in this field have been made to probe this mechanism, most of them using two different wavelength ranges, namely 250 or 254 nm (UVC) and 300 nm (UVB), although UVC does not usually reach the earth's surface.

An increase in blood flow was detected by Lewis and Zotterman, but using a dose greatly in excess of the MED.[17] Then Van der Leun made extensive studies around the observation that 250- and 300-nm erythemas had different colors. The first one is pink-red, appears earlier, and reaches its peak sooner than the second, which is deeper red. Moreover, an increase in 300-nm UVR dose gave a much deeper color than a similar increase in 250-nm UVR dose. He measured skin temperature at both wavelengths using different UVR doses. However, he was unable to detect any increase in skin temperature at 250 nm even at doses such as 30 MED. His results led him to conclude that 250- and 300-nm UVR acted through two different mechanisms. He suggested that the first one has a direct action on superficial blood vessels while the second one also acts on deeper vessels.[18,19] Breit and Kligman have shown that 300-nm erythema involves deep venules.[20]

Ramsay and Challoner, using three different methods, showed results contradicting the latter results of Van der Leun: they detected an increase in skin temperature 3 h after irradiation using 250-nm radiation at 5 × MED and similar blood-flow changes after irradiation at 250 and 300 nm. They concluded that a common mechanism of erythema production exists these two wavelengths.[14]

Further investigations by Farr and Diffey using two other techniques led them to suggest that the mechanism of erythemal response to 254- and 300-nm UV radiation was compatible with a mediator diffusion theory. They concluded that although a different mediator or mechanism of mediator release may exist for these two wavelengths, the site of action of the mediator on the dermal vasculature is identical.[21]

III. PHOTOPLETHYSMOGRAPHY MEASUREMENTS IN HUMAN SKIN

Few studies have been devoted to UV radiation-induced erythema measurement using the photoplethysmography technique. This instrument detects the amount of blood present in the cutaneous vessels. Changes in blood volume in these vessels in rhythm with the pulse are measured. It is a sensitive indicator of arteriolar changes in the skin. Increase in pulse height shows an arteriolar vasoconstriction.[22]

The first study by Crockford et al. used a strain-gauge plethysmograph which didn't allow them to detect any increase in flow before erythema was well established.[23] Ramsay and Cripps[24] showed later on, using the photoelectric pulsemeter, that arteriolar dilatation was seen at all wavelengths tested: 250, 280, 290, 293, and 300 nm, using a maximum of 2 × MED. It could be detected as early as 2.5 h after irradiation and showed an increase in

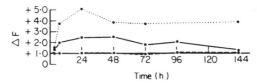

Time (h)

Figure 1 Changes in pulse height in the human skin (photoelectric plethysmograph) following irradiation with a dose of 5 × MED. See text for details. (——) Subjects irradiated with 250-nm UVR; (....) subjects irradiated with 300-nm UVR; (----) nonirradiated subjects. (From Ramsay, C. A. and Challoner, A. V. J., *J. Invest. Dermatol.*, 94, 487, 1976. With permission.)

pulse height of at least 50% in comparison to the control. By 90 h the 250-nm erythema was no longer visible and no increase in pulsation over the control site was detected. In contrast, the 300-nm erythema was still visible at 120 h and arteriolar dilatation was still present. Intradermal epinephrine was found to produce marked pallor of UV erythema, which could be due to venular constriction. However, arteriolar constriction was confirmed with the pulsemeter.[24]

A further study by Ramsay and Challoner[14] utilized three different methods: thermal conductance to measure changes in blood flux, mostly in the superficial part of the dermis, and skin thermometry and photoelectric plethysmography to assess overall changes in both superficial and deeper vessels. Vascular changes were evaluated after UV irradiation at 250 and 300 nm with a dose of 5 × MED. (See Figures 1 and 2.) Pulse height, which reflects changes in blood flow, increased after 6 h of irradiation using a 250- or 300-nm source. ΔF (pulse height of test site/pulse height of control) reached a maximum between 24 and 48 h for the 250-nm wavelength and by 24 h for the 300-nm wavelength. Again, the rise in ΔF at 300 nm persisted through 144 h and was significant at all time intervals to 72 h. Blood volume, represented by ΔV (corrected voltage at control site/corrected voltage at test site), showed the same rise between 24 and 48 h for both wavelengths. However, no statistically significant difference was found between exposed and unexposed skin due to wide variation in both groups.

As stated above, the authors' conclusion to the overall study was that erythema from both parts of the spectrum may be mediated by identical mechanisms, whether direct or via the release of a vasodilator material.[14]

IV. LASER DOPPLER VELOCIMETRY MEASUREMENTS IN HUMAN SKIN

The measurement of microcirculatory flow obtained with the laser Doppler velocimeter is dependent on both the concentration of moving red blood cells and their mean velocity.[25]

A. UVA RADIATION-INDUCED ERYTHEMA

According to the results of Drouard et al.,[10] irradiation with a single 15 J/cm² dose of UVA alone (wavelength = 320 to 400 nm) elicited no increase in laser Doppler velocimetry (LDV)-assessed perfusion and did not cause perceptible erythema in the subjects. Repeated exposure to 10 J/cm² of UVA for 6 days followed by 10 days without exposure and then by a single exposure of 15 J/cm² caused no more perfusion changes detectable by LDV.[10]

B. UVB AND UVC RADIATION INDUCED ERYTHEMA
1. Single-Exposure Studies

Farr and Diffey assessed the vascular response of human skin to 254- and 300-nm light using reflectance measurements and the LDV technique. In a *dose-response* study they plotted the

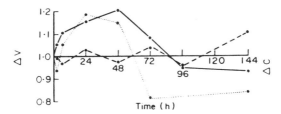

Figure 2 Changes in blood volume in the human skin (photoelectric plethysmograph) following irradiation with a dose of 5 × MED. Symbols as in Figure 1. See text for details. (From Ramsay, C. A. and Challoner, A. V. J., *J. Invest. Dermatol.*, 94, 487, 1976. With permission.)

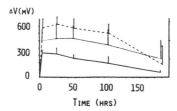

Figure 3 Results from a single exposure of human skin to a dose of 15 J/cm^2 UVA + 4 × MED UVB + UVC. Mean LDV-assessed cutaneous perfusion changes (ΔV) and subjective erythema gradings (± SD) are shown for six subjects as a function of time. (——) Capillary perfusion monitor (CPM)-assessed perfusion; (....) laser Doppler flowmeter (LDF)-assessed perfusion; (----) visual erythema grading. (From Drouard, V., Wilson, D. R., Maibach, H. I., and Guy, R. H., *J. Invest. Dermatol.*, 83, 188, 1984. With permission.)

increase in blood flux (obtained by LDV) and the increase of erythema index (obtained by reflectance) against the logarithm of the UV ray dose. An approximately linear response was obtained for both, with a considerably greater slope for 300-nm than for 254-nm radiation at all times of observation after irradiation.[21] This was confirmed by Young et al. using a UVB source with a threshold dose of 0.5 × MED.[9]

Concerning the *time course,* the maximum response was attained by Farr and Diffey between 8 and 20 h after irradiation and was followed by a gradual decline at both 254 and 300 nm.[21]

Wulf et al.,[27] on pig skin exposed to simulated sunlight, and then Young et al., on human skin submitted to 2 × MED UVB, found a peak at about 6 h postirradiation.[9] Drouard et al., using a combination of 15 mJ UVA with 4 × MED of UVB + UVC, saw a peak around 4 h after irradiation.[10] (See Figure 3.)

Relating blood flow measurements and erythema using the principles of fluid mechanics, Farr and Diffey observed that the increase in flux varied with the increase of erythema in a similar quadratic fashion in response to both 254- and 300-nm irradiation. They deduced that the mean red blood cell velocity increased significantly with both 254- and 300-nm irradiation and that there was no qualitative difference in the vascular response. This study led them, as indicated above, to the conclusion that the same blood vessels are involved in the causation of both UVB and UVC erythema.[21]

In exploring white and colored skins using UVB, Young et al.[9] showed an increase in MED with skin pigmentation by a factor of 14, similar to that previously reported by Cripps.

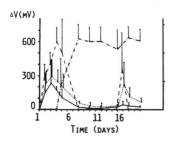

ΔV(mV)

600

300

0

1   6   11   16

TIME (DAYS)

Figure 4 Results from a multiple-exposure experiment. Details and symbols as in Figure13–3. (-.-.-.) Visual tanning grading. Laser Doppler velocimetry (LDV) blood flow values (in millivolts) are given on the left-hand ordinate, erythema and tanning ratings on the right-hand side. (From Drouard, V., Wilson, D. R., Maibach, H. I., and Guy, R. H., *J. Invest. Dermatol.*, 83, 188, 1984. With permission.)

Dose-response curves for pigmented skin have a tendency toward less steep slopes as compared to white skin.[9]

2. Repeated-Exposure Studies

Drouard et al.[10] used two different LDV devices to measure blood flow changes during repeated exposure to 10 J/cm^2 of UVA + 1 × MED UVB + C for 6 days followed by a rest of 10 days and then a single exposure of 15 J/cm^2 of UVA + 4 × MED UVB + C.[10] Results are shown in Figure 4.

It was observed that when the skin was exposed daily to 1 × MED of UVB + C the change in cutaneous perfusion reached a maximum on the third day. This implied initial cumulative damage followed by the onset of photoprotective changes in the skin with repeated exposure.

Then they showed a significantly smaller perfusion change for five daily doses of combined UVA, UVB, and UVC (10 J/cm^2 + 1 × MED) than for a single, larger exposure (15 J/cm^2 + 4 × MED). In addition, when the skin was exposed repeatedly and tanning developed, reexposure to the larger radiation produced no significant increase in cutaneous blood flow; this confirmed the well-known fact that tanning is protective.

V. CONCLUSION

Laser Doppler velocimetry and photoplethysmography are two valuable noninvasive techniques to quantify objectively ultra-violet radiations changes in human skin—even on pigmented skin. As shown in this chapter, they are two of the means by which researchers are able to separate the factors that lead to UV erythema production.

REFERENCES

1. Bastrenta, F., La Photosensibilisation Médicamenteuse: Approche des Mécanismes d'Action et Contribution à l'Étude des Moyens d'Exploration par un Test In Vitro Utilisant un Modèle Bactérien, M.D. thesis, Université Scientifique et Médicale de Grenoble, Grenoble, 1984.
2. Muel, B., Averbeck D., and Doré, J. F., Données expérimentales, in *Soleil et Mélanome: Analyse des Risques de Cancers Cutanés Moyens de Prévention,* Doré, J. F., Muir, C. S., and Clerc, F., Eds., Inserm, Nancy, 1990, chap. 3.
3. Le Bozec, P. and Belaïch, S., *Les Causes du Vieillissement Cutané,* Meunier, L., Michel, B., Duntze, F., and Meynadier, J., Eds., Sauramps Medical, Montpellier, 1990.
4. Kaoual, J., Substances Photosensibilisantes: Techniques d'Étude chez l'Homme et l'Animal, D.Ph. thesis, Faculté de Pharmacie, Nantes, 1985.
5. Wilkinson, J. B. and Moore, R. J., *Harry's Cosmetology,* Chemical Publishing Company, New York, 1982, chap. 15.

6. Girard, J., Unkovic, J., Delahayes, J., and Laffite, C., Etude expérimentale de la phototoxicité de l'essence de bergamote. Corrélations entre l'homme et le cobaye, *Dermatologica,* 158, 229, 1979.

7. Dubertret, L., Les Méfaits du soleil, *Recherche,* 223, 904, 1990.

8. Kreps, S. I. and Goldenberg, R. L., Suntan preparations, in *Cosmetics: Science and Technology,* Vol. 1, Balsam, M. S. and Sagarin, E., Eds., Interscience, New York, 1972, chap. 7.

9. Young, A. R., Guy, R. H., and Maibach H. I., Laser Doppler velocimetry to quantify UV-B induced increase in human skin blood flow, *Photochem. Photobiol.,* 42, 385, 1985.

10. Drouard, V., Wilson, D. R., Maibach, H. I., and Guy, R. H., Quantitative assessment of UV-induced changes in microcirculatory flow by laser Doppler velocimetry, *J. Invest. Dermatol.,* 83, 188, 1984.

11. Wan, S., Parrish, J. A., and Jaenicke, K. F., Quantitative evaluation of ultraviolet induced erythema, *Photochem. Photobiol.,* 37, 643, 1983.

12. Farr, P. M. and Diffey, B. L., Quantitative studies on cutaneous erythema induced by ultraviolet radiation, *Br. J. Dermatol.,* 111, 673, 1984.

13. Challoner, A. V. J., Accurate measurement of skin blood flow by a thermal conductance method, *Med. Biol. Eng.,* 13, 196, 1975.

14. Ramsay, C. A. and Challoner, A. V. J., Vascular changes in human skin after ultraviolet irradiation, *J. Invest. Dermatol.,* 94, 487, 1976.

15. Finsen, N. R., *Ueber die Bedeutung der Chemischen—Strahlen des Lichts für Medecin und Biologie,* Vogel, Leipzig, 1899.

16. Lewis, T., *The Blood Vessels of the Human Skin and their Responses,* Shaw and Sons, London, 1927.

17. Lewis, T. and Zotterman, Y., Vascular reactions of the skin to injury. VI. Some effects of ultraviolet light, *Heart,* 13, 203, 1926.

18. Van der Leun, J. C., Ultraviolet Erythema. A Study of Diffusion Processes in Human Skin, Ph.D. thesis, Utrecht, 1966.

19. Van der Leun, J. C., On the action spectrum of ultraviolet erythema, in *Research Progress in Organic, Biological and Medicinal Chemistry,* Vol. 3, Part 2, Santamaria, L. Ed., Elsevier/North-Holland, Amsterdam, 1972, 711.

20. Breit, R. and Kligman, A. M., Measurement of erythemal and pigmentary responses to ultraviolet radiation of different spectral qualities, in the *Biologic Effects of Ultraviolet Radiation,* Urbach, F., Ed., Pergamon Press, Oxford, 1969, pp. 267–275.

21. Farr, P. M. and Diffey, B. L., The vascular response of human skin to ultraviolet radiation, *Photochem. Photobiol.,* 44, 501, 1986.

22. Challoner, A. V. J. and Ramsay, C. A., A photoelectric plethysmograph for the measurement of cutaneous blood flow, *Phys. Med. Biol.,* 19, 317, 1974.

23. Crockford, G. W., Hellon, R. F., and Heyman A., Local vasomotor response to rubefacients and ultraviolet radiation, *J. Physiol.,* 161, 21, 1962.

24. Ramsay, C. A. and Cripps, D. J., Cutaneous arterioral dilatation elicited by ultraviolet irradiation, *J. Invest. Dermatol.,* 54, 332, 1970.

25. Stern, M. D., Lappe, D. L., Bowen, P. D., Chimosky, J. E., Allen Holloway, G., Jr., Keiser, H. R., and Bowman, R. L., Continuous measurement of tissue blood flow by laser-Doppler velocimetry, *Am. J. Physiol.,* 232, H441, 1977.

26. Schulze, R., *Strahlentherapie,* 86, 51, 1951.

27. Wulf, H. C., Staberg, B., and Eriksen, W. H., Laser Doppler measurement of blood flow in pig skin after sunscreens and artificial sunlight, *Photochem. Photobiophys.,* 6, 231, 1983.

Product Testing

Chapter 14

Irritancy

Edith M. De Boer and Derk P. Bruynzeel

CONTENTS

ABSTRACT: For the induction of skin irritation different schemes of open and occlusive tests, single or repeated, are being used at various test sites. Erythema as a reactive increase in skin blood flow can be used as a parameter for skin irritation. Evaluation of increase in SBF with LDF is a noninvasive, reliable, not too time-consuming method that is more sensitive than the naked eye. Pitfalls in the use of LDF are differences in local SBF, requiring exact repositioning of the probe at the test site in repeated tests without influencing the test itself. Furthermore, there is considerable regional variation in basal SBF, and there are indications that skin reactivity shows regional variation as well. Therefore, it seems advisable to vary the site of application of test substances and measure a daily control site. A clear definition of the value taken as SBF and of the response to the challenge should be given by each author.

SLS is a substance widely used to provoke irritant reactions. Unlike in most other irritants, the maximum inflammatory response to a challenge with SLS takes a long time to develop. After a single exposure of 24 h the maximum reaction is reached after about 48 h. There is a clear relationship between dose of SLS and skin response. Irritancy tests evaluated with LDF have been performed in healthy volunteers and atopics in the search for determination of ''sensitive'' skin. In studies on the irritant capacity of soaps and detergents LDF appeared to be useful. The results of studies with these kinds of irritants are highly dependent on the test method, especially the manner and duration of exposure.

LDF is of use for the evaluation of the weal and flare reaction due to intradermal injections with vasoactive drugs and the effect of previously administered histamine blockers. The irritancy of several drugs, such as tretinoin, corticosteroids, or trans-dermal drug delivery systems, free fatty acids, and moisturizers has been investigated in various tests using LDF. In the investigation on skin challenge due to industrial products LDF also proved its value.

In conclusion, in the evaluation of skin irritation due to very different stimuli, LDF has proven its value. It should be borne in mind that aspects of irritation other than increase in SBF require other methods of recording. Nowadays many investigators combine several bioengineering methods.

I. BACKGROUND OF PRODUCT EFFECTS

The skin is an important portal of entry for toxic substances and is therefore a vulnerable organ. It is our scientific assignment to find test methods that predict the possible toxic effects of substances on the skin. This is especially difficult for substances with a low irritative potential, to which the skin only reacts after repetitive contact.

The penetration of substances through the skin is thought to be a passive diffusion with a penetration rate dependent on innumerable factors, such as the concentration and molecular weight of the substance, skin thickness, temperature of the skin, humidity of the environment, duration of exposition, pH, and skin blood flow (SBF).[1]

Many tests have been proposed for the assessment of skin irritation. It is difficult to determine the sensitivity of the skin for several reasons, including the following:

1.   The interindividual susceptibility to irritants is highly divergent and not constant. Age, environmental conditions, general health, sex, and race influence personal sensitivity.[2-10] Also, intraindividual reactions to the same irritants may be widely diverse in strength depending on microclimate, sweating, perspiration insensibilis, temperature, and relative humidity.[3,5]

2.   From the intensity of a skin reaction to a primary irritant no conclusion can be drawn as to the reactivity to some other primary irritant.[3,11]

3.   Irritants damage the skin in various ways. For instance, denaturation of proteins and enzymes or damage to phospholipids of cell membranes may occur, causing changes in subepidermal blood flow as well as water loss or disturbance of electric resistance. The kind of such pathophysiological change depends on, e.g., the chemical properties of the applied irritant, its concentration, the vehicle, the manner of application, the exposure period, and the area of application. Therefore, tests for one group of irritants may be unsuitable for the evaluation of other substances.[3] Most experiments are done on humans. A problem when using animals for irritancy tests is that there is not always a clear correlation between the susceptibility of animal skin and human skin.[5,12]

Knowledge of the specific clinical reactions to substances on the skin makes interpretation of the toxicity/irritancy tests easier. The most common reaction of the skin is an inflammation, which is mostly of eczematous character. The primary irritant reactions of the skin might be divided into four main types.[1]

1. Local inflammatory response to a single exposure followed by complete healing
2. Local inflammatory response to repeated exposure with low-grade irritants
3. Chemical destruction with irreversible damage of tissue
4. Phototoxic reaction

For the evaluation of these irritant reactions laser Doppler flowmetry (LDF) has gained an important place in the set of diagnostic tools.

In the study of skin irritancy, different purposes of investigations have to be distinguished. Sometimes the overwhelming data from the literature are confusing, and some types of tests are used for more than one purpose. Principally, three topics can be distinguished *in vivo:*

1. Tests to induce irritation of the skin
2. Tests to determine the individual sensitivity to one or more irritants
3. Tests for determination of the degree of skin irritation

Examples of methods for the induction of skin irritation are open tests and closed tests. This chapter will discuss the ways these methods have to be used to elicit the kind of reaction the investigator wants to study.

Tests for the assessment of skin sensitivity (skin type assessment) of individuals include the alkali tests and the assessment of transepidermal water loss (TEWL).

Evaluating the skin reaction, whether caused by natural or artificial irritation, may be done with or without instruments, invasive and noninvasive. This chapter will be limited to macroscopic evaluation of erythema and LDF. LDF is nowadays an appreciated, widely used, noninvasive, sensitive method for the determination of skin irritation.

In vitro similar investigations are possible. On cultured human epidermal cells the effects of toxic stimuli can be investigated macroscopically, with instrumentation, and on the cellular level.[13] As mentioned, the product effect is highly dependent on the way the irritation is evoked. There are many ways irritation can be induced. The most commonly used methods of induction will be discussed shortly.

A. METHODS FOR INDUCING IRRITANCY
1. Open Tests

A nonocclusive way of testing irritants is generally more similar to exposure to irritants in normal life and during work than a patch test. A standardized variation of an open test is the repeated open application test (ROAT).[14] This test, which was originally used to detect allergic reactions, can be performed to induce irritation as well. This method consists of the application of about 0.1 ml of test material twice daily for 7 days to the flexor aspect of the forearm near the cubital fossa, to an area about 5 cm × 5 cm. Follow-up (visual score) is performed after 1 week. Repeated open daily application of chemicals is a method used in tests both on humans and on animals.[6,7,12] Using daily exposure the cumulative effect of substances can be studied. Sometimes the phenomenon of accommodation occurs.

Open application of chemicals is used in the induction of skin irritation both in humans and in animals.[12,15]

2. Closed (Chamber) Tests

For testing weak irritants a method with longer exposure or more intense contact is necessary. Accordingly, the chamber tests were developed. The chambers ensure occlusion, uniform exposure, and nonleakage and are therefore useful in testing weak irritants such as soaps and synthetic detergents. Different types of chambers are used, e.g., the Duhring chamber, the big Finn chamber, and the Van der Bend square chamber.

a. Duhring Chamber

This chamber consists of a round aluminum cup with an elevated flange. The maximal capacity of the chambers is 250 μl. Frosch and Kligman[16] investigated the optimal volume for patch testing. They obtained the best results with irritants by using 100 μl of solution and 0.1 ml of ointment. With these volumes a uniform reaction in a sharply demarcated area was evoked without problems of leakage.

b. Finn Chamber

This chamber also consists of a round aluminum cup with an elevated flange, but is smaller than the Duhring chamber. The correct dose for a standard Finn chamber is 20 μl.[17] Frosch and Kligman compared the reactions of irritants with Duhring chambers and Finn chambers and found uniformly more intense reactions with the Duhring chamber. In their opinion the volume of test agent per square millimeter in a Finn chamber is too small to provide reliable results.[16] The area of contact in irritancy tests must not be too small or a reaction will fail to develop.

A variation of this chamber is the big Finn chamber, with an inner diameter of 12 mm and a volume of 86 μl. Saturation of the filter paper disk is reached with 50 μl. This increased size overcomes the problem with the too-small standard Finn chamber.

c. Van der Bend Square Chamber

This chamber consists of a square polyethylene chamber with an elevated flange, containing Whatman® filter paper (1 × 1 cm), mechanically fixed without glue in each chamber. The maximal capacity of the chambers is 100 μl and the optimal volume 30 μl (observation of author).

3. Exposure Scheme for Chamber Tests

For irritation tests repeated application is useful. Frosch and Kligman developed a scheme of application for 24 h on the first day and 6 h on the 4 following days with scoring on the 8th day. They found this a satisfying method to evaluate skin irritancy.[16,18] In a study on the irritancy of soaps, Frosch and Kligman[19] used this 8-day chamber test, applying an 8% soap solution. In off periods the skin was not protected.

Shellow and Rappaport[20] also performed tests on the irritancy of soaps. They use a variation of the test of Frosch and Kligman consisting of a 5-day soap chamber test with end scoring on day 5. Their results were in reasonable accord with the 21-day irritation patch test of Lanman et al.[21]

In a study on the toxicity of hand cleaners, van Ketel et al.[22] used a 5-day test with big Finn chambers, applying a 20% soap solution. On day 1 the time of exposure was 24 h and on the next 3 days it was 12 h. Another group of volunteers was examined with an 8-day test with an 8% soap solution; the protocol consisted of exposure of 24 h on day 1 and 6 h on the following 4 days and scoring on day 8. There was a reasonable accord between the two test methods. A disadvantage of an 8-day test is that it is more inconvenient to the volunteers since they have to keep the test area dry in the weekend as well and have to return for one more examination.

B. EVALUATION OF SKIN TESTS

For the determination of contact allergy an internationally accepted system of application and scoring has been established.[23] The situation is quite the opposite for the determination of skin irritation. Wide variations in the duration of exposure, frequency of exposure, manner of application, and evaluation of the results have been tried out in an effort to develop a reliable method for the investigation of irritants. For each test endless variations have been attempted.

Minor insults of the skin resulting in subclinical damage are difficult to quantify visually. Mild irritation presents clinically as a dry and rough aspect of the skin with fine cracks (chapping), erythema, and sometimes edema, papules, and wrinkling of the skin.

Frosch and Kligman[18,19] evaluated tests by a visual score, grading for erythema, scaling, and fissuring on the 8th day. The scores of erythema, scaling, and fissuring can be summed and used in a final score. A disadvantage of a visual score is the subjectivity of the observer.

In experiments of short duration mild reactions usually occur, showing predominantly erythema. For the evaluation of subtle, mildly irritant reactions the degree of erythema, as a reactive increase in SBF, can be used as a parameter.[24-26] The erythema can be evaluated by a visual score, as shown above, and by measuring SBF by means of LDF, which is a very direct method. A later section of this chapter will be devoted to one aspect of skin irritation: erythema.

LDF, a noninvasive, direct technique for recording SBF, is a reliable, easy-to-handle procedure for cutaneous blood-flow measurements. The method is not too time-consuming. It is important to realize that there are factors that can affect the interpretation of the signal, such as signal processing, bandwidth, calibration, movement artifacts, the use of different types of lasers, and the effect of the probe (holder) on the skin. Recently, these hazards have been reviewed[27] and guidelines for measurement of cutaneous blood flow by LDF given.[28] In a publication it should be clearly indicated in which way the LDF has been used. A standardized method should be used for LDF in irritancy tests.[29] With this method repeated measurements can be performed in such a way that recording does not influence the test itself. The method is not sensitive to slight changes in ambient conditions, does not require extremely accurate repositioning of the probe, and prevents the influence of possible differences in skin reactivity at different skin sites.

II. PRODUCT TESTING STRATEGIES USING THE LASER DOPPLER FLOWMETER

In a review of data from the literature about investigations on skin irritation evaluated by LDF it is striking that all observers use various ways of application of the irritants (location and duration, open or patch tests, single and repeated application). The duration and way of recording SBF with the LDF differs among investigators, as does the interpretation of the thus acquired readings. In repeated measurements faults can be made easily with the replacing of the probe in the conventional probe holder. Comparison with controls is likewise variable. It is therefore difficult to compare the results of different investigators and maybe even the results of different experiments by the same investigators.

A. PITFALLS IN IRRITANCY TESTS

Because of considerable local variation of SBF and the small area of recording, interpretation of results must be performed cautiously. Recordings on different spots are highly influenced by this local variation. Repeated measurements on the same spots require accurate positioning of the probe at exactly the same site.

The surface area of a standard probe commonly used for LDF is 1 mm^2, and one special laser Doppler flowmeter is equipped with a probe with surface area of 23 mm^2 (Perimed, Stockholm, Sweden).

Besides this local variation in cutaneous blood supply in adjacent areas there is also a regional variation. The face and fingertips are the most highly perfused areas.[30] Regional variation in SBF is usually encountered;[31,32] differences between the wrists and the proximal parts of the forearms can even be considerable.[33] According to our findings, proximal and distal parts of the arms show significant differences in blood flow, which is of importance when a series of substances is tested.[29] Furthermore, skin reactivity on the forearm skin is different in different areas, as has recently been demonstrated again on corticosteroid-induced skin blanching.[34] The skin of the forearm is very suitable for irritancy measurements because of its accessibility, whereas the skin of the back is thicker and less sensitive to irritation. The LDF is less accurate in measuring SBF in strong reactions, so the irritants tested should cause only a mild irritation of the skin.

It has been suggested[35,36] that pigmentation of the skin may absorb an appreciable amount of light from the LDF. The presence of coincidental flat pigmented elements such as freckles and nevi did not influence the test results.[29] Elevated (pigmented) nevi, however, may show a clear influence on SBF and should be avoided.

Reliable measurements for quantification of irritancy tests can be expected if the following procedure is used:

1. Use a probe holder that makes easy repositioning possible— e.g., one made of transparent silicone rubber.
2. The attachment of the probe holder to the skin should not influence the test area.
3. In case of measurements on the volar aspects of the forearms, a row of test sites along the longitudinal axis is advised. The upper arm and the back are also common test sites.
4. The site of application of the test substances should be varied, e.g., according to a Latin square.
5. A nontreated control site, for daily comparison, should be included.

The response of a test reaction can be defined as the flow value measured at a test site minus the flow value at the corresponding control site at the same time. In a publication it should always be clearly indicated which values are used and how the response to an irritant is defined.

B. SODIUM LAURYL SULFATE (NONANOIC ACID AND HYDROCHLORIDE)

Sodium lauryl sulfate (SLS) is a very common test substance used by virtually all workers in the field of skin irritation. SLS is widely used as an anionic detergent and wetting agent in shampoos, toothpastes, cosmetics, and pharmaceutical agents.[2] Because it does not seem to sensitize it is suitable for irritancy tests.[37] SLS causes a short but intense stimulus of the skin. The ensuing reactions are variable depending on the site of application and the concentration. The mechanism of action is not exactly known. Suggested mechanisms are a direct action on the stratum corneum components, such as protein denaturation, removal of lipids, or interaction with living epidermal cells to alter their proliferative capacity, and/or action on dermal components to release mediators of inflammation.[38]

Epicutaneous tests with irritants usually evoke skin reactions which slowly subside after removal of the patch at 24 h. The decrease in intensity is one of the features commonly used to distinguish an irritant test reaction from sensitization. Some irritants, however, show an optimal inflammatory response at a later time. Bruynzeel et al.[2] applied patch tests with aqueous solutions of low SLS concentrations for 24 h. Evaluation was done after removal of the patches and at 48 and 72 h using a visual score. These threshold irritant concentrations of SLS induced stronger reactions at 48 h than at 24 or 72 h, so the course of these reactions could not be distinguished from allergic reactions. Dahl et al.[10] performed similar threshold tests with SLS and found a great variation in individual and intraindividual site-to-site reactivity. Almost all investigators use SLS. Some of these publications are mentioned by the authors. In other experiments SLS is often used as a control.

Gabard[38] provoked irritation in five male volunteers by applying 120 µl of 5% aqueous SLS on the forearms under occlusion for 24 h and measuring SBF for the next 10 days. The LDF values increased, attaining their maximum between the first and the third day after removal of the patches, and then decreased slowly. Simultaneously, the color reflectance, TEWL, and skin capacitance were determined. TEWL showed a rapid increase and then a rapid decline. The color measurement was not in agreement with the visible erythema. The skin capacitance decreased much later and persisted longer.

Van Neste and co-workers performed many studies using SLS. They introduced the term "rough dermatitic skin" for the condition of the skin after experimental induction of irritation by occlusive application of SLS.[39] In six male volunteers 1%, 5%, and 10% aqueous solutions of SLS were applied in a fixed order for 48 h. Simultaneously, a clinical evaluation, TEWL, and SBF were recorded. No difference between right and left arms was observed. No significant difference was observed between 5% and 10% SLS. At 7 days SBF had normalized completely, before recovery of TEWL. There was a linear correlation between LDF and TEWL.

In another group of 21 volunteers[40] Van Neste performed a similar study. SBF measurements only reflect part of the secondary inflammatory changes. The combination of TEWL and SBF measurements gave a more complete recording of the skin damage. (A close relationship between the irritation and the applied concentration was found.) A significant linear regression between TEWL or SBF and log of SLS concentration was demonstrated. The pH of the test solution did not influence the SBF changes. A clear relationship between the applied dose of SLS and the corresponding increases in skin blood flow and erythema has been noted by others as well.[2,41,42] However, sometimes an inconsistent reaction pattern or even unpredictable vagaries in responses to irritants such as SLS have been reported.[10]

In a study on the monitoring of skin response to SLS Van Neste and De Brouwer[43] patch tested ten healthy volunteers with 5% aqueous SLS on the neck and volar forearm for 48 h. Monitoring was performed on each of the next 10 days with a visual score, LDF, TEWL, and skin capacitance. LDF values increased after the insult and gradually normalized in the test period. No significant influence of body site was seen, and there was no time-site interaction. TEWL was equal at both sites, and skin capacitance measurements showed significant differences.

Lammintausta et al.[44] performed repeated open tests on the backs of volunteers with several concentrations of SLS followed by a patch test. Assessment was done with a visual score, LDF, and TEWL. Open exposure to SLS prior to patch tests decreased the reactivity as measured with the visual score and LDF. TEWL values were inconsistent. A considerable variation at different test sites and test times was noted. They found that repeated open application produced a hyporeactive state of the skin which influenced the patch tests that followed.

Many studies on SLS have been performed by the group of Serup and Agner. In one study 12 volunteers were patch tested for 24 h with 60 µl of an aqueous solution of SLS (0.12, 0.25, 0.50, and 1.00%) on systematically varied sites on the flexor side of the upper arms.[45] Evaluation was done by visual scoring, LDF, and TEWL 1 h and 24 h after removal. A significant linear relationship between dose of SLS and skin response was found. Also, a delayed time course of irritation in threshold reactions was observed. In another study[46] 20 volunteers were patch tested on the upper arms with 60 µl of 2.5% aqueous SLS, 20% nonanoic acid in propanol, and 4% aqueous HCL for 24 h. Evaluation was performed the next day, 1 h after removal, and 72 h later by a visual score, LDF, TEWL, electrical conductance, and skin thickness. A significant increase in SBF was observed for each substance compared to the vehicle and an empty chamber. With LDF and skin thickness measurements no significant differences for the three irritants were seen, except for the recording 24 h after removal, when SLS scored highest. However, significant differences were found using TEWL and electrical conductance.

A topic of interest is, of course, individual susceptibility to chemical and physical stimuli. Is it possible to evaluate the risk of future dermatitis? Does a "sensitive skin" really exist? It is well known that a reaction to one irritant does not necessarily predict a reaction to another,[3,11] but efforts to designate individuals with sensitive skin have been made.[47] UV sensitivity shows a correlation with reactivity to some chemical irritants.[47] In the search for determination of sensitive skin SLS seems a useful irritant.

For this purpose[48] Agner and Serup patch tested 60 µl of 1.0% aqueous SLS and 10% nonanoic acid in propanol on each arm for 24 h in 19 volunteers. Evaluation was done 30 min and 24 h after removal by visual scoring, LDF, TEWL, electrical conductance measurements, and skin thickness. Significant correlations between right and left arm tests were found for visual scoring, LDF, and TEWL. A significant interindividual variation was found, probably due to true biological differences in skin susceptibility to irritants between different subjects. This may be useful in the differentiation between individuals with delicate and less sensitive skin. The intraindividual variation, comparing contralateral and symmetrical sites, was always less than the interindividual variation and may be attributed to biological "site-to-site" variation in skin conditions which are also present in unexposed skin. Electrical conductance measurements were not helpful for this purpose.

In another study on the same subject[49] 70 volunteers (nonatopics) were patch tested with 60 µl of 0.50% aqueous SLS for 24 h. Test reactions were evaluated 1 h after removal by a visual score, LDF, TEWL, skin thickness, and skin color. LDF was useful for the recording of irritation, but did not have predictive value. Basal transepidermal water loss (before exposure) appeared to be a predictor of susceptibility of skin barrier function to SLS, with a high basal TEWL indicating a higher susceptibility to SLS. Increased light reflection from the skin, indicating fair skin, pointed in the same direction. These data confirm previous studies suggesting that a high basal TEWL is associated with a higher probability of developing irritant contact dermatitis.[50,51]

In a study on the susceptibility of individuals with atopic dermatitis to SLS 28 patients were compared to 28 controls.[52] On the unaffected flexor side of the upper arms 60 µl of 0.5% aqueous SLS was applied for 24 h. Evaluation was performed 1 h after removal by visual score, LDF, TEWL, skin thickness, and skin color. Neither basal SBF nor postexposure SBF as significantly different in atopics and controls. Basal TEWL was increased in patients with atopic dermatitis. The skin response to SLS was statistically increased in atopics when evaluated by visual score and increase in skin thickness, but not by increase in TEWL, SBF, or skin color.

Repeat patch tests are less commonly performed. Freeman and Maibach[53] patch tested 11 volunteers with 100 µl of 2% aqueous SLS for 24 h on several anatomic sites. After 1 week the patch tests were repeated. Evaluation was done with a visual score, LDF, and TEWL on each of the next 4 days. An augmented response after the second challenge was seen. LDF proved to be of limited use since it produced widely fluctuating values in the presence of spotty erythema due to mild irritation.

Elsner et al.[54] performed irritancy tests with 3% SLS for 24 h not only in the forearm but also in the vulva. LDF showed a higher rise in forearm skin than in vulvar skin. At the same time TEWL, capacitance, pH, and color reflectance were determined. It was concluded that bioengineering techniques are less suitable to quantify irritant dermatitis in the vulva than in the forearm.

C. SOAPS AND SYNTHETIC DETERGENTS

Skin irritancy due to weak irritants such as soaps and synthetic detergents (syndets) is common in all kinds of professions. Estimation of the irritant capacity of cleaners on the skin is difficult. Exposure during work is usually different from the artificial exposure during irritancy tests. The most-used test sites are often not the areas exposed at work. Furthermore, the test provides for exclusive exposure to the irritant investigated. During work, additional factors—mechanical, chemical, and climatological—play a role. This limitation of the value of tests is also important in the investigation of other irritants, such as metalworking fluids.

Two types of liquid hand cleaners may be distinguished: soaps, made by the saponification of animal or vegetable fats, having an alkaline nature, and synthetic detergent cleaners, with a wide pH range. Detergents can also be classified according to chemical properties: anionic, cationic, amphoteric, and nonionic.[55] Soap is an example of an anionic detergent. Many investigators have performed endlessly varying tests on the effects of soaps and detergents on the skin. The purpose of most studies has been to find a "mild soap". It is impossible to

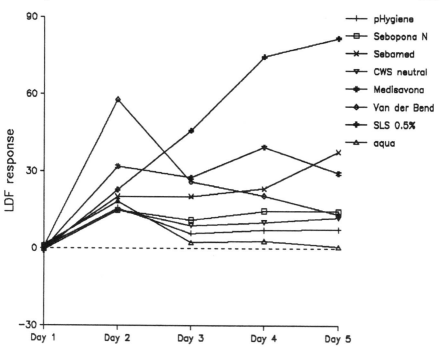

Figure 1 The mean laser Doppler flowmetry (LDF) response of all individuals for each substance on each day. See text for details of study. (The tested substances are liquid alkaline soaps and synthetic detergents.)

mention all these experiments and their results. Reviews are given by Frosch and Kligman[19] and Smeenk[56,57], who have performed extensive studies on soaps. Smeenk introduced immersion tests with soaps and other detergents used in housekeeping. Tupker[55] also did elaborate studies on the influence of detergents on human skin, assessing irritation by TEWL.

De Boer et al.[58] studied the irritancy of six liquid alkaline soaps and synthetic detergents on the Dutch market, especially advertised for "mildness". The test concentration was 20% *in aqua;* water, as a control, was tested as well. On the forearms 60 µl of the solution was applied according to a random scheme of positioning. Exposure was for 24 h on day 1 and 12 h on each of the following 3 days. Evaluation with a visual score for erythema, wrinkling, and shine and LDF was performed on days 2, 3, 4, and 5. The mean LDF response of all individuals, calculated each day for each of the test substances, is presented in Figure 1. Only mild reactions occurred.

LDF appeared to be a reliable method in the quantitation of irritant reactions due to repeated application of detergents. Discrimination of irritancy of various detergents could be achieved. LDF seemed somewhat more sensitive for small gradations of erythema than the naked eye. The visual score for erythema and the SBF values correlated well. However, other effects of soap-like irritants, such as shine and wrinkling (soap effect), are not detected by LDF, in contrast with the total visual score. These changes, though, are important in the development of irritant contact dermatitis. The test system itself and the methods of evaluation of the results used for irritancy tests with soaps appear to be questionable. It becomes clear that the results of each "soap study" are ambiguous and that at least incongruous, or perhaps even opposite, results can be obtained by changes in the scheme of application or the time of scoring. This necessitates the exact mentioning of test procedure and circumstances in each article on this subject. A standardization would be appreciable and would make comparison of different studies possible. Unnecessary overlap of efforts of investigators could be prevented.

D. VASOACTIVE DRUGS

After extensive studies on the reproducibility of skin prick tests with histamine in volunteers Van Neste[59] has drawn the conclusion that it is mandatory to state the precise place of measurement and time after challenge when reporting on instrumental evaluation of the skin response to histamine.

In a later study the effect of oral intake of two different H_1 blockers on the skin response to histamine was determined using LDF.[60] A decrease of response was seen and a difference in the effect of the two drugs could be demonstrated. A similar study on the effects of an orally taken H_1 antagonist and of pretreatment with a topical glucocorticoid on the response of SBF to histamine was done by Hammarlund et al.[61] SBF recordings were unaffected by the topical corticoid.

The vasodilator methyl nicotinate has been used to investigate differences in percutaneous absorption of the skin at different sites.[62] Measuring the increase of SBF after topical application a correlation between the appendageal density of the skin and the penetration rate was demonstrated. This model also can be used to evaluate the effect of barrier creams.[63]

E. TOPICAL DRUGS, COSMETICS, AND TOILETRIES
1. Tretinoin

On the forearms of five healthy volunteers twice a day during 4 weeks a cream containing 0.05% tretinoin (all-*trans*-retinoic acid) was applied and compared to the cream base.[38] LDF, done several times a week during 6 weeks, showed an increase in SBF after 2 to 3 weeks, but this increase was not significant. Interindividual variations were very large. Simultaneously, skin color reflectance, TEWL, and skin capacitance were recorded. The color measurements did not have the same time course as the erythema and SBF, probably due to the induced scaling. TEWL first showed a rise and then came back to normal values. The combination of the different measuring devices is helpful in the characterization of the various aspects of skin reactivity.

2. Sodium Bicarbonate

LDF has proven its value in the evaluation of the irritancy potential of chemicals in vulvar skin as well. In a study on the irritant potential of menstrual pads containing sodium bicarbonate as a deodorant the effect of the pads and of sodium bicarbonate alone on the skin was monitored by a visual score, LDF, TEWL, skin surface pH, and capacitance.[64] The pads did not significantly affect the vulva of 50 healthy volunteers with either method in short-term use.

3. Transdermal Delivery System

Nowadays more attention is drawn to the topical application of drugs to produce a systemic effect after diffusion through the skin. In six male volunteers an acrylic copolymer adhesive patch containing 20% timolol free base was applied on the forearm for 10 h.[65] Evaluation was done by a visual score, LDF, and hemoglobin index (measured by reflectance spectrophotometry), and additionally the plasma concentration of timolol was measured. A significant increase in SBF was recorded, related to the amount of timolol which diffused through the skin as determined in the plasma. Plasma concentrations in subjects who did not develop erythema were very low. LDF was of value for the determination of the variation of diffusivity among individuals.

4. Corticosteroids and Anti-Inflammatory Drugs

Vaanaanen and Hannuksela[66] studied the photoprotective and antierythematous effects of topical creams containing corticosteroids and anti-inflammatory drugs such as acetylsalicylic acid, indomethacin, butyl hydroxytoluene, and diphenhydramine on the erythema induced by UVB in healthy volunteers. For evaluation a visual score and LDF were done. When applied before irradiation acetylsalicylic acid and indomethacin inhibited UV erythema. When applied after irradiation indomethacin, diphenhydramine, and potent corticosteroids reduced UVB erythema. LDF appeared to be a useful tool.

5. Moisturizers

LDF is suitable in the study on the course of mild irritant reactions of the skin and the influence of moisturizers on the healing rate. In a study of the effect of eight moisturizers in preventing and treating irritant dermatitis, Hannuksela and Kinnunen[67] had 12 healthy female volunteers wash their arms with a dishwashing detergent for 1 min twice a day during 1 week. On one upper arm eight moisturizers were applied after each washing. The next week the moisturizers were applied on the other arm. Evaluation was done with LDF and TEWL. The untreated arm showed an increase in TEWL and LDF values. The simultaneously treated arm showed no change in SBF and a minor increase in TEWL, demonstrating that emollients prevented the development of irritant dermatitis. In the second week all values normalized. The differences, though statistically significant, were only slight. No discrimination could be made between the various moisturizers. It was concluded that the regular use of emollients prevented the development of irritant dermatitis from detergent usage. Previously, Blanken et al.[68] studied the influence of four after-work emollients on the healing of irritant skin reactions. On the forearms of 30 volunteers 5 patch tests with 2.5% aqueous SLS were performed (12, 24, and 48 h). Afterwards emollients were applied twice daily for 5 days. Skin irritation was assessed with LDF, TEWL, and a visual score. No significant effect on the regeneration rate could be demonstrated with either method.

Other studies have been performed on moisturizers and their possible irritant effect on the skin. In studies on the moisturizing effect of urea-containing creams, Serup[69] used several noninvasive techniques and found an efficient improvement of hydration and a reduction of scaling. In an irritancy study on urea 17 volunteers were patch tested for 24 h with 20% urea in petrolatum and in water.[70] Evaluation was done with LDF, TEWL, and ultrasound A-scan. It was concluded that the irritant effect of urea in petrolatum was significantly higher than that of urea in water. The latter was not more irritant than water alone. The three bioengineering methods were in agreement.

6. Free Fatty Acids

De Haan et al.[13] performed a comparison study of *in vitro* toxicity tests using cultured epidermoid cells (trypan blue test and mitochondrial metabolization test) and *in vivo* tests assessed with LDF and a visual score. Free fatty acids with chain length C6, C7, C9, C10, C11, C13, and C18 were used. Patch tests were done for 20 h on the forearms. Skin irritation increased with increasing chain length from C6 through C11 and decreased after C13. A good correlation between *in vitro* and *in vivo* tests was found except for the C13 fatty acid. LDF appeared to be useful in the discrimination of toxic effects on the skin. In a similar *in vitro* experiment an increase of toxicity of alcohols with increasing chain length was demonstrated.[71]

F. INDUSTRIAL PRODUCTS

Many industrial products may affect the skin. In most cases the repeated and accumulated minor insults of mechanical, chemical, or thermal origin are responsible for the development of occupational dermatitis. The cases of irritant contact dermatitis induced in this manner far outnumber the cases of contact sensitization.

1. Rock Wool

Rock wool is a man-made mineral fiber used for insulation which can cause mechanical skin irritation. In a study on the irritant potential of rock wool of different diameters, 2 mg of test substance was applied under occlusion for different time periods.[72] Evaluation was done with a clinical score and LDF. It was concluded that LDF was a useful experimental tool for the evaluation of irritant patch test responses to mechanical irritants.

2. Metalworking Fluids

Irritant contact dermatitis is common in metalworkers exposed to metalworking fluids (MWF). The irritancy of three commercially available water-based fluids (cutting fluids) in maximal user's concentration and two neat oils was investigated in 23 healthy volunteers.[73] Additionally, the irritancy of several groups of components of the cutting fluids was studied. After cellophane tape stripping, repeated patch tests were performed during 5 days and evaluated by a visual score and by LDF according to a standard method.

In general, the MWF caused marginal skin irritation. In a comparison of the MWF and water, the cutting fluids were ranked higher in irritancy than the neat oils, but the effect of only one cutting fluid could be distinguished from that of water. The components of the cutting fluids also caused no strong reactions. The LDF response showed that one emulsifier was more irritant than water, and according to the visual score one corrosion inhibitor was more irritant.

From this experiment it was concluded that MWF are generally only mildly irritant when used in proper concentrations. Irritant contact dermatitis, however, develops as a result of repeated and cumulative subclinical insults in which MWF, especially those that are water based, play a crucial role because exposure at work is often extensive.

G. MISCELLANEOUS
1. Nickel Sulfate vs. Chloride

LDF can be used as a sensitive method to determine the optimal test preparation for diagnosing contact sensitivity. The least irritant concentration, chemical variant, or vehicle can be demonstrated.

For routine patch testing, nickel sulfate is used for diagnosing nickel contact sensitivity. As an alternative, nickel chloride has been suggested. Wahlberg[74] used LDF to demonstrate in healthy volunteers that the chloride was more irritant to the skin than the sulfate, with petrolatum as vehicle. The naked eye could not discern this difference. LDF is useful for the demonstration of minor skin irritation.

2. Water

Water is often used as a solvent for substances tested, and therefore patch tests with water can be used as a control. The results of tests with water are incongruous. After 24 or 48 h of occlusion no significant changes in SBF have been reported,[40,46] but on the other hand positive reactions to patch tests with water *have* been reported.[58,69] For the evaluation, in order to get reproducible values it is very important to perform measurements some time after removal of the occlusive material.

3. PUVA Inhibition of Immediate Contact Reactions

LDF can be used for the evaluation of blood flow in studies on the effect of UV therapy on contact reactions. Larmi[75] studied the effect of trioxsalen bath psoralen-UVA (PUVA) therapy on nonimmunologic immediate contact reactions. In 12 dermatologic patients the effect of various concentrations of benzoic acid and methyl nicotinate on PUVA-treated and nontreated sites on the back was studied. Evaluation of edema and erythema was done with a visual score and LDF. PUVA therapy appeared to suppress the nonimmunologic immediate contact reactions on the PUVA-exposed area and occasionally on the nonexposed test sites.

III. COMMENTS

LDF is a reproducible, noninvasive method for measuring SBF, being more sensitive than the naked eye. It is a useful tool for the assessment of increases of SBF as a measure of inflammation—e.g., due to irritation. Other aspects of irritation are not detected by LDF, and therefore many investigators use it in combination with other bioengineering methods.

Since there are many ways in which investigators use this apparatus, one should be very alert in comparing results of different studies. The values given as SBF as a result of LDF recordings vary. Van Neste and associates[25,39,40] record for 60 s and take the mean level as the SBF value. Agner and Serup[45,48] use the mean of three recordings of unknown duration after stabilization of the level. Gabard[38] records for 30 s and calculates the average. Lammintausta et al.[6] calculate the mean between the highest and lowest values measured after 60 s of stabilization. De Boer and co-workers[29,42,58] calculate the mean of a recording period of 60 s after about 60 s of stabilization.

The reaction to an irritant is measured by the change in SBF. The response to an irritant might be defined as the SBF value measured at the test site minus the value the same day at a control site.[42,58] It is also possible to compare to the vehicle. Since SBF is not constant in time it is advisable to take a control value each measuring day.

As mentioned before, the exact way an irritant challenge is performed can influence the result. Furthermore, the sensitivity to irritants may show racial differences. There is some evidence that blacks are less susceptible to cutaneous irritants than whites, but the reverse also has been demonstrated. Reviews are given by Berardesca and Maibach[76] and Wilhelm and Maibach.[77] In a study on white and Hispanic subjects similar microcirculatory responses to irritation were seen, but other bioengineering methods demonstrated differences in barrier function.[78]

A putative difference between irritant reactivity in males and females was investigated by Lammintausta and associates.[6] Neither repeated daily open applications of SLS nor patch tests could demonstrate a significant sex-related difference. No age-dependent susceptibility to primary irritants seems to be present.[11] For a good interpretation of irritancy tests the examined group should be described carefully. Because there is a marked regional difference in reactivity to irritants it is also mandatory that the exact site of investigation is indicated.

It would be of common interest to all investigators working on skin irritancy to come to a consensus regarding the way these experiments should be performed. In research on contact sensitization such a consensus was reached years ago. Although the mechanism of skin irritation is of course far more complicated, we think that it might be possible to come to some agreement on the assessment of skin irritation. Several different types of experiments will always be necessary since different substances cause irritation via various mechanisms. In the case of LDF it does not seem to be too complicated to come to an agreement on its use in these experiments—for instance, which value should be taken as the SBF value and which value should be noted as the response to a challenge.

A problem when investigating mild irritants is provoking sufficient irritation for evaluation. Pretreatment of the test area to provide for increased exposure to the irritant can be assessed in several ways. For instance, cellophane tape stripping of the stratum corneum is used. This causes a nonstandardized, variable effect on the skin. Ghadially et al.[79] introduced a more controlled pretreatment. The test site is rubbed with acetone-soaked cotton until a certain rise in TEWL is reached. Unfortunately, we could not reproduce these results. It might be interesting to induce in the same manner a certain rise of SBF prior to exposure to the mild irritant to be tested.

The conclusion can be drawn that LDF is an adequate tool in the evaluation of skin irritation of various origins and that immense fields are waiting to be explored. Cooperation of investigators on the subject will probably have a positive effect on future developments. Combinations of methods for the determination of several aspects of skin irritation will provide more insight into this complicated matter.

REFERENCES

1. Suskind, R. R., Short-term toxicity tests for non-genotoxic effects, in *The skin: Predictive Value of Short-Term Toxicity Tests,* Bourdeau, P., Ed., John Wiley & Sons, New York, 1990, chap. 11.

2. Bruynzeel, D. P., Van Ketel, W. G., Scheper, R. J., and Von Blomberg-van der Flier, B. M. E., Delayed time course of irritation by sodium lauryl sulfate: observations on threshold reactions, *Contact Dermatitis,* 8, 236, 1982.

3. Björnborg, A., Skin Reactions to Primary Irritants in Patients with Hand Eczema, Ph.D. thesis, Göteborg, 1968.

4. Bruynzeel, D. P., Angry Back or Excited Skin Syndrome, Ph.D. thesis, Free University Amsterdam, 1983, chap. 5.

5. Gloor, M., *Pharmakologie dermatologischer Externa,* Springer-Verlag, Berlin, 1982, chap. 10.

6. Lammintausta, K., Maibach, H. I., and Wilson, D., Irritant reactivity in males and females, *Contact Dermatitis,* 17, 276, 1987.

7. Lammintausta, K., Maibach, H. I., and Wilson, D., Susceptibility to cumulative and acute irritant dermatitis. An experimental approach in human volunteers, *Contact Dermatitis,* 19, 84, 1988.

8. Berardesca, E. and Maibach, H. I., Sodium-lauryl-sulphate-induced cutaneous irritation. Comparison of white and hispanic subjects, *Contact Dermatitis,* 19, 136, 1988.

9. Goh, C. L. and Chia, S. E., Skin irritability to sodium lauryl sulphate—as measured by skin water vapour loss—by sex and race, *Clin. Exp. Dermatol.,* 13, 16, 1988.

10. Dahl, M. V., Paso, F., and Trancik, R. J., Sodium lauryl sulfate irritant patch tests. II. Variations of test responses among subjects and comparison to variations of allergic responses elicited by Toxicodendron extract, *J. Am. Acad. Dermatol.,* 11, 474, 1984.

11. Coenraads, P. J., Bleumink, E., and Nater, J. P., Susceptibility to primary irritants, *Contact Dermatitis,* 1, 377, 1975.

12. Anderson, C., Sundberg, K. and Groth, O., Animal model for assessment of skin irritancy, *Contact Dermatitis,* 15, 143, 1986.

13. De Haan, P., Heemskerk, A. E. J., Gerritsen, A., De Boer, E. M., Sampat, S., Van der Raay-Helmer, E. M., and Bruynzeel, D. P., Comparison of toxicity tests on human skin and epidermoid (A431) cells using free fatty acids as test substances, *Clin. Exp. Dermatol.,* 18, 428, 1993.

14. Hannuksela, M. and Salo, H., The repeated open application test (ROAT), *Contact Dermatitis,* 14, 221, 1986.

15. Frödin, T. and Anderson, C., Multiple parameter assessment of skin irritancy, *Contact Dermatitis,* 17, 92, 1987.

16. Frosch, P. J. and Kligman, A. M., The Duhring chamber. An improved technique for epicutaneous testing of irritant and allergic reactions, *Contact Dermatitis,* 5, 73, 1979.

17. Andersen, K. E., Burrows, D., and White, I. R., Allergens from the standard series, in *Textbook of Contact Dermatitis,* Rycroft, R. J. G., Menné, T., Frosch, P. J., and Benezra, C., Eds., Springer-Verlag, Berlin, 1992, chap. 13.

18. Frosch, P. J., Irritancy of soaps and detergents in bars, in *Principles of Cosmetics for the Dermatologist,* Frost, P. and Horwitz, S. N., Eds., C. V. Mosby, St. Louis, 1982, chap. 1.

19. Frosch, P. J. and Kligman, A. M., The soap chamber test. A new method for assessing the irritancy of soaps, *J. Am. Acad. Dermatol.,* 1, 35, 1979.

20. Shellow, W. V. R. and Rapaport, M. J., Comparison testing of soap irritancy using aluminum chamber and standard patch methods, *Contact Dermatitis,* 7, 77, 1981.

21. Lanman, B. M., Elvers, W. B., and Howard, C. S., The case of human patch testing in a product development program, in *Proc. Joint Conf. Cosmetic Science,* The Toilet Goods Association, Washington, D.C., 1968, 135.
22. Van Ketel, W. G., Bruynzeel, D. P., Bezemer, P. D., and Stamhuis, H. I., Toxicity of handcleaners, *Dermatoligica,* 168, 94, 1984.
23. Fregert, S., *Manuel of Contact Dermatitis,* 1st ed., Munksgaard, Coppenhagen, 1974.
24. Staberg, B. and Serup, J., Allergic and irritant skin reactions evaluated by laser Doppler flowmetry, *Contact Dermatitis,* 18, 40, 1988.
25. Van Neste, D., Masmoudi, M., Leroy, B., Mahmoud, G., and Lachapelle, J. M., Regression patterns of transepidermal water loss and cutaneous blood flow values in sodium lauryl sulfate induced irritation: a human model of rough dermatitic skin, *Bioeng. Skin,* 2, 103, 1986.
26. Kligman, A. M. and Wooding, W. M., A method for the measurement and evaluation of irritants on human skin, *J. Invest. Dermatol.,* 49, 78, 1967.
27. Obeid, A. N., A critical review of laser Doppler flowmetry, *J. Med. Eng. Technol.,* 14, 178, 1990.
28. Bircher, A., De Boer, E. M., Agner, T., Wahlberg, J. E., and Serup, J., Guidelines for measurement of cutaneous blood flow by laser Doppler flowmetry, *Contact Dermatitis,* 30, 65, 1994.
29. De Boer, E. M., Bezemer, P. D., and Bruynzeel, D. P., A standard method for repeated recording of skin blood flow using laser Doppler flowmetry, *Dermatosen,* 37, 58, 1989.
30. Tur, E., Tur, M., Maibach, H. I., and Guy, R. H., Basal perfusion of the cutaneous microcirculation: measurements as a function of anatomic position, *J. Invest. Dermatol.,* 81, 442, 1983.
31. Sundberg, S., Acute effects and long-term variations in skin blood flow measured with laser Doppler flowmetry, *Scand. J. Clin. Lab. Invest.,* 44, 341, 1984.
32. Tenland, T., Salerud, E. G., Nilsson, G. E., and Öberg, P. A., Spatial and temporal variations in human skin blood flow, *Int. J. Microcirc. Clin. Exp.,* 2, 81, 1983.
33. Tur, E., Maibach, H. I., and Guy, R. H., Spatial variability of vasodilatation in human forearm skin, *Br. J. Dermatol.,* 113, 197, 1985.
34. Meyer, E., Smith, E. W., and Haigh, J. M., Sensitivity of different areas of the flexor aspects of the human forearm to corticosteroid-induced skin blanching, *Br. J. Dermatol.,* 127, 379, 1992.
35. Mustakallio, K. K. and Kolari, P. J., Irritation and staining by Dithranol (Anthralin) and related compounds, *Acta Derm. Venereol.,* 63, 513, 1983.
36. Serup, J. and Kristensen, J. K., Blood flow of morphoea plaques as measured by laser Doppler flowmetry, *Arch. Dermatol. Res.,* 276, 322, 1984.
37. Prater, E., Goering, H. D., and Schubert, H., Sodium lauryl sulphate: a contact allergen, *Contact Dermatitis,* 4, 242, 1978.
38. Gabard, B., Appearance and regression of a local skin irritation in two different models, *Dermatosen,* 39, 11, 1991.
39. Van Neste, D., Mahmoud, G., and Masmoudi, M., Experimental induction of rough dermatitic skin in humans, *Contact Dermatitis,* 16, 27, 1987.
40. Van Neste, D., Evaluation de l'État Fonctionnel de la Peau par des Méthodes non Invasives, Ph.D. thesis, Catholic University Louvain, Belgium, 1990.
41. Nilsson, G. E., Otto, U., and Wahlberg, J. E., Assessment of skin irritancy in man by laser Doppler flowmetry, *Contact Dermatitis,* 8, 401, 1982.
42. De Boer, E. M., Occupational Dermatitis by Metalworking Fluids, Ph.D. thesis, Free University Amsterdam, 1989.
43. Van Neste, D. and De Brouwer, B., Monitoring of skin response to sodium lauryl sulphate: clinical scores versus bioengineering methods, *Contact Dermatitis,* 27, 151, 1992.

44. Lammintausta, K., Maibach, H. I., and Wilson, D., Human cutaneous irritation: induced hyporeactivity, *Contact Dermatitis*, 17, 193, 1987.

45. Agner, T. and Serup, J., Sodium lauryl sulphate for irritant patch testing—a dose-response study using bioengineering methods for determination of skin irritation, *J. Invest. Dermatol.*, 95, 543, 1990.

46. Agner, T. and Serup, J., Skin reactions to irritants assessed by non-invasive bioengineering methods, *Contact Dermatitis*, 20, 352, 1989.

47. Frosch, P. J. and Wissing, C., Cutaneous sensitivity to ultraviolet light and chemical irritants, *Arch. Dermatol. Res.*, 272, 269, 1982.

48. Agner, T. and Serup, J., Individual and instrumental variations in irritant patch-test reactions—chemical evaluation and quantification by bioengineering methods, *Clin. Exp. Dermatol.*, 15, 29, 1990.

49. Agner, T., Basal transepidermal water loss, skin thickness, skin blood flow and skin colour in relation to sodium-lauryl-sulphate-induced irritation in normal skin, *Contact Dermatitis*, 25, 108, 1991.

50. Coenraads, P. J. and Pinnagoda, J., Dermatitis and water vapour loss in metal workers, *Contact Dermatitis*, 13, 347, 1985.

51. Pinnagoda, J., Tupker, R. A., Coenraads, P. J., and Nater, J. P., Prediction of susceptibility to an irritant response by transepidermal water loss, *Contact Dermatitis*, 20, 341, 1989.

52. Agner, T., Susceptibility of atopic dermatitis patients to irritant dermatitis caused by sodium lauryl sulphate, *Acta Derm. Venereol.*, 71, 296, 1990.

53. Freeman, S. and Maibach, H. I., Study of irritant contact dermatitis produced by repeat patch test with sodium lauryl sulfate and assessed by visual methods, transepidermal water loss, and laser Doppler velocimetry, *J. Am. Acad. Dermatol.*, 19, 496, 1988.

54. Elsner, P., Wilhelm, D., and Maibach, H. I., Multiple parameter assessment of vulvar irritant contact dermatitis, *Contact Dermatitis*, 23, 20, 1990.

55. Tupker, R. A., The Influence of Detergents on the Human Skin, Ph.D. thesis, Ryksuniversiteit Groningen, The Netherlands, 1990.

56. Smeenk, G., De Invloed van Detergentia op de Huid, Ph.D. thesis, Ryksuniversiteit Leiden, The Netherlands, 1968.

57. Smeenk, G., The influence of detergents on the human skin, *Arch. Clin. Exp. Dermatol.*, 235, 180, 1969.

58. De Boer, E. M., Scholten, R. J. P. M., Van Ketel, W. G., and Bruynzeel, D. P., Quantitation of mild irritant reactions due to repeated application of liquid cleaners. A laser Doppler flowmetry study, *Int. J. Cosmet. Sci.*, 12, 43, 1990.

59. Van Neste, D., Skin response to histamine dry skin prick test: influence of duration of the skin prick on clinical parameters and on skin blood flow monitoring, *J. Dermatol. Sci.*, 1, 435, 1990.

60. Rihoux, J. P. and Van Neste, D., Quantitative time course study of the skin response to histamine and the effect of H1 blockers. A 3-week crossover double-blind comparable trial of cetirizine and terfenadine, *Dermatologica*, 179, 129, 1989.

61. Hammarlund, A., Olsson, P., and Pipkorn, U., Blood flow in histamine- and allergen-induced weal and flare responses, effects of an H1 antagonist, alpha-adrenoceptor agonist and a topical glucocorticoid, *Allergy*, 45, 64, 1990.

62. Tur, E., Maibach, H. I., and Guy, R. H., Percutaneous penetration of methyl nicotinate at three anatomic sites: evidence for an appendageal contribution to transport?, *Skin Pharmacol.*, 4, 230, 1991.

63. Wilhelm, R. P., Do barrier creams protect against irritant contact dermatitis in experimental studies? *Proc. ACDS Meeting*, San Francisco, 1992, p.19.

64. Wilhelm, D., Elsner, P., Pine, H. L., and Maibach, H. I., Evaluation of vulvar irritancy potential of a menstrual pad containing sodium bicarbonate in short-term application, *J. Reprod. Med.*, 36, 556, 1991.

65. Kubota, K., Koyoma, E., and Yasuda, K., Skin irritation induced by topically applied timolol, *Br. J. Clin. Pharmacol.,* 31, 471, 1991.
66. Vaananen, A. and Hannuksela, M., UVB erythema inhibited by topically applied substances, *Acta Derm. Venereol.,* 69, 12, 1989.
67. Hannuksela, A. and Kinnunen, T., Moisturizers prevent irritant dermatitis, *Acta Derm. Venereol.,* 72, 42, 1992.
68. Blanken, R., Van der Valk, P. G. M., Nater, J. P., and Dijkstra, H., After-work emollient creams: effects on irritant skin reactions, *Dermatosen,* 35, 95, 1987.
69. Serup, J., A double-blind comparison of two creams containing urea as the active ingredient, *Acta Derm. Venereol. Suppl.,* 177, 34, 1992.
70. Agner, T., An experimental study of irritant effects of urea in different vehicles, *Acta Derm. Venereol. Suppl.,* 177, 44, 1992.
71. De Haan, P., Sampat, S., and Bruynzeel, D. P. B., Toxicity of alcohols of different chain length towards keratinocytes in comparison with other compounds with different chain length, poster session at Int. Sympo. Irritant Contact Dermatitis, Groningen, The Netherlands, 1991.
72. Eun, H. C., Lee, H. G., and Paik, N. W., Patch test responses to rockwool of different diameters evaluated by cutaneous blood flow measurement, *Contact Dermatitis,* 24, 270, 1991.
73. De Boer, E. M., Scholten, R. J. P. M., Van Ketel, W. G., and Bruynzeel, D. P., Irritancy of metalworking fluids: a laser Doppler flowmetry study, *Contact Dermatitis,* 22, 86, 1990.
74. Wahlberg, J. E., Nickel chloride or nickel sulfate? Irritation from patch test preparations as assessed by laser Doppler flowmetry, *Dermatol. Clin.,* 8, 41, 1990.
75. Larmi, E., PUVA treatment inhibitis nonimmunologic immediate contact reactions to bensoic acid and methyl nicotinate, *Int. J. Dermatol.,* 28, 609, 1989.
76. Berardesca, E. and Maibach, H. I., Racial differences in sodium lauryl sulphate induced cutaneous irritation: black and white, *Contact Dermatitis,* 18, 65, 1988.
77. Wilhelm, K. P. and Maibach, H. I., Factors predisposing to cutaneous irritation, *Dermatol. Clin.,* 8, 17, 1990.
78. Berardesca, E. and Maibach, H. I., Sodium-lauryl-sulphate induced cutaneous irritation. Comparison of white and hispanic subjects, *Contact Dermatitis,* 19, 136, 1988.
79. Ghadially, R., Halkier-Sorensen, L., and Elias, P. M., Effects of petrolatum on stratum corneum structure and function, *J. Am. Acad. Dermatol.,* 26, 387, 1992.

Chapter 15

A Preliminary Comparison of Laser Doppler Velocimetry and Visual Ranking of Corticosteroid-Induced Skin Blanching

Eric W. Smith and Howard I. Maibach

CONTENTS

ABSTRACT: The human skin blanching assay, using visual assessment, is still widely used as a reliable, qualitative, comparative indicator of topical corticosteroid availability and potency. Theoretically, the blood perfusion rate of the skin as measured by laser Doppler velocimetry should provide a more objective technique for quantifying the vasoconstriction induced by topical corticosteroids. A preliminary bioassay using optimized methodology was conducted to compare visual assessments with instrumental perfusion data of skin blanching induced by betamethasone 17-valerate and hydrocortisone. The visual assessment data were similar to that observed previously and clearly demonstrated significant differences between the response profiles of the two corticoids in both application modes. In contrast, the laser Doppler velocimetry assessment was slow to conduct and yielded relatively imprecise data that did not demonstrate any significant differences between the two formulations in either application mode. From these results it would appear that laser Doppler velocimetry is not a viable alternative to visual assessments in topical corticosteroid bioassays.

I. INTRODUCTION

In the last four decades topical corticosteroids have been the most important family of drugs used to treat dermatological conditions. With the advances that have been made recently in quantifying the bioavailability and bioequivalence of drug formulations in general, researchers have similarly strived to improve the methods used to assess the topical availability of anti-inflammatory corticoid products. The vasoconstrictor assay first introduced by Wells[1] and refined by McKenzie and Stoughton[2] has been the basis of the visual assessment of corticosteroid potency for the last 30 years. The methodology of this bioassay has been refined and optimized[3,4] to eliminate many of the *in vivo* variables, thereby producing a reliable and highly reproducible technique[5] for assessing drug potency, ranking formulations, and comparing topical drug availability. The assay utilizes the skin whitening side effect following corticosteroid absorption; the intensity of this skin blanching is directly proportional to the potency of the drug molecule or the ease with which the drug is delivered to the stratum corneum from the applied vehicle. The assay has repeatedly demonstrated that the topical availability of the same drug incorporated in the same concentration into different delivery vehicles may be markedly different.[6]

Recently, the human skin blanching assay has been criticized[7] for the apparent subjectiveness of the visual method of assessing the intensity of the induced vasoconstriction, although researchers have found, without exception, that there is a direct correlation between visual scores and the quantitative measurements of optical instruments when comparative evaluations have been conducted.[8] However, quantitative assessment of the intensity of vasoconstriction would allow more advanced statistical analysis of the data than is currently possible with the qualitative, visual rankings. Toward this goal, scientists have evaluated a number of instrumental techniques with which the assessment of skin pallor following controlled corticoid application may be carried out.[3]

One technique that, theoretically, is applicable in this regard is laser Doppler velocimetry,[9,10] which measures the magnitude of tissue perfusion by moving blood elements, especially red blood cells. Single-frequency light from a helium-neon laser source is transmitted to the surface of the skin via an optical fiber cable. The light beam entering the skin is refracted and reflected by the superficial tissues and by moving blood cells within the vessels. Some of the reflected radiation is captured by the optical fiber and transmitted back to the instrument. Much of this radiation is reflected by the stationary structures within the skin and is of the same frequency as the incident radiation. On the other hand, radiation reflected by the moving red blood cells is frequency or Doppler shifted in proportion to the number and velocity of the moving elements. The instrument analyzes the reflected radiation, separates the incident from the Doppler-shifted frequency, and reports the relative magnitude of the latter, either as a voltage signal or as a percentage difference when compared to a reference standard. The magnitude of the result is therefore dependent on the number and speed of the blood cells passing through the incident field. This instrumental technique has been shown to be useful in documenting the vasodilation induced by methyl nicotinate[11] and other vasodilators.[12] Theoretically, this instrument should be applicable to measuring the difference in the intensity of the vasoconstriction induced by corticoid application when compared to adjacent, unmedicated skin or to adjacent application sites that have had other corticosteroids or delivery vehicles applied to them.

On this premise, a preliminary investigation was carried out to compare the results from a laser Doppler velocimeter with visual observations in an *in vivo* skin blanching assay using optimized methodology and thereby assess the applicability of the instrumental technique in this regard.

II. PRODUCT TESTING PROTOCOLS

A. HUMAN SKIN BLANCHING ASSAY

The skin blanching assay was conducted according to the optimized protocol as reported previously.[3,4] Two healthy volunteers who had previously been screened for blanching response were used in this preliminary study. Six adhesive labels, from which two 7 × 7 mm squares had been punched, were applied to the flexor aspects of each arm to demarcate a total of 48 application sites. The topical corticosteroid formulations to be compared were applied in a double-blind, randomized manner to these sites by extrusion from a 1-ml syringe (with the needle cut to 5 mm to facilitate expulsion) and were spread evenly over the sites with a glass rod. The sites of one arm of each volunteer were occluded with impervious tape, thereby preventing evaporation of moisture and delivery vehicle components. The sites of the other arm remained unoccluded, but were covered with a porous guard to prevent accidental removal of the formulations. The formulations remained in contact with the skin for 6 h, after which time the guards, occlusive strips, and demarcating labels were carefully removed and any residual formulation was gently washed from the application sites. Thereafter, one trained observer assessed the degree of induced blanching at each site at 8, 9.5, 11, 12.5, 14, 15.5, 17, and 32 h after initial application by assigning a number between 0 and 4 to represent the perceived pallor at each site. The blanching response is usually expressed as percentage of total possible score (%TPS) which could have been recorded for that preparation.[5] After decoding, these data are used to generate two blanching response vs. time profiles (occluded and unoccluded data) from which an area under the blanching curve (AUBC) may be calculated. Qualitative statistical analysis (χ^2 test) may be applied to the results to examine the significance of the difference between profiles.

Corticosteroid formulations from different potency classifications were chosen for this comparative assessment because it was known from several previous studies that these would produce very different response profiles from the visual assessment data. It was decided that, for laser Doppler velocimetry to show promise as a viable alternative to visual assessment methodology, the instrumental evaluation must have the ability to discern these gross differences in skin blanching induced by two corticoids of different potency. If the instrument was unable to distinguish between the divergent responses of these two drugs, then it certainly would not be able to discriminate between the very subtle differences in blanching that are elicited by different formulations containing the same drug in equivalent concentration. The two cream formulations chosen for comparison were Persivate® (Lennon, South Africa), which contains 0.12% betamethasone 17-valerate, and Hytone® (Dermik, Pennsylvania), which contains 2.5% hydrocortisone.

B. LASER DOPPLER ANALYSIS

The laser Doppler analysis of the blood perfusion at each application site was carried out in tandem with the visual observations. Each application site, therefore, had both visual and instrumental data recorded at each reading time. The instrumental analysis using a Periflux®PF3 instrument (Medex, Ohio) was performed by first recording the perfusion of unmedicated skin, adjacent to the application sites, on each arm of each volunteer, this value being stored in the instrument as the reference. Thereafter, the probe was moved by hand from site to site to record the difference in perfusion of the medicated regions compared to the stored reference (percentage change). The probe was held by hand perpendicularly above each application site, at a distance of approximately 1 mm from the stratum corneum surface, for a period long enough to produce a stable result from the velocimeter. These results were all expected to have a negative value as a degree of vasoconstriction was being measured at the medicated sites. After completion of the study the application patterns were decoded and the instrumental data analyzed in terms of percentage change in perfusion for each preparation and application mode (occluded or unoccluded) vs. the time after corticoid application. These data could then be compared to the appropriate visual assessment data to estimate the correlation between the two methods.

III. RESULTS

A. UNOCCLUDED APPLICATION MODE

The results of the visual assessment of skin blanching for the unoccluded application mode are depicted in Figure 1 as the blanching response (%TPS) vs. the time after drug application. These profiles are typical of those observed in previous trials and clearly show that the skin blanching elicited by betamethasone 17-valerate (Persivate®) is markedly greater (AUBC = 1000) than that generated by the hydrocortisone (Hytone®; AUBC = 356).

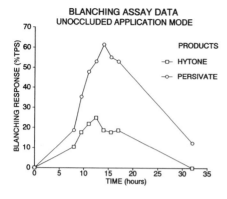

Figure 1 Visually assessed blanching response following application of betamethasone 17-valerate (Persivate®) and hydrocortisone (Hytone®) without occlusion. %TPS = percentage of total possible score.

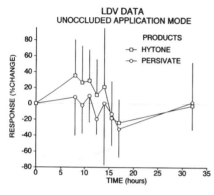

Figure 2 Laser Doppler velocimetry assessment of blanching response following application of betamethasone 17-valerate (Persivate®) and hydrocortisone (Hytone®) without occlusion.

Qualitative statistical evaluation (χ^2 tests) of the profiles indicates that the values for the two products at each time interval, except at 8 h, are significantly different. These results follow the expected trend; the more potent betamethasone 17-valerate generating skin blanching of significantly greater intensity and longer duration than that caused by hydrocortisone, with maximum blanching observed at approximately 12 to 16 h after initial drug application.[6]

The laser Doppler velocimetry data for the unoccluded application mode are depicted in Figure 2 as the average percentage change in blood perfusion (plus or minus 1 SD) vs. time after drug application. In contrast to the distinctive visual results, the instrumental data are much less definitive. Unexpectedly, there is an initial increase in perfusion recorded for both corticoids, especially for the hydrocortisone, which is evident up to the 14 h reading. The increase in perfusion is of lesser magnitude for the betamethasone 17-valerate formulation and declines after 11 h. The readings at 15.5 and 17 h are clearly negative and indicative of vasoconstriction. The magnitude of these negative results appears greater in rank order for the betamethasone 17-valerate formulation, although these minima occur later (17 h) than the corresponding period of maximum blanching observed visually. Furthermore, quantitative statistical analysis of these profiles (Student's *t*-distribution test, null hypothesis of equal means, variance unknown)[13] indicates that there are no significant differences between the betamethasone 17-valerate and hydrocortisone mean values at any of the recording times. In the unoccluded application mode, therefore, the laser Doppler velocimeter is unable to distinguish between the intensities of the skin blanching induced by the two different corticosteroids.

B.   OCCLUDED APPLICATION MODE

The results of the visual assessment of skin blanching for the occluded application mode are similarly depicted in Figure 3. These profiles are again typical of the occluded methodology

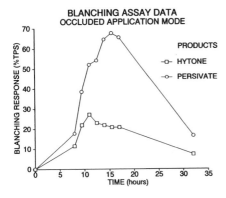

BLANCHING ASSAY DATA
OCCLUDED APPLICATION MODE

PRODUCTS

-□- HYTONE
-○- PERSIVATE

Figure 3 Visually assessed blanching response following application of betamethasone 17-valerate (Persivate®) and hydrocortisone (Hytone®) under occlusion.% TPS = percentage of total possible score.

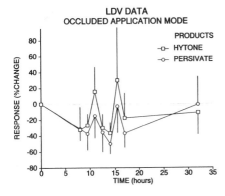

LDV DATA
OCCLUDED APPLICATION MODE

PRODUCTS

-□- HYTONE
-○- PERSIVATE

Figure 4 Laser Doppler velocimetry assessment of blanching response following application of betamethasone 17-valerate (Persivate®) and hydrocortisone (Hytone®) under occlusion.

and show greater response at peak blanching and a more rounded profile when compared to the unoccluded data. Here, too, the skin blanching elicited by the betamethasone 17-valerate is markedly greater (AUBC = 1166) than that generated by the hydrocortisone (AUBC = 453) and is significantly different by χ^2-analysis at all recording times except 8 and 32 h.

The laser Doppler velocimetry analysis of the occluded application sites (Figure 4) is more distinctive than that of the unoccluded sites, with all but two of the average perfusion values being negative with respect to the reference standard. The instrumental values for the betamethasone 17-valerate are all negative with a minimum at 14 h, which is 1.5 h earlier than the corresponding visually observed maximum response for this formulation. Data for the hydrocortisone formulation are also negative at all recording times except at 9.5 and 15.5 h. As with the unoccluded data, the variance about the mean values for each corticoid at each recording time is of such magnitude that there are no statistically significant differences between the means at any of the observation times. However, the instrumental values for the betamethasone 17-valerate formulation appear to be smaller in rank order than those of the hydrocortisone product, as would be expected.

Noteworthy from both the occluded and unoccluded data is the magnitude of the variance about the mean values recorded by the velocimeter. This variance, exemplified by the length of the standard deviation bars on the graphs, is smaller for the occluded data, presumably because the vasoconstriction is more intense in this application mode. However, in both application modes the variance of the data is of such magnitude that the means are not significantly different at any of the reading times.

IV. CONCLUSIONS

It is evident from the data that the precision of the laser Doppler velocimetry analysis is relatively low when compared to the visual ranking of vasoconstriction. There may be several factors contributing to this imprecision; the hand manipulation of the probe is probably one of the major problems. Optimally, the probe and optical fiber should be fixed to the tissue under assessment (using adhesive tape or glue) and be completely stationary during the assessment. This is not possible in the optimized blanching assay because a large number of application sites are required along the entire length of the forearm to overcome the physiological variations in vasoconstrictive response to corticosteroids at different forearm locations.[14] It is therefore essential to move the sensor from site to site at each observation time, holding it as motionless as possible at a constant distance from the skin. Any pressure from the probe on the stratum corneum will slow blood perfusion through that region and yield inaccurate data. Undoubtedly the ability on the part of the operator to hold the probe stationary would improve with experience. However, it was evident from this study that hand manipulation was extremely difficult; in some cases several minutes were required before a stable perfusion reading was obtained. This was considerably longer than the 30 s or so required for visual ranking of the 12 application sites on each arm. Routine analysis with this instrument would certainly require some mechanical device for holding and positioning the probe and fiber.

Generally, it would appear from the results of this study that laser Doppler velocimetry is unsuitable as an objective, instrumental replacement for visual observations of the intensity of skin blanching in topical corticosteroid assessment. It must be borne in mind that the data reported here were generated from a preliminary study using only 2 volunteers and 1 observer as opposed to 12 and 3, respectively. However, even with this contracted study the visual comparison of the two corticoids produced the expected, definitive separation of weak from potent drug response. Conversely, the instrument was unable to make a significant distinction between the products, although the rank order of response appeared to follow the expected sequence. It is highly unlikely, therefore, that the instrument could be used in bioequivalence testing of similar formulations that are chemical equivalents.

The instrument is clearly not as capable of distinguishing different degrees of vasoconstriction as well as it is able to discern degrees of vasodilation. These observations are in agreement with those of other researchers.[15] It has been suggested[16] that the postcorticoid application of a vasodilator such as methyl nicotinate and subsequent assessment of the inhibition of vasodilation by the corticosteroid may be an improved method for using the laser Doppler velocimeter. However, an earlier study[15] has questioned the usefulness of reactive hyperemia in this regard. Furthermore, the use of additional chemicals to attempt quantification of corticosteroid-induced blanching, which itself is not yet fully understood,[17] merely complicates the bioassay and is therefore unwarranted.

REFERENCES

1. Wells, G. C., The effect of hydrocortisone on standardized skin-surface trauma, *Br. J. Dermatol.,* 69, 11, 1957.
2. McKenzie, A. W. and Stoughton, R. B., Method for comparing percutaneous absorption of steroids, *Arch. Dermatol.,* 86, 608, 1962.
3. Haigh, J. M. and Kanfer, I., Assessment of topical corticosteroid preparations: the human skin blanching assay, *Int. J. Pharm.,*19, 245, 1984.
4. Smith, E. W., Meyer, E., Haigh, J. M., and Maibach, H. I., The human skin blanching assay as an indicator of topical corticosteroid bioavailability and potency: an update, in *Percutaneous Absorption. Mechanisms—Methodology—Drug Delivery,* Bronaugh, R. L. and Maibach, H. I., Eds., Marcel Dekker, New York, 1989, 443.

5.   Smith, E. W., Meyer, E., and Haigh, J. M., Accuracy and reproducibility of the multiple-reading skin blanching assay, in *Topical Corticosteroids,* Maibach, H. I. and Surber, C., Eds., S. Karger, Basel, 1992, 65.

6.   Smith, E. W., Meyer, E., and Haigh, J. M., Blanching activities of betamethasone formulations. The effect of dosage form on topical drug availability, *Arzneim. Forsch.,* 40, 618, 1990.

7.   Shah, V. P., Peck C. C., and Skelly, J. P., Vasoconstriction-skin blanching assay for glucocorticoids—a critique, *Arch. Dermatol.,* 125, 1558, 1989.

8.   Haigh, J. M. and Smith, E. W., Topical corticosteroid-induced skin blanching measurement: eye or instrument?, *Arch. Dermatol.,* 127, 1065, 1991.

9.   Nilsson, G. E., Tenland, T., and Oberg, P. A., Evaluation of a laser Doppler flowmeter for measurement of tissue blood flow, *IEEE Trans. Biomed. Eng.,* 27, 597, 1980.

10. Nilsson, G. E., Salerud, E. G., Tenland, T., Oberg, P. A., and Wahlberg, J. E., Laser Doppler flowmetry. New technique for noninvasive assessment of skin blood flow, *Cosmet. Toilet.,* 99, 97, 1984.

11. Guy, R. H., Wester, R. C., Tur, E., and Maibach, H. I., Non-invasive assessments of the percutaneous absorption of methyl nicotinate in humans, *J. Pharm. Sci.,* 72, 1077, 1983.

12. Stevenson, J. M., Maibach, H. I., and Guy, R. H., Laser Doppler and photoplethysmographic assessment of cutaneous microvasculature, in *Models in Dermatology,* Vol. 3, Maibach, H. I. and Lowe, N. J., Eds., S. Karger, Basel, 1987, 121.

13. Dunn, O. J., *Basic Statistics: A Primer for the Biomedical Sciences,* 2nd ed., John Wiley & Sons, New York, 1977, chap. 8

14. Meyer, E., Smith, E. W., and Haigh, J. M., Sensitivity of different areas of the flexor aspect of the human forearm to corticosteroid-induced skin blanching, *Br. J. Dermatol.,* 127, 379, 1992.

15. Amantea, M., Tur, E., Maibach, H. I., and Guy, R. H., Preliminary skin blood flow measurements appear unsuccessful for assessing topical corticosteroid effect, *Arch. Dermatol. Res.,* 275, 419, 1983.

16. Kristensen, J. K., Bisgaard, H., and Søndergaard, J., A new quantitative technique for ranking vascular corticosteroid effects in man using laser Doppler velocimetry, *J. Invest. Dermatol.,* 84, 445, 1985.

17. Ahluwalia, A. and Flower, R. J., Investigations into the mechanism of vasoconstrictor action of the topical corticosteroid betamethasone 17-valerate in the rat, *Br. J. Pharmacol.,* 108, 544, 1993.

Chapter 16

Sunscreens

Véronique Epstein-Drouard

CONTENTS

I. INTRODUCTION

Human skin possesses its own natural protection against sunburn. This consists of the thickness of the stratum corneum and the pigmentation of the skin. However, this natural protection is often not sufficient to protect the skin from acute or chronic damage. It is interesting to know that in ancient times people already used various recipes with sunscreening properties. For religious or other purposes some equatorial tribes still paint themselves with mud or plant mixtures, which help protect their skin. More closely, our grandparents wore covering clothes and used umbrellas for protection against sun rays.

After World War I the first sunscreens were created, but their absorbing capacity for solar radiation was poor.[1] Nowadays, extensive sunscreen preparations are being developed to help prevent the harmful effects of sun rays or to allow safe tanning.

II. ULTRAVIOLET RAYS AND SUNSCREENS

Sunscreen formulations are composed of a vehicle and a sunscreen agent. Many factors can influence the effectiveness of such a preparation:

- pH
- Solvent used
- Thickness of the film applied on the skin
- Stability of the product in the sun

The most important constituent is, of course, the sunscreen agent. Sunscreen agents usually absorb at least 85% of ultraviolet (UV) radiation within the wavelength region from 290 to 320 nm: they are chemical UVB sunscreens.

Opaque sunblock agents form a physical barrier against light in the UV and visible ranges, scattering wavelengths from 290 to 777 nm. The most commonly used are titanium dioxide and zinc oxide.[2] Another class of sunscreen agents allowed in cosmetic products in the European Economic Community (EEC), but not yet in over-the-counter (OTC) sunscreen products in the U.S., is a range of chemical sunscreens with a maximum absorption in the UVA wavelength, the so-called UVA sunscreens.

A variety of substances which significantly potentiate tanning under the stimulus of UVA radiation have been used to enhance protection; for example, psoralen increases the appearance of a self-protective tanning. However, scientific opinion is far from unanimous with regard to the usefulness and hazards of these sunscreen formulations.[3,4]

III. SUNSCREEN PROTECTANT EFFECT ASSESSMENT

The degree of attenuation needed for a particular quantity of erythematogenic radiation reaching the skin first depends on the dosage of the formulation to be used and on the type of protection desired.[5] To be able to classify a sunscreen preparation it is necessary to assess its efficiency. The general practice is to examine *in vivo* skin sensitivity to erythemal radiation (UVB) with and without sunscreens, using an artificial sun-ray lamp with a filter: this is the determination of the sun protective factor (SPF).

The SPF of a product is calculated from the exposure time interval required to produce the minimal erythema dose (MED) for protected skin and from the exposure time interval required to produce the MED for unprotected skin (control site):[2]

$$SPF = \frac{\text{protected skin exposure time}}{\text{unprotected skin exposure time}}$$

Several methods have been described to measure SPF, including the two official ones following:

- The DIN (Deutsches Institut für Normung) norm method, currently used in Europe[6]
- The method recommended by the Over-the-Counter Drug (OTC) Panel of the Food and Drug Administration (FDA) in the U.S.[7]

Both of these are based on the determination of the MED (see Chapter 13), and both of them involve visual assessment of the erythema. The differences between the two methods lie mainly in the quantity of sunscreen preparation applied on the skin (1.5 mg/cm² or µl/cm² for the DIN method vs. 2 mg/cm² for the FDA method) and the type of lamp used.

A method to evaluate the effect of UVA sunscreens through their tanning potential is actually being discussed.

IV. LASER DOPPLER VELOCIMETRY MEASUREMENTS

The aim of the laser Doppler velocimetry (LDV) technique is to measure concentration and mean velocity of red blood cells.[8] LDV was applied to blood flow measurements following UV-induced changes in the skin, with and without sunscreen, after a single or repeated exposure.

A. SINGLE-EXPOSURE STUDIES

Wulf et al. measured pig skin blood flow after exposure to artificial sunlight from a solar simulator for 30 to 240 s using two different sunscreen preparations, one with a SPF of 5, and the other with a SPF of 6. They found an enhancement of cutaneous blood flux in nonprotected and in protected areas. Using the SPF 5 sunscreen the blood flow was 77% of the blood flow enhancement of the unprotected area, and using the SPF 6 sunscreen it was 62% of the unprotected area.[9]

Then Drouard et al. used a single dose of combined UVA, UVB, and UVC (15 J/cm² + 4 × MED) on either untreated human skin or skin treated with a UVB radiation filter alone (octyl methoxycinnamate), with a psoralen alone (5-MOP), with both of them, and with their vehicle alone. Their data confirmed the enhancement of blood flow even in protected areas with a significant difference between untreated areas and those treated with the UVB sunscreen. Those results were reasonably coincidental with the subjective visual erythema assessment. They showed that, among the constituents tested, octyl methoxycinnamate was the only ingredient active against erythema. Neither the vehicle nor the 5-MOP appeared to perform any function in screening erythematogenic radiation.[10]

B. REPEATED-EXPOSURE STUDIES

Wulf et al. repeated the trial described in Section IV.A three days later. It resulted in only a reduction of the increase in pig skin blood flow compared with the first exposure; almost no effect of the sunscreens was seen. This illustrates that previous sunlight exposure may give some protection against sunlight, especially UVB.[9]

Another observation of interest was that blood flow measurements in skin burn resulting from a single or repeated exposure were seen to be lower than in corresponding unburned areas. A total stop of blood flow was even recorded in some areas.

Drouard et al.[10] measured blood perfusion changes during repeated exposure to 10 J/cm² of UVA + 1 × MED UVB + C for 6 days, followed by 10 days without exposure and

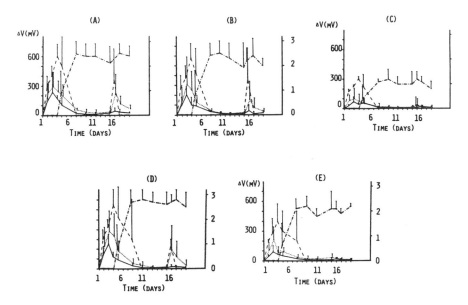

Figure 1 Results from the multiple-exposure experiments. Mean laser Doppler velocimetry (LDV)-assessed cutaneous perfusion changes (ΔV) and subjective erythema and tanning gradings (± SD) for five subjects are shown as a function of time for five treatment categories: (A) control; (B) sunscreen lotion vehicle alone; (C) sunscreen lotion with a UVB sunscreen agent, 5% 2-ethylhexyl p-methoxycinnamate; (D) sunscreen lotion with 30 ppm 5-methoxypsoralen; and (E) sunscreen lotion with both cinnamate and 5-methoxypsoralen. (——) Capillary perfusion monitor (CPM)-assessed perfusion; (....) laser Doppler flowmeter (LDF)-assessed perfusion; (----) visual erythema grading; (-.-.) visual tanning grading. LDV blood flow values (in millivolts) are given on the left-hand ordinate.

then a single exposure of 15 J/cm^2 of UVA + 4 × MED UVB + C, on human skin treated as described in Section IV.A. Results are shown Figure 1.

When skin was exposed to five small daily doses and tanning had developed, reexposure to the larger irradiation 10 days later produced no significant increase in cutaneous perfusion in any of the treatment categories, including the control.[10] This confirms the conclusion of Wulf et al. regarding the protective effect of a previous exposure to sunlight and the well-known fact that tanning itself is protective.

The data also showed that, on the basis of a daily exposure to 1× MED UVB + C radiation, the change in skin blood flow passes through a maximum on day 3. This confirmed the conclusions of Kaidbey and Kligman that initial cumulative damage was followed by the onset of photoprotective changes in the skin with continued repeated exposure.[11]

No significant photo effect of 5-MOP on cutaneous perfusion was observed.

V. CONCLUSION

Laser Doppler velocimetry may have application in the quantitative assessment of sun protectant efficacy either on a fundamental basis or in the objectivation of the sun protective factor. As shown in this chapter, little research has been done on the subject, leaving room for extensive work either testing other sunscreen agents or using photoplethysmography.

REFERENCES

1. Greiter, F., Bilk, P., and Doskoczil, S., History of sunscreens and the rationale for their use, in *Principles of Cosmetics for the Dermatologist,* Part 8, Frost, P. and Horwitz, S. N., Eds., C. V. Mosby, Saint Louis, 1982, 187.
2. Wilkinson J. B. and Moore, R. J., *Harry's Cosmetology,* 7th ed., Godwin, G., Ed., Chemical Publishing Company, New York, 1982, 222.
3. Gshnait, F., Brenner, W., and Wolff K., Photoprotective effect of a psoralen UVA-induced tan, *Arch. Dermatol. Res., 263,* 181, 1978.
4. Brenner, W. and Gshnait F., The value of topical sunscreens containing psoralens, *Arch. Dermatol. Res., 267,* 189, 1980.
5. Kreps, S. I. and Goldemberg, R. L., Suntan preparations, in *Cosmetics: Science and Technology,* 2nd ed., Balsam, M. S. and Sagarin, E., Eds., Interscience, New York, 1972, chap. 7.
6. Experimentelle dermatologische Bewertung des Erythemschutzes von externen Sonnenschutzmitteln für die menschliche Haut, *Deutsche Norm DIN 67501,* Beuth Verlag Gmbh, Berlin, 1.
7. Proposed monograph for OTC sunscreen drug products, *Fed. Regis.* 43 (166), 38206, 1978.
8. Stern, M. D., Lappe, D. L., Bowen, P. D., Chimosky, J. E., Allen Holloway, G., Jr., Keiser, H. R., Bowman, R. L., Continuous measurement of tissue blood flow by laser-Doppler velocimetry, *Am. J. Physiol., 232,* H441, 1977.
9. Wulf, M. C., Staberg, B., and Eriksen, W. H., Laser Doppler measurements of blood flow in pig skin after sunscreens and artificial sunlight, *Photobiochem. Photobiophys., 6,* 231, 1983.
10. Drouard, V., Wilson, D. R., Maibach, H. I., and Guy, R. H., Quantitative assessment of UV-induced changes in microcirculatory flow by laser Doppler velocimetry, *J. Invest. Dermatol., 83,* 188, 1984.
11. Kaidbey, K. H.and Kligman, A. M., Cumulative effects from repeated exposures to ultraviolet radiation, *J. Invest. Dermatol., 76,* 352, 1981.

Part III   Remittance Spectroscopy and Chromametry

General Aspects

Chapter 17

Remittance Spectroscopy: Hardware and Measuring Principles

Peter H. Andersen and Peter Bjerring

CONTENTS

I. INTRODUCTION

Since the skin is immediately present for the observer, the use of objective techniques was delayed in dermatology research compared to other fields of medicine and biophysics. This phenomenon can be explained by several factors. First, the human eye is extremely sensitive when immediately grading different color intensities. Second, a visual skin reaction very often consists of several estimates, including the size, the degree of edema, and the color of the site, partially explaining why visual estimations often are equal to objective measuring techniques. Furthermore, bioengineering equipment often measures a small area in the center of a response, whereas visual impression integrates several conditions in the cutaneous reaction, making the trained eye and finger a very powerful tool. However, in recent decades several noninvasive and objective measuring procedures have been presented and have increased our knowledge of several skin conditions.[1–3]

In general, a new noninvasive bioengineering devise for objective cutaneous measurements must be nontraumatic and should not interfere with measured skin parameters. Furthermore, to become generally accepted and widely used, the systems must be accurate and reliable and must provide highly reproducible and informative measurements. Although developed around the beginning of the 20th century and introduced into dermatology in 1939 by Edwards and Duntley,[4] reflectance spectroscopic measurements were sparsely used in scientific investigations until the last two decades. However, after the introduction of fiber optics and faster scanners, reflectance spectroscopy soon was considered a dependable and precise method for determination of cutaneous color variations, utilized to monitor dermatological treatment effects and to study skin biophysics.[1,5–7]

In contrast to *in vitro* absorbance measurements in chemistry, *in vivo* cutaneous reflectance spectroscopic data analyses are very complex.[8] The coexistence of several major absorbing molecules in the skin and the heterogeneous optical properties complicate interpretation. The

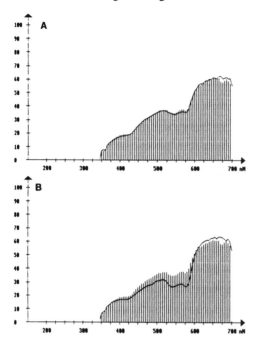

Figure 1 Reflectance measurement on normal skin. The figure shows a series of measurements on adjacent normal skin (horizontal lines) and UVB-induced erythema (solid line), with increased UVB exposure from A to D giving relative reflectance (y-axis) as a function of wavelength (x-axis). Generally, skin reflectance increases from the near-ultraviolet range and throughout the visible part of the light spectrum. (Please see text for further details.)[5]

broad and multiple absorption bands of skin chromophores and the diffuse scattering of incoming light make it difficult to describe exactly the cutaneous optical events mathematically.[7–10] Early studies have been using pigment indexes to quantitate changes in skin color. Recently, an improved method for data analysis, was suggested, using a multiple regression method for calculations of the optical *amounts* of different chromophores present in the skin.[11] The technique was found reliable in *in vitro* systems and at experimentally induced variations in pigment content caused by venous congestion or ultraviolet (UV) light irradiation.[11,12] In the present chapter general aspects of the reflectance spectroscopic technique, different hardware systems, and various methods of data analysis will be discussed.

II. REFLECTANCE SPECTROSCOPY EQUIPMENT

Reflectance spectroscopy of the skin can be performed in the wavelength range from UV to infrared. Most studies have been performed in the near-UV and visible ranges, and generally the skin is irradiated with light ranging from the upper portion of the UV region (350 nm) to the beginning of the infrared (800 nm) part of the spectrum[13] (Figure 1). The irradiation may be either polychromatic (white) with monochromatic detection or monochromatic with polychromatic light sampling. In order to reduce the influence of ambient light, most systems today use polychromatic irradiation with monochromatic detection.

The skin is illuminated through a flexible light guide (fiber), and the backscattered light reflected off the skin surface from the different epidermal and dermal layers is collected by a light-integrating system. Through a similar fiber the reflected light is guided back to a wavelength scanning and detector device. Combined with wavelength signals from the scanner light signals are converted to electrical signals, which are digitized and further processed by a computer[8,9,13–15] (Figure 2).

Several of the commercially available hardware systems used today in dermatology are not developed specifically for *in vivo* cutaneous measurements; however, often only very small changes of the hardware part physically in contact with the skin make these instruments very useful for clinical and experimental investigations. These standard systems generally meet high standards of stability and accuracy.

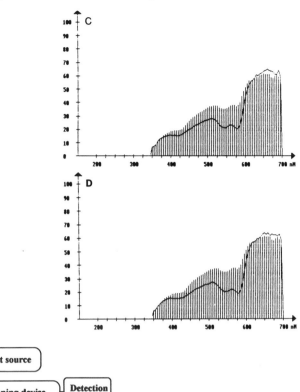

Figure 1 Continued

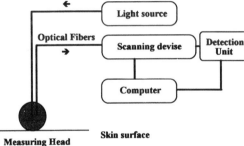

Measuring Head Skin surface

Figure 2 Schematic reflectance unit. The different parts of a reflectance spectroscopic unit are shown.

A. SCANNING SYSTEMS

In principle, three different types of scanners have been used for *in vivo* spectroscopy of the skin: slow-turning prisms or gratings,[13] fast-turning gratings (e.g., Mono Light®, Surrey KT15 2SN, England), and wedge filters combined with diode array systems (Minolta®, Higashi-Ku, Osaka 541, Japan; Carl Ziess®, Optische und Elektronische System Komp., Oberkochen, Germany).

Using a prism or a holographic grating on a reflecting surface, the slowly turning spectrophotometer splits the backscattered light from the skin into a fan-shaped beam slowly swept in front of narrow slits, behind which a detector is placed. Normally, only a single sweep through the wavelength spectrum is performed, allowing several measurements on every single wavelength followed by computerized averaging. The measuring time may last minutes. The fast-turning holographic grating systems incorporate a stabilized motor-driven prism which rotates several times per second. At each rotation one spectrum is registered. A computer system connected to the scanner averages a series of individual spectra in order to increase precision. Accurate and selective wavelength positioning and light detection sensitivity are of utmost importance. Wavelength stability depends highly on the mechanical construction of the prism- or grating-turning mechanism, and the selectivity is mainly influenced by the size of the inlet and outlet slit opening placed in front of the detector. Narrow

A

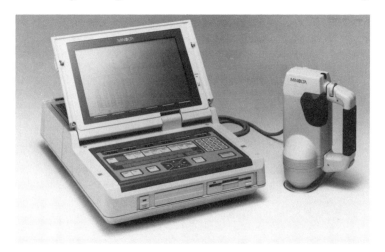

B

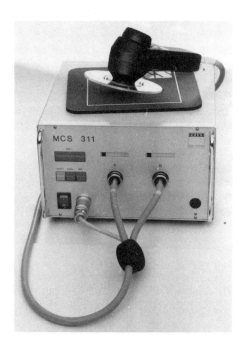

Figure 3   The Minolta (A) and Ziess (B) systems. The figure shows two different reflectance spectroscopic units based on advanced techniques, both using dielectric filters and semiconductor arrays.

slits reduce bandwidth and increase wavelength precision, but also decrease the amount of detectable light. Detector accuracy and sensitivity depend on low dark-current values, temperature stabilizing systems, low-noise amplifiers, and high-resolution A/D converters.

Recently developed spectrophotometers are designed with dielectric filters and an array of semiconductor detectors containing no mechanical, moving parts (see Figure 3). A linear array of, e.g., 1024 closely packed individual photodiodes is covered by a wedge filter, which allows the passage of a specific narrow wavelength range to each underlying photodiode.

Chromophore in vitro absorbance

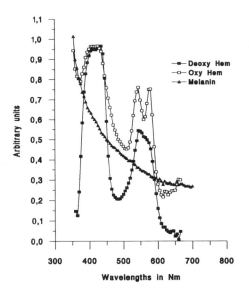

Figure 4 Chromophore *in vitro* absorbance. As a function of wavelength, *in vitro* absorbance of melanin and oxygenized (Oxy Hem) and deoxygenized hemoglobin (Deoxy Hem) is shown. Melanin has the strongest absorption in the near-UV range, with a gradually diminished attenuation. The two forms of hemoglobin absorb specifically. (Please see text for details.)

The skin is irradiated by a short quasi-continuous xenon flash lamp, and the reflected light is recorded by the wedge filter-covered photodiode array. The wavelength specificity is highly dependent on the quality of the light filter and the number of individual photodiodes comprising the array.

B. CHROMAMETRY (THREE-COLOR MEASUREMENTS)

Color measurements using the Commission Internationale de l'Eclairage (CIE) principle and more dedicated tristimulus systems have also been used to evaluate skin color. Tristimulus methods using the CIE three-dimensional color coordinate system measure skin brightness and utilize two color coordinates to express green/red and blue/yellow relations[27] (a detailed description can be found in Chapter 19). Other systems determine erythema and melanin pigmentation based on the work of Diffey, B. L. et al. in 1984,[24] who used red green photodiodes to calculate the logarithmic relation between green and red reflectance to describe skin erythema and red reflectance to measure melanin pigmentation.[14]

III. SKIN OPTICS

In the skin several major absorbing molecules coexist in a heterogeneous way, making the optical properties of skin very complicated. Furthermore, broad and multiple absorption bands of skin chromophores and diffuse scattering of the incoming light make it complex optically to define the events.[7–9,16] When the skin surface is irradiated with light, entering photons may be either absorbed or scattered by different cutaneous molecules (see Figure 4).

In vivo the stratum corneum predominantly induces diffuse forward scattering and only influences the total skin reflectance minimally.[8,9] The epidermal melanin heavily absorbs all wavelengths, but the strongest attenuation occurs in the near-ultraviolet range of the spectrum.[8,9,17] Both oxygenized and deoxygenized hemoglobin absorb the incoming light specifically and influence *in vivo* reflectance spectrograms accordingly.[1,8,9] Oxygenized hemoglobin absorbs strongest at 405 nm, with a relatively low absorbance from 430 until the two characteristic absorbance maxima at 550 and 575 nm, giving a double-peak appearance. Deoxygenized hemoglobin absorbs strongest at 430 nm, followed by relatively low attenuation until 550 nm. Both hemoglobins absorb minimally from 620 nm[8] (Figure 4). Although the

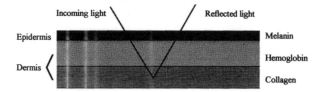

Figure 5   Using the simplified three-layer skin model it is assumed that incoming light passes the epidermal melanin and the nonabsorbed part, which penetrates the epidermis, reaches the hemoglobin containing papillary dermis before being linearly reflected by dermal collagen. It eventually reaches the skin surface, passing the hemoglobin- and melanin-containing layers twice.

hemoglobins absorb potently from 405 to 430 nm, *in vivo* hemoglobin is analyzed most easily from 550 to 575 nm[1,7,9] (Figure 1). This is partially due to a strong melanin influence in the near-ultraviolet range, but is also caused by cutaneous scattering properties of the epidermis especially, allowing diminished penetration of shorter wavelengths.[8,17] Furthermore, dermal collagen, which when purified appears almost 100% white, will reflect all nonabsorbed wavelengths, but since shorter wavelengths scatter more easily, the longer wavelengths tend to penetrate deeper into the dermis and thus are more likely to be absorbed by hemoglobin. Therefore, light of shorter wavelengths is less well represented in the papillary dermis and, hence, is sparsely absorbed by hemoglobin molecules. *In vivo* skin reflectance gradually increases from the UV range until the near-infrared part of the spectrum; this is caused by melanin, with a minor hemoglobin influence from 405 to 430 nm and more pronounced effects seen from 550 to 575 nm[1] (Figure 17–1).

A.   DATA ANALYSIS TECHNIQUES

In order to perform chromophore comparisons several authors have suggested a simplified three-layer model allowing the introduction of pigment indexes as a measure for chromophore content in the skin.[7,9,11,16] Optically, the human skin is then described in terms of two different heavily absorbing and only mildly scattering layers on top of a nearly absolutely reflecting collagen layer in the dermis [7,9,11,16] (Figure 5). The outer stratum corneum, which mainly induces diffuse forward scattering, allows penetration of all wavelengths into the stratum corneum. The incoming light is intensely absorbed here by the epidermal melanin, and the nonabsorbed part, which penetrates the epidermis, reaches the hemoglobin-containing papillary dermis. [7–9,17,18] This layer has the most pronounced dermal influence on skin color. The remaining part of the light is then diffusely reflected by dermal collagen almost linearly, reaching the skin surface after passing through the hemoglobin- and melanin-containing layers twice.[8,9,19]

Using this three-layer optical skin model, skin reflectance is given by the equation[1,7,9,17]

$$R_{tot} = T_1^2 \times T_2^2 \times R_c \qquad (17\text{–}1)$$

Where

R_{tot} = remittance from skin surface minus regular reflectance caused by skin air interface (n=1.55),
T_1 = transmittance through the melanin layer in the epidermis,
T_2 = transmittance through the hemoglobins in the dermis, and
R_c = reflectance from the collagen matrix in the dermis.

Using this very simple Equation 17–1, several groups have developed very useful pigment indexes, calculating the reflectance in a specific part of the spectrum to estimate cutaneous chromophore content. It is assumed that "optical windows" exist which allow different chromophore amounts to be calculated by mathematical analysis of specific and separate parts of the measured *in vivo* skin reflectance. Pigment indexes were proposed to describe melanin pigmentation, bilirubin quantities, and the degree of erythema.[1,7,9,17] Due to the strong UV absorbance of melanin, it has been proposed that the slope of the *in vivo* reflectance spectrogram at 365 to 395 nm correlates with melanin content in the skin.[1,8]

The melanin index (MI) equation is

$$(MI) = 1 \,/\, \text{Slope}_{(365 \text{ nm}-395 \text{ nm})} \times 100$$

Some authors have also suggested using the integrated slope from 705 to 800 nm to measure melanin pigmentation.[8,17,18] Theoretically, it is most reasonable to determine cutaneous melanin concentrations in the range of the spectrum, where it is the major absorbing molecule.[8] However, some hardware systems have low reproducibility in the UV range, and since hemoglobins have a very low absorbance in the near-infrared range this region of the spectrum may serve as an alternative range in which to calculate melanin concentrations with minimal hemoglobin influence.[8,17]

Most groups, however, agree that the determination of erythemas should be done in the area of the spectrum around the double peak caused by oxygenized hemoglobin. The most widely used "erythema index" (EI) was originally defined by Dawson et al.:[9]

$$(EI) = 100 \, (\log 1/R_{560} + 1.5(\log 1/R_{540} + \log 1/R_{575})) - 2 \, (\log 1/R_{510} + \log 1/R_{610})$$

Using these parameters reflectance spectroscopy have been used successfully to describe different dermatological treatment effects objectively including ranking the relative potency of corticosteroids and *in vivo* effects of UV irradiation on both melanin pigmentation and development of erythema.[7,14,20,21] Since none of the skin chromophores absorbs in narrow bands, both high melanin pigmentation and severe erythema could influence the calculations of other chromophore contents.[1,8] In order to improve data analysis further and allow transcutaneous chromophore concentration calculations, the following assumptions were made and mathematical transformations were performed. Since absorption by melanin and hemoglobins was much more prominent than the scattering induced by these layers, the transmittance was approximated according to the Lambert-Beer law.[7,8,9] Then Equation 17–1 can be expressed as

$$-1/2 \, \ln R_c/R_{tot} = {}_{melanin}e^{c_1 d_1} + {}_{ox.hem}e^{c_2 d_2} + {}_{deox.hem}e^{c_3 d_3} \qquad (17\text{–}2)$$

where **e** is the extension coefficient, **c** the concentration and **d** the optical thickness of the chromophore layers. Since Equation 17–2 is defined as a sum, the Choleski decomposition method of multiple regression analysis can be applied.[22,23] By this mathematical method the optical amount of the skin chromophores (**cd**) can be calculated if **e** (the extension coefficient—wavelength-dependent absorbance) is known *a priori*.[22] The mathematical computer estimate of **cd** may be either positive or negative and is given in arbitrary units.[22] In order to perform the estimation of the optical chromophore amount (**cd**), the absorption characteristics of *in vivo* melanin, carotene, and oxygenized and deoxygenized hemoglobin and the reflectance (Rc) of collagen were measured.[11] Essentially, the Choleski decomposition method of multiple regression analysis allows the use of **n** chromophores, and the model can be extended to cover bilirubin or exogene-induced pigments also.[22]

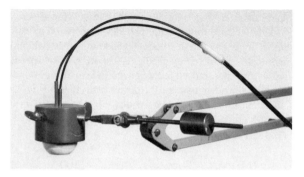

Figure 6   Measuring head. Our detection system consists of a light-integrating sphere and a balanced weight to reduce and standardize the pressure onto the measuring site. The setup furthermore allows small movements by the subject due to respiration, etc.

After measuring *in vivo* skin reflectance, chromophore amounts are detected using only parts of the spectrum and are subsequently incorporated into the calculations of the following pigments. *In vivo* melanin concentration is calculated from 360 to 390 nm, which corresponds to its main absorbing area in the recorded spectrum and is also the area with minimal absorbance of other chromophores.[11] Hemoglobins is then analyzed from 515 to 610 nm, incorporating the previously determined amount of melanin.[11] This means that the degree of erythema is calculated after the computer has "depigmented" the skin in order to reduce chromophore covariation to a minimum. For most practical purposes other chromophores are only included if these pigments are present in relevant amounts.[11]

B. *IN VIVO* REFLECTANCE SPECTROSCOPY— STATE OF THE ART

1. Equipment
Most available systems have been developed for industrial use in the painting industry and generally show inter-measurement variation far below 1% *in vitro*, allowing several measurements in the entire spectrum in seconds. This precision generally exceeds the need in most biological systems.[1,13] However, depending on the design of the detecting devise, some loss of accuracy may be induced in *in vivo* experiments. In most investigations the vascular status is of importance, making it essential to standardize and reduce the weight of the measuring equipment which is placed on the skin surface[12–14,24] (Figure 6). Especially in units with xenon flash systems built into the measuring head the large physical size may impede accurate repositioning onto the measuring site (Figure 3). Previous investigations in the authors' laboratory comparing the reproducibility of repeated measurements on the same anatomical site showed an inter-measurement variation of 5 to 10% (data not published internationally). This difference was induced by small inaccuracies when applying the detection devise to the exact same anatomical location, but also by small physiological time-dynamic changes in vascular status of the test site. With small adaptations for *in vivo* measurements most commercial systems will be able to meet these requirements, but it is very important to describe the nature of the technical alternations exactly to allow data comparisons between laboratories.

2. Data Comparison
When several bioengineering techniques are used to describe the same biological phenomena, authors very often compare the sensitivity of the different types of equipment. For example,

erythema of the skin can be assessed by scanning reflectance spectroscopy, tristimulus color measurements, laser Doppler blood flowmetry, and temperature measurement. It is important to realize that although the different types of measurements may directly or indirectly correlate with identical physiological values, vast differences exist between related measuring techniques.[1,14,25] Tristimulus systems using the CIE three-dimensional color coordinate system measure skin brightness and utilize two color coordinates to express green/red and blue/yellow relations[27] (a detailed description can be found in Chapter 19). Methods based on red/green photodiodes use the logarithmic relation between green and red reflectance to describe skin erythema and use red reflectance to measure melanin pigmentation.[14] Since reflectance spectroscopy measures skin reflectance within a continuous spectrum, differences between various techniques may be observed, e.g., when erythema or vasoconstriction are investigated using reflectance spectroscopic measurements or laser Doppler flowmetry. Although laser Doppler flowmetry and reflectance spectroscopy often show good correlation it is important to understand the differences between the two measured optical phenomena.[2,9] Laser Doppler flowmetry utilizes a monochromatic high-intensity laser beam directed onto the skin surface, determining the Doppler shift caused by moving erythrocytes as a measure of cutaneous blood *flow*.[2] Therefore, laser Doppler flowmetry only includes *moving* hemoglobin, but reflectance spectroscopy measures both arterial and venous hemoglobin absorbance independently of cutaneous blood flow.[9] Since laser Doppler flowmetry can be performed at different wavelengths with moderate to high tissue penetration depths, these measurements may or may not be influenced by the blood flow of the deeper dermal vessels, which is normally not measured using scanning reflectance spectroscopy.[2,12] Specificity and sensitivity of laser Doppler flowmetry or reflectance spectroscopy are often dependent on the nature of the vascular phenomena in question and probably only seldom of different equipment sensitivity.

3. Data Analysis

Originally, reflectance spectroscopic measurements were analyzed by comparing the raw spectra measured at exposed and unexposed sites[4,20] (Figure 17–1). Due to the difficulty of reporting changes by a single parameter, pigment indexes were introduced. These allowed more convenient comparisons of measurements and statistical calculations.[9] Using these methods the effects of several dermatological treatments were objectively shown.[6,21,26] However, due to complicated skin optics some conclusions based on pigment indexes could be questioned theoretically.[1,11] Therefore, it was essential to improve data analysis techniques even further. Computerized calculation of cutaneous chromophore content allows accurate determinations of melanin pigmentation and separation of erythemas into venous and arterial components *after* computerized extraction of the melanin pigmentation component from the *in vivo*-obtained reflectance spectrum.[11] Since the introduction of computerized chromophore analysis, new and important information on the cutaneous vascular physiology has been achieved both in normal and diseased skin and in several dermatologic models.[11,12,15]

IV. SUMMARY

In vivo reflectance spectroscopic measurements allow exact analysis of skin pigmentation and skin color changes. In several studies the method has proven both sensitive and reliable; dermatological models and treatment modalities have been investigated, giving new information about vascular and pigmentary changes. Since many reflectance spectroscopic units are modified for dermatological use, the nature of these alterations must be stated clearly to allow data comparisons between laboratories.

REFERENCES

1. Bjerring, P., and Andersen, P. H., Skin reflectance spectrophotometry, *Photodermatology,* 4, 167, 1987.
2. Nilsson, G. E., Tenland, T., and Oberg, P. Å., Evaluation of a laser Doppler flowmeter for measurement of tissue blood flow, *IEEE Trans. Biomed. Eng.,* 27, 597, 1980.
3. Nilson, G. E., Measurements of water exchange through skin, *Med. Biol. Eng. Comput.,* 15, 209, 1987.
4. Edwards, E. A. and Duntley, S. Q., The pigments and color of human living skin, *Am. J. Anat.,* 65, 1, 1939.
5. Feather, J. W., Ryatt, K. S., and Dawson, J. B., Reflectance spectrophotometric quantification of skin colour changes induced by topical corticosteroid preparations, *Br. J. Dermatol.,* 106, 436, 1982.
6. Ryatt, K. S., Feather, J. W., Dawson, J. B., and Cotterill, J. A., The usefulness of reflectance spectrophotometric measurements during psoralens and ultraviolet A therapy for psoriasis, *J. Am. Acad. Dermatol.,* 85, 558, 1985.
7. Wan, S., Jaenicke, K. F., and Parrish, J. A., Comparison of the erythemogenic effectiveness of ultraviolet-B (290–320 nm) and ultraviolet-A (320–400 nm) radiation by skin reflectance, *Photochem. Photobiol.,* 37, 547, 1983.
8. Anderson, R. R. and Parrish, J. A., Optical properties of human skin, in *The science of Photomedicine, Reagen, J. D., Parrish, J. A., Eds., Plenum Press, New York,* 1982, 147.
9. Dawson, J. B., Barkar, D. J., Ellis, D. J., Grassam, E., Cotterill, J. A., and Fisher, G. W. A theoretical and experimental study of light absorption and scattering by in vivo skin, *Phys. Med. Biol.,* 25, 695, 1980.
10. Diffey, B. L., A mathematical model for ultraviolet optics in skin, *Phys. Med. Biol.,* 1983.
11. Andersen, P. H. and Bjerring, P., Non invasive computerized analysis of skin chromophores in vivo by reflectance spectroscopy, *Photodermatol. Photoimmunol. Photomed.,* 7, 249, 1990.
12. Andersen, P. H., and Abrams, K., Bjerring, P., and Maibach, H., A time correlation study of UVB induced erythema measured by reflectance spectroscopy and laser Doppler flowmetry, *Photodermatol. Photoimmunol. Photomed.,* 8, 123, 1991.
13. Bjerring, P. and Andersen, P. H., Skin reflectance spectrophotometry, *Photodermatology,* 4, 167, 1987.
14. Feather, J. W., Hajizadeh-Saffar, M., Leslie, G., and Dawson, J. B., A portable scanning reflectance spectrophotometer using visible wavelengths for the rapid measurements of skin pigments, *Phys. Med. Biol.,* 34, 807, 1989.
15. Frank, K. H., Kessler, M., Appelbaum, K., and Dummler, W., The Erlangen micro-light-guide spectrophotometer EMPHO 1, *Phys. Med. Biol.,* 34, 1883, 1989.
16. Wan, S., Parrish, J. A., and Jaenicke, K. F., Quantitative evaluation of ultraviolet induced erythema, *Photochem. Photobiol.,* 37, 643, 1983.
17. Kollias, N. and Baqer, A., Spectroscopic characteristics of human melanin in vivo, *J. Invest. Dermatol.,* 85, 593, 1985.
18. Kollias, N. and Baqer, A. H., Quantitative assessment of UV-induced pigmentation and erythema, *Photodermatology,* 5, 53, 1988.
19. Findlay, G. H., Blue skin, *Br. J. Dermatol.,* 83, 127, 1970.
20. Tang, S. and Gilchrest, B., Spectrophotometric analysis of normal, lesional and treated skin of patients with port wine stains (PWS), *J. Invest. Dermatol.,* 78, 340, 1982.
21. Ryatt, K. S., Feather, J. W., Dawson, J. B., and Cotterill, J. A., the usefulness of reflectance spectrophotometric measurements during psoralens and ultraviolet A therapy for psoriasis, *J. Am. Acad. Dermatol.,* 9, 558, 1983.

22. Makridakis, S. G., Multiple regression, in *Forecasting, Methods and Applications,* John Wiley & Sons, New York, 1983, 246.

23. Martin, R. S., Peters, G., and Wilkinson, J. H., Symetric decomposition of a positive definity matrix, *Numeriche Math.,* 7, 362, 1965.

24. Diffey, B. L., Oliver, R. J., and Farr, P. M., A portable instrument for quantifying erythema induced by ultraviolet radiation, *Br. J. Dermatol.,* 111, 663, 1984.

25. Farr, P. M. and Diffey, B. L., A quantitive study of the effect of topical indomethacin on cutaneous erythema induced by UVB and UVC radiation, *Br. J. Dermatol.,* 115, 453, 1986.

26. Queille-Roussel, C., Poncet, M., and Schaefer, H., Quantification of skin colour changes induced by topical corticosteroid preparations using Minolta Chroma Meter, *Br. J. Dermatol.,* 124, 264, 1991.

27. Robertson, A. R., The CIE 1976 color difference formulas, *Color Res. Applic.,* 2, 7, 1977.

Chapter 18

Standardization of Measurements

Peter H. Andersen

CONTENTS

I. INTRODUCTION

The present chapter describes important variables needing attention before and during *in vivo* bioengineering experiments, but with special emphasis on the reflectance spectroscopic technique, suggesting guidelines for human laboratory experiments. Although studies to outline standardization routines probably are performed in most bioengineering laboratories, only a few publications on this topic exist. Recently, however, an article titled "Guidelines for Transepidermal Water Loss (TEWL) Measurements" has been printed bringing new attention to the importance of generally accepted data sampling routines.[1] Despite the fact that most TEWL measurements are performed using identical equipment constructed by the same manufacturer, especially developed for cutaneous investigations, relatively high data discrepancy is still seen between laboratories.[1] Since most commercial reflectance spectroscopic units are made specifically for industrial use in the painting industry and are only modified for *in vivo* human measurements, it becomes even more meaningful to introduce general guidelines for *in vivo* reflectance spectroscopic measurements.

The present chapter is based on previous publications on variables shown to influence skin biophysics in general, and skin color specifically, and the author's own experiments performed through several years working with reflectance spectroscopic skin measurements. Some of the data have not been published earlier internationally.

II. WHAT TO STANDARDIZE

A. EQUIPMENT

In order to control equipment accuracy and prevent any y- or x-axis drift it is important to follow the recommendations given for calibrations, warming up periods, and maintenance. Some reflectance spectrophotometers are modified locally; these systems especially must also include procedures for the added or modified units. Using fiber optics it is significant to realize that some fabrications may darken over time if ultraviolet (UV) light is transmitted. Furthermore, when several users are handling the equipment it may be advisable to keep a log book in which to summarize equipment recommendations and calibration routines to ensure that proper procedures are followed. In a tense situation it may be necessary to repair special sensitive spare parts in order to keep an experiment running when technical help may not be immediately available. Several systems require daily calibration on a 100% white tile made of magnesium oxide.[2-6] This standard should be kept in a clean place and renewed regularly. To perform the measurement as uniformly as possible, physical contact between the measuring unit and the skin should be standardized.[5,6] The weight of the measuring head and the detection angle are the most important variables, and some groups have developed a balanced system minimizing the pressure on the skin and allowing small movements of the panelist due to respiration, etc.[6,7] (please see Chapter 17, Figure 6).

B. PANELISTS
It must be emphasized that included subjects must be healthy; they should be screened for illness by clinical laboratory testing and physical examination. Furthermore, informed consent must be obtained and the experiment must be approved by the local ethics committees. To perform inter-individual comparisons, data must be obtained on a uniform group and both age and sex must be controlled. Furthermore, it is very important to realize that intra-individual variation may be influenced by the menstrual cycle, seasonal variation, and regional cutaneous sensitivity.[1,6,8]

C. SURROUNDINGS
The optimal room climate exists in only a few laboratories, but for practical purposes a well-ventilated, humidity- and temperature-controlled facility may be acceptable.[1,6-9] However, it is extremely important to control these variables exactly, since both ambient temperature and humidity may influence skin biophysics in general but particularly modify skin color of the panelist.[9] Skin reflectance spectroscopic measurements require artificial room illumination to standardize the influence of ambient light and allow an accurate and unchanging "zero" calibration.[9] Finally, it is very important to allow the panelist to adjust to the room; however, in the author's experience acclimatization periods longer than 20 min did not improve measurement accuracy.[9]

D. EXPERIMENTAL PLANNING
To plan an experiment with several panelist arriving at certain time intervals for measurements, strict scheduling is needed to calculate time periods spent with each subject.[10] To do this accurately, pilot experiments are recommended. If several measures are obtained on each subject, time sheets and appointment charts must be prepared beforehand, and computer-controlled data collection will limit sampling errors by a stressed scientist or technician, who may be 1 h behind with a waiting room full of impatient panelists.

E. EXPERIMENTAL SETUP
One of the major advances in cutaneous experiments is the possibility of immediate comparisons between diseased and healthy skin or matching exposed sites to uninvolved areas. Using the subject as his own control, very sensitive paired statistical methods may be included.[9,10] In most cutaneous models either the arm—especially the underarm—or the back is used for investigations. The underarm is very often chosen for the convenience of both the panelist and the investigator. However, this region is probably the anatomical test site with the highest physiological skin biophysical variation per square centimeter.[11,12] Furthermore, panelist movements are often much greater when they are sitting up, able to follow every activity in the investigation facility. Therefore, for most experiments it can be recommended that the subject be placed in a supine position, laying on the stomach and using the back as the test site, applying the compounds in a randomized manner in order not to induce any methodological error.[9,10]

III. DISCUSSION

The present chapter describes general guidelines for reflectance spectroscopic measurements, suggesting useful guidelines to standardize procedures, thus allowing easier comparisons between data obtained in different bioengineering facilities.

To perform accurate and reliable *in vivo* measurements, several factors and sources of variation need attention. General experimental variation is related to panelist homogeneity, the measuring facility, and the reflectance spectrophotometer. Furthermore, skin color may change owing to variations in cutaneous blood flow, skin temperature, mental state, and pressure of the measuring devise on the test site, making it essential to control as many variables as possible through the experimental setup. However, when proper standardization procedures are followed, skin reflectance spectroscopy data are easily obtained, and statistically significant data can be obtained with panelist groups smaller than ten.[10]

REFERENCES

1. Pinnagoda, J., Tupker, A., Agner, T., and Serup, J., Guidelines for transepidermal water loss (TEWL) measurements, *Contact Dermatitis,* 22, 164, 1990.
2. Diffey, B. L., Oliver, R. J., and Farr, P. M., A portable instrument for quantifying erythema induced by ultraviolet radiation, *Br. J. Dermatol.,* 111, 663, 1984.
3. Feather, J. W., Hajizadeh-Saffar, M., Leslie, G., and Dawson, J. B., A portable scanning reflectance spectrophotometer using visible wavelengths for the rapid measurements of skin pigments, *Phys. Med. Biol.* 34, 807, 1989.
4. Mendelson, Y., Kent, J. C., Yocum, B. L., and Birle, M. J., Design and evaluation of a new reflectance pulse oximeter sensor, *Med. Instrum.* 22, 167, 1988.
5. Frank, K. H., Kessler, M., Appelbaum, K., and Dummler, W., The Erlangen micro-light-guide spectrophotometer EMPHO 1, *Phys. Med. Biol.,* 34, 1883, 1989.
6. Bjerring, P. and Andersen, P. H., Skin reflectance spectrophotometry, *Photodermatology,* 4, 167, 1987.
7. Dawson, J. B., Barkar, D. J., Ellis, D. J., Grassam, E., Cotterill, J. A., and Fisher, G. W., A theoretical and experimental study of light absorption and scattering by in vivo skin, *Phys. Med. Biol.,* 25, 695, 1980.
8. Agner, T., Damm, P., and Skouby, S. O., Menstrual cycle and skin reactivity, *J. Am. Acad. Dermatol.,* 24, 566, 1991.
9. Andersen, P. H. and Bjerring, P., Spectral reflectance of human skin in vivo, *Photodermatol. Photoimmunol. Photomed.,* 7, 5, 1990.
10. Andersen, P. H., Abrams, K., Bjerring, P., and Maibach, H., A time correlation study of UVB induced erythema measured by reflectance spectroscopy and laser Doppler flowmetry, *Photodermatol. Photoimmunol. Photomed.,* 8, 123, 1991.
11. Tur, E., Maibach, H., and Guy, R. H., Spatial variability of vasodilation in human forearm skin, *Br. J. Dermatol.,* 113, 197, 1985.
12. Andersen, P. H., unpublished data.

Chapter 19

Chromametry: Hardware, Measuring Principles, and Standardization of Measurements

Peter Elsner

CONTENTS

I. INTRODUCTION

Skin color is a clinical parameter of utmost importance for the dermatologist. Indeed, the experienced clinician may base a number of diagnoses mainly on specific colors of skin lesions. The slate-gray/bluish color of lichen planus and the orange red of pityriasis rubra pilaris are but two examples of the diagnostic value of colors. On the other hand, the intensity of a color may provide useful information on the intensity of a pathological process. Redness is a classical sign of inflammation, and the degree of inflammation may be graded on the degree of redness.

The perceived color of the skin depends on a number of variables, including pigmentation, blood perfusion, and desquamation. Since color perception is a subjective sensory and neurophysiological process, the evaluation of color is highly observer dependent. This has been a concern not only in dermatology, but especially in industries such as dye production and application, printing, etc., where highly consistent colors are necessary. In order to measure color objectively instead of having it judged by subjective observers, color measuring devices (colorimeters) have been developed, of which the Minolta colorimeter (or chromameter) has become especially popular.

While the various instruments in the Minolta chromameter series differ in technical details adjusted to special measuring situations, their measuring principle is identical.

II. HARDWARE

The Minolta chromameters from the 200 and 300 series are lightweight, portable instruments.[1] The measuring unit of the CR 300 has a length of 201 mm with a diameter of 60 mm. It is connected through a 1300-mm cable to a 220 × 200 × 50 mm control unit (DP-301;

Figure 1   The Minolta Chromameter CR 300 with the measuring probe and the control unit.

Figure 1). Whereas all measuring is performed with the measuring unit, calibration, data storage, printing of results, and transfer of data through the serial RS232 interface are done with the control unit.

III. MEASURING PRINCIPLE

A. ILLUMINATION OF THE OBJECT

The perceived color of an object as the spectral distribution of remitted light depends on the light with which the object is illuminated. This refers to both the color of the lighting and the angle at which the light hits the object. Since the perceived color of an object may be expressed as the ratio of the remitted to the absorbed light, equal lighting over the entire visible spectrum is essential. Selective lighting results in limited color information; e.g., a red object illuminated with a red light would appear as black when perceived under a green lamp.

In the Minolta chromameter, lighting is achieved by a xenon flash lamp that emits an intensive white light covering the whole visible spectrum. In order to control for variation in illumination, part of the emitted xenon light is sent to a set of color sensors, whereas the rest illuminates the object (dual-beam system; Figure 2). The color sensors analyze the illuminating light, and through a microprocessor the light remitted from the object is controlled for variations in the illuminating light.

Through the construction of the measuring unit as a hollow chamber the light emitted from the flash lamp hits the object surface from all angles. However, only the light remitted at 0° to the axis of the instrument (90° to the object surface) is collected for color measurement (d/0 measurement principle; Figure 3).

B. SPECTRAL SEPARATION AND MEASUREMENT

The remitted light is transferred to three photodiodes. Before each diode, a color filter ensures a specific spectral sensitivity with peaks at 450 nm (blue), 550 nm (green), and 610 nm (red).

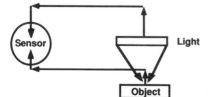

Figure 2   Dual-beam system designed to control for differences in lighting of the object.

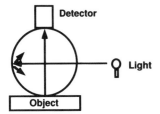

Figure 3   Principle of the d/0 measurement system.

The sensitivity of the photodiodes is adjusted according to a "standard observer" as defined by the Commission Internationale de l'Eclairage (CIE) in 1931. These photodiodes therefore simulate the human eye with its three blue-, green-, and red-sensitive cones in the central foveola. The light reaching the sensors is transformed into electrical signals that are used by the microprocessor for the calculation of the color values in the chosen color space.

The tristimulus chromameter with its three-color-channel design is different from a spectrophotometer that uses multiple sensors to "scan" the remitted light over the whole visible spectrum (for details see Chapter 17). A spectrogram may be transformed into color values analogous to the tristimulus results. However, since these calculations are based on a higher number of values, the spectrophotometric color measurements tend to be more accurate than those obtained with tristimulus instruments.

C. NUMERICAL EXPRESSION OF THE COLOR VALUE

Colors can be described by their hue (color position in the color wheel), their lightness, and their saturation.[2] The first color quantification system was developed in 1905 by the American artist A. H. Munsell, who compared a given color specimen with a collection of paper color clips with different hue, lightness (called value), and saturation (called chroma). In a revised version, this color system is still in use. Any color is expressed as a combination of "H V/C" (hue, value, chroma) in relation to the Munsell color charts.

In 1931 the CIE introduced the XYC color space based on the sensory physiology of vision, where each value corresponds to the intensity of primary color receptor (red, green, and blue) stimulation. The Yxy color space is based on the XYC space, but allows a better visualization for colors with identical lightness.

In 1976 the CIE defined the L*a*b* color space (CIELAB) that is easily visualized and that has been used in practically all skin bioengineering studies. The CIELAB space is defined by three axes vertical to each other (Figure 4): a* is the red-green axis with −a* green and +a* red, and b* is the yellow-blue axis with −b* blue and +b* yellow. The third axis is L* for lightness.

The L*C*H° color space is based on the CIELAB space, but uses cylindrical coordinates instead of rectangular coordinates.[2] The Hunter Lab color space is another modification of the CIE Yxy color space. The Minolta chromameters display color values according to the following systems: XYZ, Yxy, L*a*b*, L*C*H°, and Hunter Lab.

D. CALCULATION OF COLOR DIFFERENCES

One important advantage of the CIELAB system is the easy calculation of distance between points in the color space, which follows simple vector geometry. The distance between points

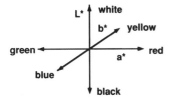

Figure 4   CIE L*a*b* color system.

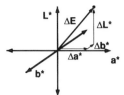

Figure 5 Color difference in the CIE L*a*b* space.

on the axes is calculated as the arithmetic difference; e.g., on the red-green axis a* a color difference may be expressed as

$$\Delta a^* = a^*_1 - a^*_2$$

The distance between two points anywhere in the color space has to include all three axes, and it is given by the CIELAB formula (Figure 5):

$$\Delta Ea^*b^* = \mathrm{sqrt}(\Delta L^{*2} + \Delta a^{*2} + \Delta b^{*2})$$

The distinction of colors by a standard observer is not equally distributed over the a*b* plane; it is more accurate in the center (toward a* = 0 and b* = 0) and less precise in the periphery.[3] Therefore, the color distance given by the CIELAB formula is more precise than the human eye for highly positive values of a* and b*. This has been a concern for quality control in industry, where imperceptible color differences lead to the rejection of materials. Therefore, other formulas for the calculation of color differences were developed. In dermatological applications the high precision of the chromameter in distinguishing extreme colors would be welcome and the CIELAB formula should be applied.

IV. STANDARDIZATION OF MEASUREMENTS

In skin bioengineering the chromameter has been used for the quantification of erythema in the study of irritant dermatitis,[4–8] and, in the context of dermatopharmacological studies, for estimating topical corticoid activity in the vasoconstriction test[9–11] and measuring percutaneous penetration in the nicotinate model.[12] In photodermatology the instrument has been used for the quantification of pigmentation in the assessment of skin type,[13] photosensitivity,[14,15] and pigmenting reaction to psoralen-ultraviolet A (PUVA) therapy.[16]

Although no formal guidelines for the standardization of chromametry in skin bioengineering have been published, a number of procedures seem to be reasonable and are practiced by most researchers (Table 1).

Table 1 Recommendations for chromameter measurements in skin bioengineering studies

- Control ambient environmental conditions, including room temperature and humidity, since they will influence skin perfusion.
- Have the volunteer or patient rest and unocclude the measurement site for a minimum of 15 min before taking measurements.
- Take measurements at the same time of the day to reduce day-to-day variation.
- Always include the measurement of an untreated reference skin site.
- Avoid applying pressure on the measurement site since this will cause blanching of the skin.
- Repeat measurements at the same site three times and take the mean.

Table 2   **Correlation of reflectance parameters with visual scores**

Luminance	(ΔL^*)	-0.45
Redness	(Δa^*)	0.80
Yellow hue	(Δb^*)	-0.45
Total	(ΔEa^*b^*)	0.85

Note: Data from Reference 6. All correlation coefficients are highly significant.

A. SENSITIVITY

In several studies it was shown that the color as measured with the chromameter is proportional to visual scoring.[4,5,7,8] Correlation coefficients between reflectance parameters and visual scores are given in Table 2. However, "sensitivity" of the chromameter, meaning the detection of a minimal redness compared to normal skin, is not superior to the human eye. This is to be expected since for low values of redness the instrument is adjusted to the "standard human observer." The situation is different for the assessment of erythema intensity, where the chromameter is superior to the human eye.

B. REPRODUCIBILITY

In the measurement of standard color plates the reproducibility of the instrument is very high. As expected, the reproducibility is much lower in *in vivo* situations in man and even lower when considering day-to-day variation. Therefore, chromameter data should be preferably used as relative data in comparison to an intraindividual standard of "normal skin".

Serup and Agner reported an excellent technical reproducibility of measurements with the chromameter, although less so on the a* axis compared to the other axes.[8] In a recent study Lahti et al. compared the Minolta chromameter with a spectroradiometer, a two-channel erythema meter, and a laser Doppler flowmeter (LDF) in the assessment of erythema induced by benzoic acid, methyl nicotinate, and UV irradiation.[4] The *in vivo* repeatability of the measurements as lowest for the LDF, but it detected clinically invisible blood flow changes. The reproducibility was highest for the two-channel erythema meter and the Minolta chromameter.

REFERENCES

1. Chroma-Meter CR-300, CR-310, CR-331, Minolta Camera Co. Ltd., 1991.
2. Precise Color Communication. Color Control from Feeling to Instrumentation, Minolta Camera Co. Ltd., 1993.
3. McDonald, R., European practices and philosophy in industrial colour-difference evaluation, *Color Res. Applic.*, 15, 249, 1990.
4. Lahti, A., Kopola, H., Harila, A., Myllälä, R., and Hannuksela, M., Assessment of skin erythema by eye, laser Doppler flowmeter, spectroradiometer, two-channel erythema meter and Minolta chroma meter, *Arch. Dermatol. Res.*, 285, 278, 1993.
5. Wilhelm, K. P., Surber, C., and Maibach, H. I., Quantification of sodium lauryl sulfate irritant dermatitis in man: comparison of four techniques: skin color reflectance, transepidermal water loss, laser Doppler flow measurement and visual scores, *Arch. Dermatol. Res.*, 281, 1989.
6. Wilhelm, K. P. and Maibach, H. I., Skin color reflectance measurements for objective quantification of erythema in human beings, *J. Am. Acad. Dermatol.*, 21, 1989.
7. Babulak, S. W., Quantitation of erythema in a soap chamber test using the Minolta Chroma (Reflectance) Meter: comparison of instrumental results with visual assessment, *J. Soc. Cosmet. Chem.*, 37, 475, 1986.

8. Serup, J. and Agner T., Colorimetric quantification of erythema—a comparison of two colorimeters (Lange Micro Color and Minolta Chroma Meter CR-200) with a clinical scoring scheme and laser-Doppler flowmetry, *Clin. Exp. Dermatol.*, 15, 267, 1990.

9. Chan, S. Y. and Po, A. L., Quantitative skin blanching assay of corticosteroid creams using tristimulus colour analysis, *J. Pharm. Pharmacol.*, 44, 371, 1992.

10. Queille, R. C., Poncet, M., and Schaefer, H., Quantification of skin-colour changes induced by topical corticosteroid preparations using the Minolta Chroma Meter, *Br. J. Dermatol.*, 124, 264, 1991.

11. Pershing, L. K., New approaches to assess topical corticosteroid bioequivalence: pharmacokinetic evaluation, *Int. J. Dermatol.*, 1, 14, 1992.

12. Berardesca, E., Cespa, M., Farinelli, N., Rabbiosi, G., and Maibach, H., In vivo transcutaneous penetration of nicotinates and sensitive skin, *Contact Dermatitis*, 25, 35, 1991.

13. Andreassi, L., Casini, L., Simoni, S., Bartalini, P., and Fimiani, M., Measurement of cutaneous colour and assessment of skin type, *Photodermatol. Photoimmunol. Photomed.*, 7, 20, 1990.

14. Seitz, J. C. and Whitmore, C. G., Measurement of erythema and tanning responses in human skin using a tri-stimulus colorimeter, *Dermatologica*, 177, 70, 1988.

15. Westerhof, W., Estevez, U. O., Meens, J., Kammeyer, A., Durocq, M., and Cario, I., The relation between constitutional skin color and photosensitivity estimated from UV-induced erythema and pigmentation dose-response curves, *J. Invest. Dermatol.*, 94, 812, 1990.

16. Marrakchi, S., Decloquement, L., and Pollet, P., et al., Variation in 8-methoxypsoralen profiles during long-term psoralen plus ultraviolet A therapy and correlations between serum 8-methoxypsoralen levels and chromametric parameters, *Photodermatol. Photoimmunol. Photomed.*, 8, 206, 1991.

Chapter 20

Erythema Measurements in Diseased Skin

Enzo Berardesca

CONTENTS

I. INTRODUCTION

The noninvasive quantification of skin color and, in particular, erythema is a useful tool that can help the dermatologist to assess skin disease and evaluate the efficacy of treatment. One of the first approaches to the instrumental quantification of erythema was done by Daniels and Imbrie in 1958.[1] They used a tristimulus reflectance meter to check the reliability of four filters (red, green, blue, and amber) in assessing erythema and tanning, reporting that the green filter was more sensitive to the effects of erythema than the amber and the blue. On the other hand, the red filter was only slight sensitive to erythema; therefore, it is not useful in quantifying red color of the skin. All four filters were capable of measuring melanin, even though the red one is more sensitive in measuring pigmentation in the absence of erythema. The overall study confirmed previous findings[2] and recommended for an appropriate analysis of skin color a blue filter with maximum transmittance at 460 nm, a filter with transmittance at 542 or 576 nm for maximum response to oxyhemoglobin, and a filter with maximum transmittance between 620 and 660 nm for sampling the region in the red end of the visible range, which depends primarily on melanin.

Since then several instruments and analytical methods of skin spectra have been developed and used to investigate skin disease and to quantify the effects of topically applied drugs.[3-5]

II. SKIN DISEASE

Psoriasis has been evaluated by Ryatt et al.[6] using remittance spectroscopy. Their study investigated erythema and melanin indexes in subjects undergoing psoralen-ultraviolet A (PUVA) treatment. Measurements were done on psoriatic lesions and unaffected skin. The mean initial erythema index of uninvolved skin rose from 33 to 54 units during 4 weeks of treatment (44.6 J/cm^2); accordingly, erythema of psoriatic lesions decreased from 108 to an average value of 70 in the same period. Initial responses were dose dependent, and the technique was shown to be very useful in monitoring the dose/benefit ratio. Indeed, higher doses were needed to produce the same therapeutic efficacy after 4 weeks when the protective effect of pigmentation was present. The final erythema index 1 month after the last dose of UVA was greater than that of uninvolved skin, confirming that an abnormal vascular pattern was still present even though the lesion was apparently cleared. In conclusion, the study showed that disease activity could be mirrored by the vascular compartment of psoriatic skin.

0-8493-8371-4/95/$0.00+$.50
© 1995 by CRC Press, Inc.

The changes in erythema response, measurable by remittance spectroscopy but not by the naked eye, can be used as a basis to discontinue PUVA therapy despite the presence of some degree of psoriasis, clinically reducing the cost of treatment and side effects for patients.

Patients with icterus or anemia and those undergoing hemodialysis were investigated by Deleixe-Mauhin et al.[7] using a chromameter. The skin color of patients undergoing hemodialysis was different from controls for L* (low) and b* (high) values, but not for a* values. No relationship was found between skin color and duration of hemodialysis or treatment with erythropoietin. Significant differences were detected in erythema (a* values) between controls (7.9) and patients with anemia (5.8) and icterus (6.2).

One of the most fascinating areas of investigation using remittance spectroscopy techniques is the early detection of malignant melanoma and its differentiation from benign nevi. Using a sophisticated mathematical approach, Marchesini et al.[8] derived from spectra ranging from 420 to 780 nm four variables that significantly differed when comparing melanoma and nevi. Two of these variables are related to the mild presence of hemoglobin in melanoma due to a different vascularization: indeed, in melanoma, during vertical growth, blood vessels surround a closely packed neoplastic proliferation, whereas in nevi the vascular network is diffuse and growing between variously sized nets of nevus cells. In addition, benign nevi lack aggressive infiltration of the epidermis, and light can penetrate into the dermis where blood vessels are located. These data confirm changes in blood flow recorded in malignant pigmented lesions using laser Doppler velocimetry.[9] However, using this technique, blood flow is reported to be increased in malignant melanoma. It is clear that even though the two techniques (remittance spectroscopy and laser Doppler velocimetry) are both used to quantify erythema, they do not give the same information. The study in remittance spectroscopy[8] also shows that quantification of the total amount of melanin is not discrimant in the differential diagnosis: the data agree with the clinical criteria that the appearance of the lesion (dark or light) is not critical to the diagnosis.

III.   IRRITATION

The instrumental evaluation of erythema in the course of irritant reactions has been proposed by several investigators and since the development of cheaper and portable instrumentation has become a practice widely used in many labs.

Wilhelm and Maibach,[10] using a tristimulus colorimeter, correlated the degree of erythema (a* value) with the concentration of the irritant applied (from 0.125 to 3.0% sodium lauryl sulfate [SLS]). Differences in red color reflectance (Δa* value) and change in total skin color (ΔE*ab) were highly significantly correlated with the applied dose of SLS ($p < 0.0001$) and to visual scores. Tristimulus reflectance spectroscopy showed low variability in measuring different surfaces repeatedly, but a considerable variation in measuring untreated skin from day to day.[10] In a further study[11] the same group compared color reflectance with transepidermal water loss (TEWL) and laser Doppler velocimetry in the evaluation of surfactant-induced irritation. All techniques detected significant irritation after the application of the lowest dose of the irritant (0.125% SLS), with the highest level of significance for TEWL. The study concludes that although TEWL is the more sensitive technique to measure irritation, color reflectance is a helpful complementary tool to evaluate skin redness.

Patch test reactions have been quantified by Mendelow et al.[12] using remittance spectroscopy: the technique was highly correlated with visual scoring, and within the reactions classified with the same score it was possible to differentiate allergen formulations. Color reflectance also has been compared to laser Doppler velocimetry in the quantification of patch test reactions:[13] both techniques have very good reproducibility and reliability. These methods appear to some extent complementary in the evaluation of erythema since laser Doppler velocimetry measures total blood flow, mainly determined by the arteriolar tone, whereas colorimetry by the Commission Internationale de l'Éclairage (CIE) system, which

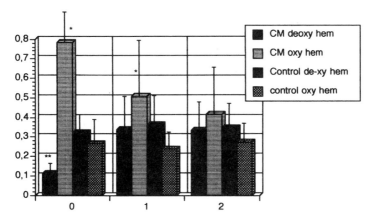

Figure 1 Relative changes in oxygenized (oxyhem) and deoxygenized hemoglobin (deoxyhem) after skin delipidization with chloroform/methanol (CM) compared to control skin. CM only produced subjective and objective changes with a significant increase of oxygenized hemoglobin at 0 h and 1 h * and reduction of deoxygenized hemoglobin at 0 h (**) compared to control skin (*, ** $p < 0.01$). (From Berardesca, E., Andersen, P. H., Bjerring, P., and Maibach, H., *Contact Dermatitis*, 27, 8, 1992. With permission.)

takes into account the non-linear color perception of the human eye, mainly reflects the capillary accumulation of skin blood flow.

Erythema induced by organic solvents has been evaluated recently by Berardesca et al.[14] using remittance spectroscopy with an improved data analysis, allowing the quantification of the amount of oxygenized and deoxygenized hemoglobin after chemical irritation of the skin with organic solvents. Erythema caused by application of a chloroform/methanol mixture (2:1) induced a significant increase of oxygenized hemoglobin (Figure 1) due to dilatation of arterioles in the dermal subpapillary plexus. Chloroform/methanol extracts polar lipids, sphingomyelin, and ceramides and causes skin irritation associated with erythema and painful sensation. On the other hand, delipidization with ether/acetone did not result either in visible changes or changes in oxygenized hemoglobin. This computerized approach to the investigation of erythema and blood flow allows a better understanding of skin pathophysiology, with specific regard to the dynamics of blood vessels during contact dermatitis.

A comparison between color reflectance readings for quantifying erythema and cellular infiltrate during irritant reactions induced by anthralin has been done by Westerhof et al.[15] Visual estimation of erythema may be misleading since the redness observed in pigmented skin is known to be comprised of genuine erythema and the red component of brown color. The study showed more pronounced changes of skin redness in patients with vitiligo not detectable by the naked eye; these changes in erythema were correlated with the presence of granulocytes and monocytes in the epidermis of vitiliginous skin. Chromameter readings indicated that the erythematous reaction was higher in vitiliginous skin, confirming that the human eye is not accurate in the quantitative assessment of complex colors.

IV. URTICARIA

Weal formation has been evaluated in prick test reactions by remittance spectroscopy and compared to skin thickness measurements and visual scoring;[16] discrimination between negative and positive (+/++) reactions is significant. Baseline skin shows an erythema index of 36, whereas the index for positive reactions (+) is 71.3 (Figure 2). Negative skin sites have a slightly higher (but not significant) erythema index compared to basal skin due to a

Visual score	Mean	± SD
+*	71.3	9.9
++°	56.8	12.8
+++^	60.2	13.3
negative	47.7*^	12.4
basal	36.6*°^	11.4

Figure 2 Evaluation of prick test reactions with remittance spectroscopy. Erythema index scores were compared to visual assessment. *°^: significant differences were found between groups. (From Berardesca, E., Gabba, P., Nume, A., Rabbiosi, G., and Maibach, H., *Acta Derm. Venereol.,* 72, 261, 1992. With permission.)

dermographic reaction related to the procedure of prick testing itself. Remittance spectroscopy is of questionable value in assessing stronger reactions (++/+++) because of the reduction of blood flow and hemoglobin content associated with skin whitening in the center of the lesion and because of the probable maximization of skin blood flow which occurs already at ++ levels. Ultrasound for measurements of skin thickness can be used in combination with remittance spectroscopy since it is fast and easy to perform, even though it detects mainly macroscopic changes during weal formation[16] and, therefore, can quantify only important reactions.

REFERENCES

1. Daniels, F. and Imbrie, J. D., Comparison between visual grading and reflectance measurements of erythema produced by sunlight, *J. Invest. Dermatol.,* 30, 295, 1958.
2. Lasker, G. W., Seasonal changes in skin color, *Am. J. Phys. Anthropol.,* 12, 553, 1954.
3. Andersen, P. H. and Bjerring, P., Spectral reflectance of human skin in vivo, *Photodermatol. Photoimmunol. Photomed.,* 7, 5, 1990.
4. Anderson, R. R. and Parrish, J. A., The optics of human skin, *J. Invest. Dermatol.,* 77, 13, 1981.
5. Weatherall, I. L. and Coombs, B. D., Skin color measurements in terms of CIELAB color space values, *J. Invest. Dermatol.,* 99, 468, 1992.
6. Ryatt, K. S., Feather, J. W., Dawson, J. B., and Cotterill, J. A., The usefulness of reflectance spectrophotometric measurements during psoralens and ultraviolet A therapy for psoriasis, *J. Am. Acad. Dermatol.,* 9, 558, 1983.
7. Deleixe-Mauhin F., Krezinski, J. M., Rorive, G., and Pierard, G. E., Quantification of skin color in patients undergoing hemodialysis, *J. Am. Acad. Dermatol.,* 27, 950, 1992.
8. Marchesini, R., Cascinelli, N., Brambilla, M., Clemente, C., Mascheroni, M., Pignoli, E., Testori, A., and Venturoli, D. R., In vivo spectrophotometric evaluation of neoplastic and nonneoplastic skin pigmented lesions. II. Discriminant analysis between nevus and melanoma, *Photochem. Photobiol.,* 55, 515, 1992.
9. Tur, E. and Brenner, S., Cutaneous blood flow measurements for the detection of malignancy in pigmented skin lesions, *Dermatology,* 184, 8, 1992.
10. Wilhelm, K. and Maibach, H., Skin color reflectance measurements for objective quantification of erythema in human beings, *J. Am. Acad. Dermatol.,* 21, 1306, 1989.

11. Wilhelm, K., Surber, C., and Maibach, H., Quantification of sodium lauryl sulphate irritant dermatitis in man: comparison of four techniques: skin color reflectance, transepidermal water loss, laser Doppler flow measurement and visual scores, *Arch. Dermatol. Res.*, 281, 293, 1989.

12. Mendelow, A. Y., Forsyth, A., Feather, J. W., Baillie, A. J., and Florence, A. T., Skin reflectance measurements of patch test responses, *Contact Dermatitis*, 15, 73, 1986.

13. Serup, J. and Agner, T., Colorimetric quantification of erythema—a comparison of two colorimeters (Lange Micro Color and Minolta Chroma Meter CR 200) with a clinical scoring scheme and laser Doppler flowmetry, *Clin. Exp. Dermatol.*, 15, 267, 1990.

14. Berardesca, E., Andersen, P. H., Bjerring, P., and Maibach, H., Erythema induced by organic solvents: in vivo evaluation of oxygenized and deoxygenized haemoglobin by reflectance spectroscopy, *Contact Dermatitis*, 27, 8, 1992.

15. Westerhof, W., Buehre, Y., Pavel, S., Bos, J. D., Das, P. K., Krieg, S. R., Siddiqui, A. H., Increased anthralin irritation response in vitiliginous skin, *Arch. Dermatol. Res.*, 281, 52, 1989.

16. Berardesca, E., Gabba, P., Nume, A., Rabbiosi, G., and Maibach, H., Objective prick test evaluation: non invasive techniques, *Acta Derm. Venereol.*, 72, 261, 1992.

Chapter 21

Skin Pharmacology

Ethel Tur

CONTENTS

I. INTRODUCTION

The determination of the presence and concentration of pharmaceutical substances in the skin is of paramount importance to many aspects of general and skin pharmacology, such as transdermal penetration of drugs. Since the skin is partially transparent in a fairly wide wavelength range, beginning in the ultraviolet and ending in the infrared, qualitative as well as quantitative information about cutaneous concentration and pharmacodynamics of drugs can be deduced from noninvasive spectroscopic studies of the skin. When light illuminates the skin, part of it will be specularly reflected due to the refractive index mismatch at the air-skin interface. Another part enters the skin and, depending on the wavelength, is then subject to a series of absorption, scattering, and reflection processes that take place at the stratum corneum, epidermis, dermis, and subcutaneous fat layer.[1] The scattered light which comes out from the skin carries spectral signatures of the skin chromophores and of other pharmaceutical agents when present. This scattered light is also called "diffusely reflected" or "remitted" light and these three terms will be used interchangably.

A variety of spectroscopic techniques can be used to measure this remittance, from large complex, laboratory-oriented spectrophotometers down to hand-held instruments that provide only limited spectral information. These techniques can be broadly divided into two classes:

1. Detailed spectroscopic studies—here one characterizes the diffuse reflection at each wavelength in a given spectral range. Common ranges include the ultraviolet (UV), visual, and infrared. Practically, finite resolutions must be used; i.e., the detector always "sees" a few nanometers around the wavelength of interest. The number of independent spectral channels collected is given by the ratio (covered spectral range)/(resolution bandwidth). This number is quite large so that such studies produce a very detailed spectral signature of the skin. However, they require the use of relatively expensive, complex, and large measuring instruments.

2. Studies based on the measurement of only a few spectral degrees of freedom. Using one to three standardized spectral bands, two instruments have recently gained popularity. Chromameters are relatively compact instruments originally developed for color measure-

ments (chromametry, colorimetry); they use a white light source and measure the remitted light through three standardized filters (hence the name tristimulus color analysis). They are moderately expensive but easy to operate. With the melanin/hemoglobin spectrophotometer just the green and red wavelength bands are of importance, corresponding to the absorption spectra of hemoglobin and melanin. Green and red light-emitting diodes are used, and their readings can be converted to melanin and erythema indices. These instruments are very simple to operate, relatively inexpensive, and handy.

Specialized devices which include only a single spectral channel to monitor one characteristic of the scattered light have also been proposed.

Spectroscopic techniques in skin pharmacology have many advantages. They are sensitive, reproducible, and noninvasive, while providing instant *in vivo* objective assessments. There are very few limitations as to the choice of body region and site of examination. Since certain spectral signatures are uniquely associated with the chemistry of the drug or one of its constituents, these methods, using different wavelengths for different purposes, can trace the metabolism of the drug or its release from its vehicle. Obviously, special markers, having very distinct spectra, also can be used to advantage in following pharmacological processes in the skin.

Indeed, reflectance spectrophotometry was able to record erythema at lower dilutions of red color than the human eye, thus proving more sensitive at red color perception.[2] The same was shown *in vivo* when the spectrophotometric instrument detected erythema at UV radiation doses below the minimal visually detectable erythema.[2,3] It was also more sensitive in perceiving brown ink as well as *in vivo* tanning.[3] In addition to being more sensitive than judgments made by the eye, it is also more reproducible.[1,4]

As for reproducibility, both colorimetry and direct spectroscopy were found to be highly reproducible by Deleixhe-Mauhin et al.,[5] who compared the skin color of healthy controls to patients with uremia, anemia, or icterus with a chromameter, and by Berardesca et al.,[6] who monitored prick test reactions with a UV spectrophotometer. These latter investigators also found the technique to be very sensitive to small erythematous changes.

This chapter reviews applications of remittance spectroscopy and chromametry to skin pharmacology. Oddly enough, the potential of spectroscopy in the study of skin pharmacology has not been fully and systematically investigated. More often than not, the importance of spectroscopic techniques in skin pharmacology appears in the literature only as a by-product of the main objectives of the investigators.

The following sections survey the pharmacological issues which were addressed using remittance (reflectance) spectroscopy and chromametry (colorimetry). The main topics to be covered are direct and indirect evaluation of percutaneous penetration, hydration of the skin and barrier function, action of topical corticosteroids, transdermal delivery-drug release from patches, measurements of systemically administered drugs which affect the skin color, and systemic drugs which affect skin diseases that can be evaluated by the method.

II. PENETRATION OF DRUGS THAT HAVE A CHARACTERISTIC OPTICAL ABSORPTION

Remittance spectroscopy allows *in vivo* detection of topically applied drugs, whose penetration into skin can be evaluated from the time dependence of the remittance. Kolmel et al.[7] studied the penetration behavior of UV-absorbing drugs (sunscreens) after topical application, utilizing spectral remittance in the UV region. Since the penetration depth of the incident light in the UV spectral region is restricted to the stratum corneum and epidermis, the method allows the determination of drug distribution only in these regions. After topical application the drug shades the skin; this lowers the remittance. When released from the vehicle the drug slowly penetrates into the stratum corneum, which is the main diffusion barrier, and then diffuses into the epidermis and finally reaches the dermis. As the drug is transported into deeper skin layers the effect on the remittance signal, which at UV wavelengths results from scattering of light in the stratum corneum and epidermis, decreases and the remittance approaches that of untreated skin. The time dependence of remittance reflects variations of drug distribution in the areas of skin penetrated by the incident light. Thus, a quantitative description of the pattern of remittance after topical application of a light-absorbing drug can be obtained by taking into account the optical properties of untreated skin and the variations due to the time-dependent distribution of the light-absorbing drug. The spectral remittance of the skin is lowered after topical application of a vehicle film containing the light-absorbing drug since the light has to pass through the film. Therefore, measurements were also done on skin treated with unloaded vehicle. The difference between the two curves shows the *in vivo* absorption spectrum of the agent tested.

Mak et al.[8] followed drug transport across the stratum corneum, monitoring the outer stratum corneum by reflectance infrared spectroscopy. In this manner they demonstrated enhancement of percutaneous penetration. They chose a penetrant (4-cyanophenol) which has a unique absorption in the infrared spectrum, distinct from those of the stratum corneum and the vehicle components. They followed its penetration in a solution containing propylene glycol alone or propylene glycol and oleic acid as an enhancer. They also followed the solvents alone. The infrared absorption corresponding to the drug diminished faster when oleic acid was present in the solution, and an infrared absorption due primarily to propylene glycol also disappeared more quickly following application of the enhancer-containing solution. Moreover, the formulations with oleic acid were the only ones which induced a higher wave number shift in the frequency of the asymmetric C-H bond-stretching absorbance, indicating increased lipid-chain disorder—which is the assumed mechanism of percutaneous penetration enhancement by oleic acid. Thus, they could follow and study the kinetics of the penetration *in vivo* and could consider the mechanism of action of oleic acid as a penetration enhancer.

III. INDIRECT EVALUATION OF PENETRATION THROUGH MEASURABLE EFFECTS OF THE DRUG

Mendelow et al.[9] used reflectance measurements at 566 and 640 nm to evaluate patch test responses. Among other allergens, various formulations of nickel sulfate were tested in nickel-sensitive subjects. They included a formulation containing a penetration enhancer: 2% 1-dodecylazacycloheptan-2-one, which was added to the gel formulation. The addition of this penetration enhancer caused a greater mean response than the gel vehicle without the penetration enhancer. However, this difference was not statistically significant. In another study,[10] by measuring the erythemal allergic response they were able to demonstrate the inhibitory effects of a paraffin vehicle in nickel release. Thus, by measuring the erythemal results of the pharmacological process one can deduce the influence of various pharmacological factors without actually measuring them.

Reflectance measurements can be applied to the study of the time course of penetration and elimination of topical drugs. Lanigan and Cotterill[11] applied a vasodilating cream and a vasoconstricting cream to both normal skin and port wine stains and found a diminished response in the latter. Because the time to maximum vasodilation and resolution induced by the vasodilating cream was similar in both sites, it is improbable that the absorbed vasoactive agents were removed more rapidly in the port wine stain due to its increased vascularity. Therefore, one can deduce that the diminished response is due to a reduced autonomic innervation of cutaneous vasculature in port wine stains. This study demonstrates the importance of measuring not only the magnitude of the response, but also its time course. Using this technique it is possible to measure both and to arrive at conclusions which are otherwise difficult or even impossible to obtain.

Measurements of the magnitude of a hyperemic response as well as its time course were also done by Clarys and Barel,[12] who used chromametry (and laser Doppler flowmetry) to measure the percutaneous absorption of methyl nicotinate on different body regions. Regional variations in percutaneous absorption of methyl nicotinate were exhibited. Pretreatment of the skin with a plastic occlusion modified the permeability to methyl nicotinate, as was indicated by the time to onset and the magnitude of the erythema induced by the penetration of the drug.

Andersen et al.[13] utilized a reflectance spectrophotometer equipped with a computerized data analysis system to study the relationship between the dissociation constants of chemicals and skin irritation. In addition to the results related to their main objective in the study, they found a linear dose-response relationship between the oxygenized hemoglobin value as measured by the spectrophotometer and the concentration of imipramine. Since skin irritation is dependent on percutaneous absorption of the irritating drug, such measurements may give an indication of the degree of percutaneous absorption following topical application of different drug concentrations.

This method was used to study the effect of prostaglandin inhibition on anthralin- and UVB-induced inflammation using topical indomethacin.[14] Indomethacin gel application caused a small but significant reduction in the erythematous component of anthralin-induced inflammation. Maximum reduction in anthralin erythema following application of indomethacin gel was at 6 h. The effect of indomethacin was small and was greatest in the first 4 h after anthralin application. Previous studies of the effect of prostaglandin inhibition using oral aspirin and indomethacin failed to show any effect on anthralin-induced inflammation when assessed visually. Therefore, the importance of the study in this context is not only in illustrating that a pharmacological effect can be measured by the method, but also in demonstrating that subjective assessments may fail to identify small but significant differences in anthralin erythema. The slight inhibition of part of the inflammatory reaction to anthralin caused by indomethacin, as demonstrated by Ramsay et al.,[14] implies that although prostaglandins may be involved in the early phase of the response, other mediators are involved in the major portion of the inflammatory response.

Sharpe and Shuster[15] studied the effect of topical indomethacin on the wheal and flare response in chronic urticaria. Prior to pursuing their main goal, they first tried to establish whether topical indomethacin penetrated the skin and how bioactive it was. They confirmed the bioefficacy by measuring inhibition of the UVB erythema response, using a reflectance instrument to measure the effect of UVB on test sites treated with either indomethacin gel or gel base alone. The increase in erythema following UVB irradiation was lower in the indomethacin-treated sites. Thus, the reflectance measurements confirmed the penetration and bioavailability of the indomethacin gel. There was no relationship between the degree of inhibition of UVB erythema and the degree of augmentation of the wheal and flare or the degree of lowering the threshold of whealing by indomethacin.

IV. HYDRATION OF THE SKIN
AND BARRIER FUNCTION

A water-absorbant band exists in the mid-infrared region, where skin and most topical substances show no absorbance. This enables the assessment of water concentration in the uppermost layers of the stratum corneum. The advantage of this method for our topic is that water absorbance is far removed from peaks caused by commonly applied substances. Thus, Potts et al.[16] have shown that petrolatum forms a very effective occlusive barrier which allows increased hydration of the underlying stratum corneum. Removing the petrolatum, the water content returns to pretreatment values within 30–45 min. They were able to show a quantitative relationship between infrared absorbance and water concentration. With variable-angle optics it is also possible to study the water concentration at varying depths, thus sectioning the stratum corneum optically.

In vivo infrared spectroscopy was used to study the barrier function of the stratum corneum by sequential tape-stripping. Bommannan et al.[17] concentrated on spectral features which provide details on the intercellular lipids and the degree of hydration (hydrogen bonding), since these are the main determinants of barrier function of the stratum corneum. They attempted to address the subject of whether the barrier function is uniformly distributed across the layers of the stratum corneum. They found that the intercellular lipids form ordered and uniform domains in the deeper layers only, while they are less structured near the surface, the amount of lipids as well as the disorder increasing toward the surface. This may indicate that the outer layers are less effective as a barrier. Whereas infrared spectroscopy was found to be useful in studying the lipid content of the stratum corneum, the results of the evaluation of the hydration of the stratum corneum were less encouraging when compared to other methods which showed a steeper increase in water content as a function of depth. The results of their study indicate that the stratum corneum is a nonuniform membrane characterized by gradients of the measured properties. This is important for both percutaneous penetration enhancement and the development of topical therapy of damaged skin. Other studies utilizing infrared spectroscopy for stratum corneum lipids and showing that the surface lipids are highly disordered are reviewed by Potts and Francoeur.[18]

The effect of ethanol on stratum corneum lipids was studied in the same manner.[19] Immediately after treatment with ethanol the spectral parameters showed appreciable amounts of ethanol in the stratum corneum. The results suggested extraction of a disordered lipid component from the stratum corneum, returning to baseline within 24 h after treatment—indicating lipid replacement.

Serup[20] compared two topical preparations of urea cream and, among other methods, used chromametry for skin color measurements. He found a significant change in color toward yellow which was more pronounced in the 3% urea cream preparation. He suggests that this change reflects an alteration of the stratum corneum.

This chapter has concentrated on *in vivo* studies, but many investigations have been conducted with infrared spectroscopy *in vitro*, examining the stratum corneum barrier function[21] and the effect of penetration enhancers.[18,22] Such *in vivo* and *in vitro* studies provide insight into the mechanisms of transport through the stratum corneum.

V. ACTION OF TOPICAL CORTICOSTEROIDS ON THE SKIN

Chapter 23 is dedicated to potency, bioavailability, and bioequivalence of topical steroid preparations as measured by skin blanching. In short, subjective assessments of blanching correlate well with the readings of reflectance spectrophotometry.[23,24] Moreover, the method can be utilized for the evaluation of the pharmacological activity of the same formulation over time, the pharmacological activity of various concentrations of the same formulation,[23] and the influence of the vehicle.[25] Thus, it was shown that the vasoconstrictor potential of a diluted corticosteroid preparation (betamethasone-17-valerate in emulsifying ointment base) decreased with time, results which corresponded well with high-performance liquid chromatography (HPLC). Twofold dilutions were studied in the same manner; the first three dilutions gave similar results, with a rapid loss of effectiveness on the next dilution.[23] Leveque et al.[25] studied the influence of the vehicle on the potency of the same topical corticoid. They found the oil-in-water emulsion to be markedly more potent than the fatty ointment and the water-in-oil emulsion. Similar bases alone did not give any blanching effect. Thus, these results exhibit true pharmacological differences between the different formulations, differences exposed by the reflectance spectrophotometric measurements.

Feather et al.[26] used reflectance spectrophotometry to assess the relative blanching efficacy of topical corticosteroid preparations and found that these objective measurements were in agreement with those obtained by other workers using subjective assessments. However, when information on the response of an individual patient to topical corticosteroids is required, the instrument is preferable due to its greater sensitivity. They measured the response over time and showed that the relative effectiveness of the preparations did not change with elapsed time, so that if measurements were taken at the same time after application, reliable comparison could be made at any time between 4 and 24 h. In contrast, Chan and Li Wan Po,[27] who also compared the potencies of topical corticosteroid preparations, concentrating on several parameters of the pharmacological response over time, concluded that multiple-point measurements over an interval provided more information than single end-point measurements. They analyzed the slope and area under the response-time curve, which accounts for both the intensity and the duration of action of the pharmacodynamic response, and found it superior to the multiple-point comparisons used by Feather et al.[26] They concluded that the quantitative relationship between topical administration, percutaneous absorption, clearance from site of administration, and cutaneous receptor saturation of corticosteroids should be investigated further.

Such further investigations were carried out by Andersen et al.,[24,28] who used an improved reflectance spectroscopy technique which separated the arterial from the venous hemoglobin components in the cutaneous circulation. They showed that the blanching induced by topical corticosteroids was predominantly due to a decreased amount of venous hemoglobin.[28] They were able to grade corticosteroids of different potencies, applying them topically in alcoholic solutions. The relative potency assigned to the vasoconstrictor assay expresses all of the following: the ability of the formulation to penetrate the skin barrier after release from the vehicle, the binding to receptors in the skin, the activity, and the rate of clearance from the site of the reaction. The measurements, therefore, give only an overall estimation and do not identify the exact nature of the differences.

Evaluation of the therapeutic effect of topical corticosteroids was performed by Broby-Johansen et al.[29] They used colorimetry, among other methods, to rank the antipsoriatic effect of various topical corticosteroids applied for 1 week under a hydrocolloid dressing (one application only, with no changing). The redness was the last sign to disappear in accordance with skin color measurements. With increased potency of the topical corticosteroid, skin color normalized, approaching that of normal skin. Their results indicate that, using occlusion with a hydrocolloid dressing, a potent topical corticosteroid normalizes psoriatic skin in 1 week, thus demonstrating the value of short-term therapy with a potent corticosteroid in psoriatic patients (the reader is advised to consult Chapter 8).

VI. TRANSDERMAL DELIVERY

Reflectance spectrophotometry, along with visual scoring and laser Doppler flowmetry, was used for the investigation of transdermal delivery.[30] Following topical application of the beta-blocker timolol, Kubota et al.[30] studied the relationship between the magnitude of skin irritation and drug concentration in the blood which had just perfused the site of application (inner forearm). The investigators were hoping to find a "threshold" concentration critical for the development of erythema. However, the erythema was linearly correlated to the logarithm of the plasma concentration, and erythema developed whenever drug flux was sufficient. They conclude that it seems difficult to provide sufficient amounts of timolol to deliver the desired systemic effects with no concurrent erythema. However, they suggest more extensive studies where the drug is applied to other body sites, since there is regional variation in skin reactions. There is also a need for studies directed to determine whether mild erythema should exclude transdermal application as an alternate route of administration of the drug. Thus, besides addressing the main purpose of this study, it was shown that reflectance spectrometry readings were linearly correlated to the logarithm of the plasma concentration, which was itself correlated to the flux of drug through the skin. Consequently, its use as a noninvasive tool for the assessment of transdermal delivery of such drugs that cause measurable skin responses should be favorably considered.

VII. PHARMACODYNAMIC MEASUREMENTS OF SYSTEMICALLY ADMINISTERED DRUGS

Pharmacodynamic measurements of systemically administered drugs are possible with reflectance spectrophotometry. Drugs which affect skin color can be studied in the same manner as carbon monoxide poisoning.[31] In this manner, pharmacodynamic measurements of facial flushing induced by cicaprost were made using chromametry (among other methods used in this study, including measurement of plasma levels).[32] Starting at 5 µg and increasing the dose, they registered the effect-time profile for the changes in forehead skin color. Alterations in skin color occurred from the lowest dosage level. Individual differences were observed in the effect-time profile—several test subjects exhibited an early and high maximum change in color, while others showed a more sustained profile with no pronounced peak. The individual time profiles of color and plasma levels were similar. Both inhibition of platelet aggregation and changes in chromametry were dose dependent and subject to interindividual variability, a phenomenon due to differences in responsiveness and clearance. The study demonstrates a good correlation between the plasma levels and the effect, as determined by both the anti-aggregation effect and the chromametry measurements. The authors conclude that chromametric measurements of changes of forehead skin color are a valid and sensitive tool to describe prostacyclin-mimetic effects. The same was also demonstrated for iloprost administered intravenously.[33]

Prince and Frisoli,[34] studying the kinetics of beta-carotene, used remittance spectroscopy to measure the rate of its accumulation in the stratum corneum after oral administration. They obtained an absorption profile from absorption spectra derived from skin remittance data. The change in skin absorbance was calculated from the skin remittance measurements. The relative amount of beta-carotene in skin was calculated from the change in absorption at 490 nm. Remittance measurements of skin color demonstrated that the accumulation of beta-carotene in skin was delayed by up to 2 weeks compared with serum accumulation. They speculate that this delay may reflect differences in skin thickness, which can vary considerably from one individual to another. Other possible explanations include variable rates of uptake into the basal cells and subcutaneous fat. They comment that the yellow skin color may be a useful marker for evaluating compliance and determining when and which lipophilic tissues have accumulated significant amounts of pigment.

Kakizoe et al.[35] applied reflectance spectrophotometry for the study of capillary permeation of rat skin in real time. They followed norepinephrine-induced vasoconstriction and histamine-induced vasodilatation and increase in capillary permeation. Intravenous administration of norepinephrine caused a reduction in the relative absorption spectra of the reflectance spectrophotometry due to vasoconstriction. They were able to follow the time course of the effect: readings decreased immediately following injection and recovered about 3 min later. This effect was dose dependent, and so was the increase in the relative absorption spectra of the reflectance spectrophotometry due to histamine injection. Histamine was injected intradermally and measurements were done around the injection site. Here two spectra were of interest: the absorption spectrum of oxyhemoglobin and the absorption spectrum of the Evans blue dye which was concomitantly injected. Both spectra increased, suggesting the occurrence of both vasodilatation and an increase in the capillary permeation of the Evans blue dye. The method allowed the measurement of both at the same time. The increase in permeation was observed at doses lower than those that produced vasodilatation. When injecting saline intradermally they also observed changes in the relative absorption spectra, suggesting a release of vasoactive mediators from the traumatized tissue.

VIII. SYSTEMIC DRUGS FOR DERMATOLOGIC CONDITIONS

Reflectance spectrophotometry was used to study the mechanism of the effect of methotrexate on psoriasis.[36] The cutaneous response to intradermal injection of human complement split product C5a was measured before and after the intake of methotrexate. The response was significantly depressed after methotrexate intake, indicating that methotrexate is a potent inhibitor of C5a-induced skin response in patients with psoriasis. The results of the study support the idea that the anti-inflammatory effect of methotrexate may be partially responsible for its antipsoriatic effect, in addition to the inhibition of epidermal cell proliferation.

IX. CONCLUSION

Remittance spectroscopy and chromametry have been successfully used to quantify pharmacological processes *in vivo*. The method is precise, sensitive, and objective. It allows real-time analysis of penetration and other pharmacological processes, providing objective numerical data at various time points.

However, currently used techniques still suffer from quite a few disadvantages. Due to the lack of standardization and for a few other reasons the data are expressed only semi-quantitatively. In addition, it is difficult to compare values between individuals because of differences in skin color and/or the heterogeneity of the skin components. For certain purposes—for instance, for quantification of contact allergic reaction—the method is time-consuming, since only a small area can be measured at one time and multiple readings at different sites require the repositioning of the light guide between measurements.[37]

Nevertheless, with more affordable and user-friendly instruments becoming available, many of the above-mentioned current problems will be solved. Scanning techniques will simplify and shorten the length of the measurement, allowing the coverage of substantial skin areas without compromising sensitivity. Then the unique power of spectroscopic techniques will be harnessed not only for the applications which were discussed in this chapter, but also toward heretofore unexplored and novel avenues of skin pharmacology research.

REFERENCES

1. Dawson, J. B., Barker, D. J., Ellis, D. J., Grassam, E., Cotterill, J. A., Fisher, G. W., and Feather, J. W., A theoretical and experimental study of light absorption and scattering by *in vivo* skin, *Phys. Med. Biol.,* 25, 695, 1980.
2. Diffey, B. L., Oliver, R. J., and Farr, P. M., A portable instrument for quantifying erythema induced by ultraviolet radiation, *Br. J. Dermatol.,* 111, 663, 1984.
3. Seitz, J. C. and Whitmore, C. G., Measurement of erythema and tanning responses in human skin using a tri-stimulus colorimeter, *Dermatologica,* 177, 70, 1988.
4. Serup, J. and Agner, T., Colorimetric quantification of erythema—a comparison of two colorimeters with a clinical scoring scheme and laser Doppler flowmetry, *Clin. Exp. Dermatol.,* 15, 267, 1990.
5. Deleixhe-Mauhin, F., Krezinski, J. M., Rorive, G., and Pierard, G. E., Quantification of skin color in patients undergoing maintenance hemodialysis, *J. Am. Acad. Dermatol.,* 27, 950, 1992.
6. Berardesca, E., Gabba, P., Nume, A., Rabbiosi, G., and Maibach, H. I., Objective prick test evaluation: non-invasive techniques, *Acta Derm. Venereol.,* 72, 261, 1992.
7. Kolmel, K. F., Sennhenn, B., and Giese, K., Investigation of skin by ultraviolet remittance spectroscopy, *Br. J. Dermatol.,* 122, 209, 1990.

8. Mak, V. H. W., Potts, R. O., and Guy, R. H., Percutaneous penetration enhancement *in vivo* measured by attenuated total reflectance infrared spectroscopy, *Pharm. Res.,* 7, 835, 1990.
9. Mendelow, A. Y., Forsyth, A., Feather, J. W., Baillie, A. J., and Florence, A. T., Skin reflectance measurements of patch test responses, *Contact Dermatitis,* 15, 73, 1986.
10. Mendelow, A. Y., Forsyth, A., Florence, A. T., and Baillie, A. J., Patch testing for nickel allergy, *Contact Dermatitis,* 13, 29, 1985.
11. Lanigan, S. W. and Cotterill, J. A., Reduced vasoactive responses in port wine stains, *Br. J. Dermatol.,* 122, 615, 1990.
12. Clarys, P. and Barel, A. O., Percutaneous absorption of methyl nicotinate: influence of body region and occlusion, *Allergologie,* 16, S159, 1993.
13. Andersen, P. H., Nangia, A., Bjerring, P., and Maibach, H. I., Chemical and pharmacologic skin irritation in man, *Contact Dermatitis,* 25, 283, 1991.
14. Ramsay, B., Rice, N., and Lawrence, C., The effect of indomethacin on anthralin inflammation, *Br. J. Dermatol.,* 126, 262, 1992.
15. Sharpe, G. R. and Shuster, S., Topical indomethacin aggravates the weal and flare response in chronic dermographic urticaria: evidence for a new class of histamine receptors, *Acta Derm. Venereol.,* 73, 7, 1993.
16. Potts, R. O., Guzek, D. B., Harris, R. R., and McKie, J. E., A noninvasive, *in vivo* technique to quantitatively measure water concentration of the stratum corneum using attenuated total-reflectance infrared spectroscopy, *Arch. Dermatol. Res.,* 277, 489, 1985.
17. Bommannan, D., Potts, R. O., and Guy, R. H., Examination of stratum corneum barrier function *in vivo* by infrared spectroscopy, *J. Invest. Dermatol.,* 95, 403, 1990.
18. Potts, R. O. and Francoeur, M. L., Physical methods for studying stratum corneum lipids, *Semin. Dermatol.,* 11, 129, 1992.
19. Bommannan, D., Potts, R. O., and Guy, R. H., Examination of the effect of ethanol on stratum corneum *in vivo* using infrared spectroscopy, *J. Controlled Release,* 16, 299, 1991.
20. Serup, J., A double blind comparison of two creams containing urea as the active ingredient, *Acta Derm. Venereol. Suppl.,* 177, 34, 1992.
21. Lin, S. Y., Hou, S. J., Hsu, T. H. S., and Yeh, F. L., Comparisons of different animal skins with human skin in drug percutaneous penetration studies, *Methods Find. Exp. Clin. Pharmacol.,* 14, 645, 1992.
22. Hadgraft, J., Walters, K. A., and Guy, R. H., Epidermal lipids and topical drug delivery, *Semin. Dermatol.,* 11, 139, 1992.
23. Ryatt, K. S., Feather, J. W., Mehta, A., Dawson, J. B., Cotterill, J. A., and Swallow, R., The stability and blanching efficacy of betamethasone-17-valerate in emulsifying ointment, *Br. J. Dermatol.,* 107, 71, 1982.
24. Andersen, P. H., Milioni, K., and Maibach, H. I., The cutaneous corticosteroid vasoconstriction assay: a reflectance spectroscopic and laser-Doppler flowmetric study, *Br. J. Dermatol.,* 128, 660, 1993.
25. Leveque, J. L., Poelman, M. C., Legall, F., and de Rigal, J., New experimental approach to measure the skin-reflected light. Application to cutaneous erythema and blanching, *Dermatologica,* 170, 12, 1985.
26. Feather, J. W., Ryatt, K. S., Dawson, J. B., Cotterill, J. A., Barker, D. J., and Ellis, D. J., Reflectance spectrophotometric quantification of skin colour changes induced by topical corticosteroid preparations, *Br. J. Dermatol.,* 106, 437, 1982.
27. Chan, S. Y. and Li Wan Po A., Quantitative skin blanching assay of corticosteroid creams using tristimulus color analysis, *J. Pharm. Pharmacol.,* 44, 371, 1992.
28. Andersen, P. H., Broichmann, P. W., and Maibach, H. I., A corticosteroid, a non-steroidal anti-inflammatory drug and an antihistamine modulate *in vivo* vascular reactions before and during post-occlusive hyperemia, *Br. J. Dermatol.,* 128, 137, 1993.

29. Broby-Johansen, U., Karlsmark, T., Petersen, L. J., and Serup, J., Ranking of the antipsoriatic effect of various topical corticosteroids applied under a hydrocolloid dressing—skin thickness, blood flow and colour measurements compared to clinical assessments, *Clin. Exp. Dermatol.*, 15, 343, 1990.

30. Kubota, K., Koyama, E., and Yasuda, K., Skin irritation induced by topically applied timolol, *Br. J. Clin. Pharm.*, 31, 471, 1991.

31. Findlay, G. H., Carbon monoxide poisoning: optics and histology of skin and blood, *Br. J. Dermatol.*, 119, 45, 1988.

32. Hildebrand, M., Staks, T., and Nieuweboer, B., Pharmacokinetics and pharmacodynamics of cicaprost in healthy volunteers after oral administration of 5 to 20 µG, *Eur. J. Clin. Pharmacol.*, 39, 149, 1990.

33. Seifert, W., Hildebrand, M., and Knoffler, F., Facial colour by chromametry and iloprost plasma levels, in *Prostaglandins in Clinical Research*, Schroer, K. and Sinzinger, H., Eds., Alan R. Liss, New York, 125, 1989.

34. Prince, M. R. and Frisoli, J. K., Beta-carotene accumulation in serum and skin, *Am. J. Clin. Nutr.*, 57, 175, 1993.

35. Kakizoe, E., Kobayashi, Y., Shimoura, K., Hattori, K., and Jidoi, J., Real-time measurement of microcirculation of skin by reflectance spectrophotometry, *J. Pharmacol. Toxicol. Methods*, 28, 175, 1992.

36. Ternowitz, T., Bjerring, P., Andersen, P. H., Schroder, J. M., and Kragballe, K., Methotrexate inhibits the human C5a-induced skin response in patients with psoriasis, *J. Invest. Dermatol.*, 89, 192, 1987.

37. Quinn, A. G., McLelland, J., Essex, T., and Farr, P. M., Quantification of contact allergic inflammation: a comparison of existing methods with a scanning laser Doppler velocimeter, *Acta Derm. Venereol.*, 73, 21, 1993.

Product Testing

Chapter 22

Evaluation of Irritation Tests by Chromametric Measurements

Klaus-P. Wilhelm and Howard I. Maibach

CONTENTS

ABSTRACT: Human skin irritation tests of consumer products and drugs are usually evaluated by visual scores. Recently, instruments that objectively measure erythema have increasingly become commercially available. The usefulness of instrumental measurements in human irritation assays was evaluated for surfactant-induced skin irritation. The testing modalities were varied and the results compared with established parameters for evaluating irritant skin reactions, i.e., transepidermal water loss, cutaneous blood flow, and visual scoring.

I. INTRODUCTION

Assays frequently employed in order to determine skin irritation potential in humans are listed in Table 1. Irritant skin reactions are routinely evaluated by applying different scores for the degree of erythema, edema, scaling, and fissuring.[1] This is the conventional approach, but the disadvantages of lacking objectivity and parametric properties are implied. Thus, the observers' judgment of color is greatly influenced by the observation conditions, i.e., illumination, size, distribution, and shape of the reaction.[2]

Products most frequently tested for their irritation potential are listed in Table 2. For most of these products (e.g., moisturizers), significant skin irritation is routinely not expected. The reason for their testing is for product safety and marketing reasons. In order to evaluate the usefulness of objective erythema measurements, the authors' study was limited to surfactants. Because surfactants constitute a dominant proportion of all products being tested, this

Table 1   **Frequently used human skin irritation assays**

Acute irritation studies (minutes to 72 h; occluded vs. unoccluded)
21-day cumulative irritation assay (occluded vs. unoccluded)
Elbow wash test
Repeat open application test
Hand immersion test
Chamber scarification test
Soap chamber test
Stinging test

Table 2   **Products frequently tested for irritation potential**

Skin and hair cleansing products (soaps, detergents, shampoos)
Skin care products (moisturizers, after-shave lotions)
Dermatologic drugs (base preparations, active formulas)
Cosmetics (lipsticks, mascara, foundations, antiacne formulas)
Dishwashing liquids
Transdermal drug delivery systems
Deodorants
Perfumes
Tapes, bandages
Textiles

Table 3   **Example of a frequently used visual scale to quantify surfactant-induced erythema[1]**

1+	Slight erythema, either spotty or diffuse
2+	Moderate uniform erythema
3+	Intense redness
4+	Fiery redness with edema

Note: On this scale, in addition to erythema, fissuring and edema are also evaluated separately.

approach seems justified. The aim for testing surfactants is to determine the relative irritation potential of one product in comparison with another product based on differences in chemical composition.

A.   BACKGROUND OF PRODUCT EFFECTS

Skin reactions to some irritants, such as surfactants, consist of stratum corneum barrier impairment and increased cutaneous blood flow, macroscopically visible as erythema.[3-7] For visual assessment of erythema either visual analog scales or scoring systems defining each point by a characteristic morphology are used. An example of the latter type is the frequently used visual scoring system proposed by Frosch and Kligman,[1] especially designed for soap and surfactant-induced irritant skin reactions (Table 3). Besides the inherent lack of objectivity, a common problem with the use of visual scales is doubtful reactions.

B.   INSTRUMENTAL MEASUREMENTS

The degree of erythema can be evaluated indirectly by blood plethysmography,[8] or laser Doppler blood flow (LDF) measurements.[4,5,9-12] The methods described are noninvasive and correlate well with clinical grading as well as with the applied chemical dose.[4,5,9] However, until now neither technique has proved suitable for routine clinical use since each is time-

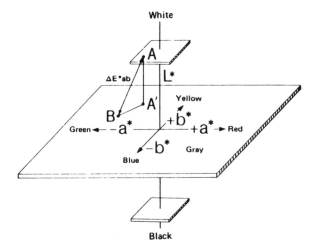

Figure 1   The L*a*b* color space (Commission Internationale de l'Éclairage, 1976), abbreviated CIELAB, where all possible colors of object are represented on the tree axis. See text for details. (Adapted from the instrument manual.)

consuming, expensive, and difficult to perform. A direct approach to instrumental erythema measurement utilizes spectrophotometry of light reflectance and absorbance by different pigments of the skin[12–18] or measurement of skin color using a standardized tristimulus system.[4,5,19–26]

Spectrophotometers used to be custom-made, stationary, and cumbersome to use on the skin. However, some portable spectrophotometers have recently become commercially available. Because hemoglobin and oxyhemoglobin contribute only a minor fraction of the absorbance in comparison with other pigments, current models calculate hemoglobin and melanin indices.[13]

Tristimulus color measurements, on the other hand, are based on standard techniques recommended by Commission Internationale de l'Éclairage (CIE).[2] A detailed description of the instrument is given in Chapter 19. Briefly, the color of any given object can be defined in different three-dimensional color spaces. For applications in skin biology, the L*a*b* color space (CIE, 1976), abbreviated as CIELAB, appears to be the most appropriate choice.[2,19,21,26] This nonlinear system mimics the eye-brain response and represents the perceived color. Hence, equal differences in perceived color differences are approximately equally represented in this three-dimensional color space. Luminance (L*) expresses the brightness or darkness of an object. The color is defined by the parameters a* (red-green axis) and b* (blue-yellow axis; Figure 1).

II. SKIN COLOR MEASUREMENTS IN SURFACTANT-INDUCED SKIN IRRITATION

A. REPRODUCIBILITY

Reproducibility of measurements is a concern in all investigative areas and is imperative in order to satisfy good laboratory practice guidelines. When comparing the results of multiple color reflectance measurements of different nonbiologic materials, namely white and red plastic plates, only minimal measurement variability was observed (Table 4). The variability, however, was greater for measurements of normal and erythematous skin. Nevertheless, the coefficient of variation of this measurement is acceptable when compared with other meas-

Table 4   Reproducibility of repeated measurements of a standard white plate, a uniformly red plate, untreated skin, and treated (erythematous) skin

	Luminance (L*)	Redness (a*)	Yellow Hue (b*)
White plate	99.2 ± 0.1 (0.2)	-0.8 ± 0.1 (0.2)	0.4 ± 0.1 (0.4)
Red plate	43.6 ± 0.1 (0.2)	53.6 ± 0.1 (0.4)	9.6 ± 0.2 (0.8)
Normal skin	63.4 ± 0.6 (2.5)	6.4 ± 0.3 (0.8)	12.2 ± 0.3 (1.2)
Erythema	61.1 ± 0.5 (1.8)	12.2 ± 0.3 (1.1)	11.0 ± 0.1 (1.5)

Note: Reflectance measurements (L*, a*, b*) demonstrate a low variability for repeated measurements, especially for nonbiologic materials. Values represent means $\pm$ SD. The range is shown in parenthesis ($n = 15-26$).

Adapted from Wilhelm, K.-P. and Maibach, H. I., *J. Am. Acad. Dermatol.*, 21, 1306, 1989. With permission.

Table 5   Change in skin color reflectance after various concentrations of sodium lauryl sulfate (SLS)

SLS Concentrations (%)	Difference in Reflectance Parameters			
	Luminance (ΔL*)	Redness (Δa*)	Yellow Hue (Δb*)	Total Color (ΔE*ab)
0.0	0.0 ± 0.0	0.0 ± 0.0	0.0 ± 0.0	0.0 ± 0.0
0.125	-0.4 ± 1.3^a (ns)	0.7 ± 0.9^a (0.05)	0.1 ± 1.0^a (ns)	1.7 ± 1.2^a (0.0001)
0.25	-0.8 ± 1.0	1.4 ± 1.3	-0.2 ± 1.6	2.5 ± 1.0
0.5	-0.7 ± 2.2^b (ns)	1.7 ± 1.6^b (ns)	-0.2 ± 1.4^b (ns)	2.9 ± 1.8^b (ns)
1.0	-0.9 ± 2.0	2.5 ± 1.9	-1.0 ± 2.0	4.0 ± 1.9
2.0	-1.4 ± 4.1^b (ns)	4.0 ± 3.5^b (0.01)	-0.9 ± 2.2^b (ns)	5.9 ± 4.0^b (0.05)
3.0	-2.5 ± 3.6	4.6 ± 3.2	-1.2 ± 2.1	6.3 ± 3.9

Note: With increasing SLS dose Δa* and ΔE*ab increased while ΔL* and Δb* decreased. Values are means $\pm$ SD ($n = 10$) and are calculated as follows: Δ reflectance = reflectance (treated) − reflectance (controls), ns = not significant; significant differences are indicated by p-values in parentheses. $\Delta E^*ab = [(\Delta L^*)^2 + (\Delta a^*)^2 + (\Delta b^*)^2]^{1/2}$.
[a]Compared to controls; [b]Compared to 1% SLS.
Adapted from Wilhelm, K.-P. and Maibach, H. I., *J. Am. Acad. Dermatol.*, 21, 1306, 1989.

urements of biologic materials. These results indicate that most of the variability of skin color reflectance measurements is due to the inherent nature of the integument *in vivo*, e.g., variability in perfusion, vasomotion, etc., rather than being attributable to the instrument.

B.   DOSE-EFFECT STUDIES

Application of the model irritant sodium lauryl sulfate (SLS) led to increases in parameter a* (redness) and total color change (ΔE*ab), whereas luminance (ΔL*) and yellow hue (Δb*) decreased (Table 5). In this study population, including both sensitive and nonsensitive individuals, the variability of the data was considerable, as demonstrated for parameter Δa* in Figure 2. Most of this variability was probably due to inherent interindividual differences in responsiveness among the subjects, as suggested by a similar variability of the visual scores (Figure 3).

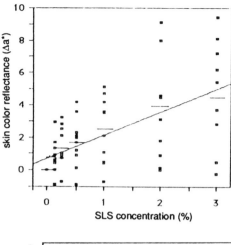

Figure 2 Differences in skin redness (Δa*) are plotted as a result of 24 h exposure to various concentrations of sodium lauryl sulfate (SLS). Means are indicated by horizontal bars (n = 10 volunteers).

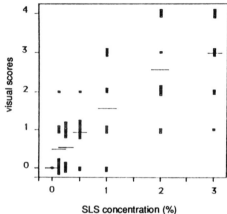

Figure 3 Erythema as evaluated by visual scores after 24 h exposure to various concentrations of sodium lauryl sulfate (SLS). Means are indicated by horizontal bars (n = 10 volunteers).

C. COMPARISON OF COLOR MEASUREMENTS WITH VISUAL SCORES AND WITH OTHER INSTRUMENTAL TECHNIQUES

The parameter Δa* was significantly linearly correlated with both visual scores and applied SLS dose (Table 6). The correlation coefficient was even stronger for the calculated parameter ΔE*ab. This, however, is by definition not only influenced by changes in redness. Therefore, Δa* is preferred as the parameter by which to quantify erythema. It is recommended, however, that all three parameters (L*, a*, and b*) are recorded even if only one parameter is given in the final report. The reasons for that recommendation will be discussed below.

Skin color reflectance measurements showed a strong correlation with laser Doppler blood flow and transepidermal water loss (TEWL) measurements (Table 7). TEWL measurements, however, were more sensitive in discriminating between various SLS concentrations than were skin color or cutaneous laser Doppler blood flow measurements (Table 7). This finding has been independently confirmed by Agner and Serup.[5]

D. DISCRIMINATING BETWEEN PRODUCTS WITH VARIABLE IRRITATION POTENTIAL

Demonstration of dose dependency is one criterion for the value of an instrumental parameter, while discrimination between the irritation potential of different products is another. Decades

Table 6   Correlation between transepidermal water loss (TEWL), lasser Doppler flowmetry (LDF), redness (Δa^*), change in total color (ΔE^*ab), visual scores (VS), and SLS dose (Conc.) 24 h after application of sodium lauryl sulfate (SLS, 0.125–3.0%) or water, comparing individual data points ($n = 70$)

	TEWL	Δa^*	ΔE^*	LDF	VS
TEWL	1	*	*	*	*
Δa^*	0.86	1	*	*	*
ΔE^*	0.83	0.92	1	*	*
LDF	0.82	0.83	0.73	1	*
VS	0.90	0.80	0.85	0.82	1
Conc.	0.76	0.59	0.63	0.68	0.79

Note: All correlation coefficients are significant ($p < 0.0001$). Values represent Pearson correlation coefficients except for VS, where Spearman rank correlation coefficients were calculated.
$\Delta E^* = [(\Delta L^*)^2 + (\Delta a^*)^2 + (\Delta b^*)^2]^{1/2}$.
Adapted from Wilhelm, K.-P., Surber, C., and Maibach, H. I., *Arch. Dermatol. Res.*, 281, 293, 1989.

Table 7   Skin irritation 24 h after treatment with sodium lauryl sulfate (SLS) as assessed by transepidermal water loss [TEWL (g/m²/h)], laser Doppler flowmetry (LDF [% of controls]), skin color reflectance (Δa^* [difference in redness]), ΔE^* (change in total color), and visual scores (VS)

% SLS	TEWL	LDF	Δa^*	ΔE^*	VS
0.0	4.7 ± 1.4	100.0 ± 0.0	0.0 ± 0.0	0.0 ± 0.0	0.0
0.125	7.9 ± 2.9^a	149.3 ± 53.0^a	0.7 ± 0.9^a	1.7 ± 1.2^a	0.5^a
	(0.0002)	(0.0005)	(0.05)	(0.001)	(ns)
0.25	9.3 ± 2.6	220.2 ± 94.4	1.4 ± 1.3	2.5 ± 1.0	0.6
0.5	12.4 ± 5.3^b	339.1 ± 182.1^b	1.7 ± 1.6^b	2.9 ± 1.8^b	0.9^b
	(0.02)	(ns)	(ns)	(ns)	(0.05)
1.0	20.2 ± 11.3	422.2 ± 264.0	2.5 ± 1.9	4.0 ± 1.9	1.6
2.0	27.5 ± 13.6^b	589.7 ± 305.1^b	4.0 ± 3.5^b	5.9 ± 4.0^b	2.5^b
	(0.0005)	(0.05)	(0.01)	(0.05)	(0.005)
3.0	41.6 ± 21.1	678.7 ± 322.6	4.6 ± 3.2	6.3 ± 3.9	3.0

Note: Values are means $\pm$ SD (means only for VS); ns = not significant ($p > 0.05$); significant differences are indicated by p values in parentheses. $\Delta E^* = [(\Delta L^*)^2 + (\Delta a^*)^2 + (\Delta b^*)^2]^{1/2}$.
[a]Compared to controls (0.0%); [b]Compared to 1% SLS.
Adapted from Wilhelm, K.-P., Surber, C., and Maibach, H. I., *Arch. Dermatol. Res.*, 281, 293, 1989.

ago Kligman and Wooding applied the Litchfield and Wilcoxon probit analysis to cumulative irritation testing with calculation of the IT_{50} (the number of days of continuous exposure which will produce a threshold reaction in 50% of the test population) and the ID_{50} (the concentration that produces a perceptible irritant reaction in 50% of the test population in 24 h).[27] This important work forms part of the basis for the 21-day cumulative irritation assay that is still widely used. The irritation potential of a homologous series of *n*-alkyl sulfates is greatly dependent on the alkyl chain length of the molecule, as indicated by the IT_{50}/ID_{50} ratio (Figure 4).

Having these data available, the authors reinvestigated the identical compounds.[30] The ranking of the surfactants according to increases in Δa^* was identical to the ranking by Kligman and Wooding (Figure 5). The main difference between the two studies was the

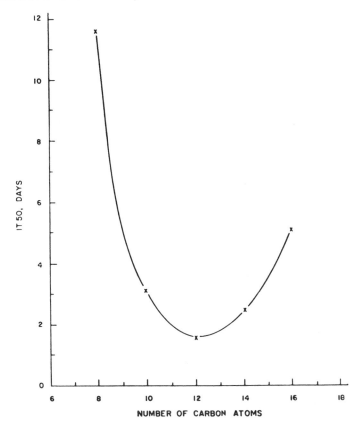

Figure 4 Evaluation of the irritation potential of a homologous series of *n*-alkyl sulfates by determination of their IT_{50} (number of days of continuous exposure that will produce a threshold reaction in 50% of the test population). Note that low IT_{50} values represent compounds with high irritation potential. (Adapted from Kligman, A. and Wooding, W. M., *J. Invest. Dermatol.*, 49, 78, 1967.)

duration (48 h vs. 21 days). There was also a good correlation between the instrumental erythema measurements and TEWL readings (Figure 6).

E. REPAIR OF IRRITANT SKIN REACTIONS

When irritant skin reactions were induced by repeated short-term (20 min) exposure to SLS, chromameter readings (Δa^*) paralleled visual scores in the induction phase of the irritant reaction as well as in the repair phase (Figure 7). In contrast, Δa^* after one single 24-h exposure to SLS remained elevated above preexposure levels for as long as 17 days after treatment, whereas visual examination revealed no erythema at that time (Figure 8). The reason for that discrepancy was significant postinflammatory hyperpigmentation. This hyperpigmentation could be verified by chromameter readings, i.e., parameter Δb^*. Parameter $\Delta b,^*$ which increases in tanning reactions, returned almost to baseline values in the chronic irritation model, while it remained elevated above preexposure levels in the acute irritation model (Figures 9 and 10). These results demonstrate the value of recording all three parameters of the CIELAB system rather than only Δa^*, especially in new experimental designs.

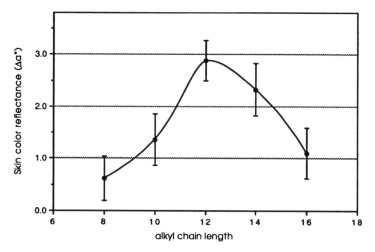

Figure 5   Evaluation of the irritation potential of a homologous series of *n*-alkyl sulfates by measuring the change in redness (Δa^*) 24-h after a single 24-h exposure to 20 mmol solutions. (Drawn are means $\pm$ SEM; $n =$ 10 volunteers.)

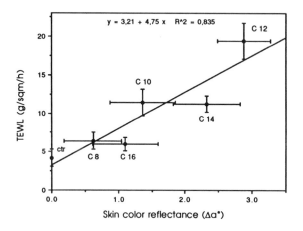

Figure 6   Correlation of parameter Δa^* with transepidermal water loss measurements (TEWL, in g/m²/h) after a single 24-h exposure to 20 mmol solutions of *n*-alkyl sulfates with variable alkyl chain length. (Means $\pm$ SEM; $n =$ 10 volunteers.)

Instrumental and visual erythema evaluation

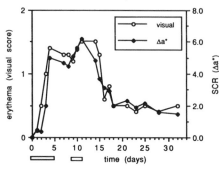

Figure 7   Induction and repair of chronic irritant skin reactions as evaluated by erythema after cumulative short-term (20 min) exposure to 7.5% sodium lauryl sulfate (SLS) for 8 consecutive days excluding the weekend. Instrumental readings (right axis) and visual scores (left axis) are shown. Bar indicates time of repeated exposure to SLS. SCR = Skin Color Reflectance. (Means; $n =$ 10 volunteers.)

Instrumental and visual erythema evaluation

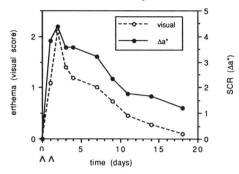

Figure 8 Induction and repair of acute irritant skin reactions as evaluated by erythema formation after 24 h exposure to 0.5% sodium lauryl sulfate (SLS). Skin color reflectance (SCR) readings (right axis) and visual scores (left axis). Arrowheads indicate the start and end of SLS exposure. (Means; n = 10 volunteers.)

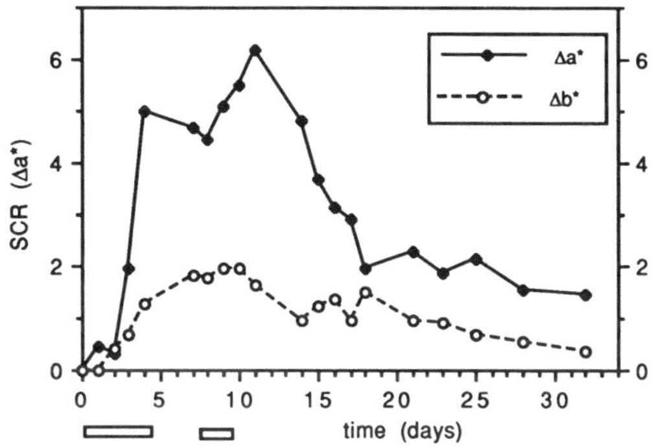

Figure 9 Drawn are SCR differences on the red-green axis (Δa^*) against color differences on the blue-yellow coordinate (Δb^*) in the chronic irritation model (see also Figure 7).

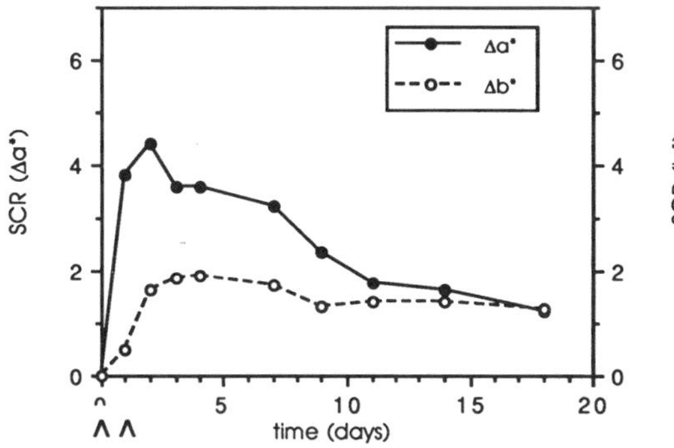

Figure 10 Drawn are SCR differences on the red-green axis (Δa^*) against color differences on the blue-yellow coordinate (Δb^*) in the acute irritation model (see also Figure 8). Note that Δa^* does not return to preexposure levels and that Δb^* remains elevated as a result of the postinflammatory hyperpigmentation.

III. PRODUCT TESTING STRATEGIES

The evaluation of human irritation tests by visual scoring alone may no longer be appropriate. Additional objective measurements are especially desirable if the assay results are to be submitted to regulatory agencies. Because various skin physiologic parameters, such as barrier function, blood flow, and erythema, are altered in irritant skin reactions, different bioengineering instruments may be used to document and quantify these reactions, e.g., evaporimeters, laser Doppler blood flow monitors, and chromameters.[4,5,9,13,22,28,29] Instead of custom-made instruments, compact, portable equipment that is commercially available can be used. Standardized instrumentation and procedures will allow interlaboratory comparability (Table 8). On-line data acquisition by increasingly powerful computer software facilitates data management.[31] However, neither combination of instruments will ever replace the

Table 8   **Advantages of visual scoring and skin color measurements**

Visual scores

- Inexpensive
- Rapid (especially clearly negative reactions)

Skin color measurements

- Objective
- Parametric, therefore statistics are more powerful
- More sensitive
- Standardized instrumentation available
- Interlaboratory comparability

expert observer—in most instances, a dermatologist. This is especially true for new experimental designs and for the evaluation of unexpected reactions. Furthermore, trained personnel are required to yield reliable results with sophisticated instruments.

The question of which parameter is best suited to quantify irritant responses remains to be definitely answered. Surfactant-induced irritant reactions are most sensitively detected by measurements of transepidermal water loss. Skin color measurements, on the other hand, can be obtained more rapidly and represent the corresponding objective parameter to visual erythema scores. Therefore, skin color measurements most certainly will be increasingly used in the evaluation of human irritation assays.

ACKNOWLEDGMENTS

This work was supported in part by Deutsche Forschungsgemeinschaft (DFG Wi-879/3–1). B. Geist provided invaluable help in the preparation of this manuscript.

REFERENCES

1. Frosch, P. J. and Kligman, A. M., The soap chamber test, *J. Am. Acad. Dermatol.*, 1, 35 1979.
2. Wyszecki, G. and Stiles, W. S., *Color Science Concepts and Methods, Quantitative Data and Formulae,* John Wiley & Sons, New York, 1982, 165.
3. Van der Valk, P. J. M., Nater, J. P., and Bleumink, E., Skin irritancy of surfactants as assessed by water vapor loss measurements, *J. Invest. Dermatol.*, 82, 291, 1984.
4. Wilhelm, K. P., Surber, C., and Maibach, H. I., Quantification of sodium lauryl sulfate irritant dermatitis in man: comparison of four techniques: skin color reflectance, transepidermal water loss, laser Doppler flow measurement and visual scores, *Arch. Dermatol. Res.*, 281, 293, 1989.
5. Agner, T. and Serup, J., Sodium lauryl sulphate for irritant patch testing—a dose response study using bioengineering methods for determination of skin irritation, *J. Invest. Dermatol.*, 95, 543, 1990.
6. Tupker, R. A., Pinnagoda, J., Coenraads, P. J., and Nater, J. P., The influence of repeated exposure to surfactants on the human skin as determined by transepidermal water loss and visual scoring, *Contact Dermatitis,* 20, 108, 1989.
7. Simion, F. A., Rhein, L. D., Grove, G. L., Wojtkowski, J. M., Cagan, R. H., and Scala, D. D., Sequential order of skin responses to surfactants during a soap chamber test, *Contact Dermatitis,* 25, 242, 1991.

8. Guy, R. H., Tur, E., and Maibach, H. I., Optical techniques for monitoring cutaneous microcirculation. Recent applications, *Int. J. Dermatol.,* 24, 88, 1985.
9. Nilsson, G. E., Tenland, T., and Öberg, P. Å., Evaluation of a laser Doppler flowmeter for measurement of tissue blood flow, *IEEE Trans. Biomed. Eng.,* 27, 597, 1980.
10. Blanken, R., Van der Valk, P. J. M., Nater, J. P., Laser-Doppler flowmetry in the investigation of irritant compounds on human skin, *Dermatosen,* 34, 5, 1986.
11. Willis, C. M., Stephens, C. J. M., and Wilkinson, J., Assessment of erythema in irritant contact dermatitis, *Contact Dermatitis,* 18, 138, 1988.
12. Bjerring, P. and Andersen, P. H., Skin reflectance spectrophotometry, *Photodermatology,* 4, 167, 1987.
13. Diffey, B. L., Oliver, R. J., and Farr, P. M., A portable instrument for quantifying erythema induced by ultraviolet radiation, *Br. J. Dermatol.,* 111, 663, 1984.
14. Feather, J. W., Ryatt, K. S., Dawson, J. B., Cotterill, J. A., Barker, D. J., and Ellis, D. J., Reflectance spectrophotometric quantification of skin colour changes induced by topical corticosteroid preparations, *Br. J. Dermatol.,* 106, 437, 1982.
15. Crowe, D. M., Willard, M. S., and Murahata, R. I., Quantitation of erythema by reflectance spectroscopy, *J. Soc. Cosmet. Chem.,* 38, 451, 1987.
16. Mendelow, A. Y., Forsyth, A., Feather, J. W., Baillie, A. J., and Florence, A. T., Skin reflectance measurements of patch test responses, *Contact Dermatitis,* 15, 73, 1986.
17. Diffey, B. L., Optical properties of skin: measurement of erythema, in *The Physical Nature of the Skin,* Marks, R. M., Barton, S. P., and Edwards, C., Eds., MTP Press, Lancaster, England, 1987, 79.
18. Pearse, A. D., Edwards, C., Hill, S., and Marks, R., Portable erythema meter and its application to use in human skin, *Int. J. Cosmet.,* 12, 63, 1990.
19. Wilhelm, K. P. and Maibach, H. I., Skin color reflectance measurements for objective quantification of erythema in man, *J. Am. Acad. Dermatol.,* 21, 1306, 1089.
20. Serup, J. and Agner, T., Colorimetric quantification of erythema—a comparison of two colorimeters (Lange Micro Color and Minolta Chroma Meter CR-200) with a clinical scoring scheme and laser-Doppler flowmetry, *Clin. Exp. Dermatol.,* 15, 267, 1990.
21. Babulak, S. W., Rhein, L. D., Scala, D. D., Simion, F. A., and Grove, G. L., Quantitation of erythema in a soap chamber test using the Minolta chroma (reflectance) meter comparision of instrumental results with visual assessments, *J. Soc. Cosmet. Chem.,* 37, 475, 1986.
22. Seitz, J. C. and Whitmore, C. G., Measurement of erythema and tanning responses in human skin using a tri-stimulus colorimeter, *Dermatologica,* 177, 70, 1988.
23. el-Gammal, S., Hoffmann, K., Steiert, P., Gaßmüller, J., Dirschka, T., and Altmeyer, P., Objective assessment of intra- and inter-individual skin colour variability: an analysis of human skin reaction to sun and UVB, in *The Environmental Threat to the Skin,* Marks R. and Plewig, G., Eds., Dunitz, London, 1991, 99.
24. Serup, J., Characterization of contact dermatitis and atopy using bioengineering techniques. A survey, *Acta Derm. Venereol., Suppl.,* 177, 14, 1992.
25. Agner, T., Basal transepidermal water loss, skin thickness, skin blood flow and skin colour in relation to sodium-lauryl-sulphate induced irritation in normal skin, *Contact Dermatitis,* 25, 108, 1991.
26. Weatherall, I. L. and Coombs, B. D., Skin color measurements in terms of CIELAB color space values, *J. Invest. Dermatol.,* 99, 468, 1992.
27. Kligman, A. M. and Wooding, W. M., A method for the measurement and evaluation of irritants on human skin, *J. Invest. Dermatol.,* 49, 78, 1967.
28. Wilhelm, K. P., Cua, A. B., Wolff, H. H., and Maibach, H. I., Surfactant-induced stratum corneum hydration in vivo: prediction of the irritation potential of anionic surfactants, *J. Invest. Dermatol.,* 101, 310, 1993.

29.   Agner, T., Noninvasive measuring methods for the investigation of irritant patch test reactions, *Acta Derm. Venereol., Suppl.,* 173, 1, 1992.
30.   Berardesca, E. and Maibach, H. I., Bioengineering and the patch test, *Contact Dermatitis,* 18, 3, 1988.
31.   Elsner, P. and Maibach, H. I., AT-based data acquisition and analysis system for the skin bioengineering laboratory, *Dermatosen,* 4, 124, 1991.

Chapter 23

In Vivo Cutaneous Assays to Evaluate Topical Corticosteroids and Nonsteroidal Anti-Inflammatory Drugs Using Reflectance Spectroscopy

Peter H. Andersen

CONTENTS

I. INTRODUCTION

The skin has several advantages as a model for compound testing. Some of the major benefits besides allowing an acceptable human experimental setup are quick and reliable visual and objective readings. As the number of improved bioengineering techniques has increased, determining different skin biophysical parameters, several cutaneous tests have been either reevaluated or redefined.[1-4]

One of the earliest skin models for *in vivo* compound testing was the cutaneous vasoconstriction assay, which is still used to rank the *in vivo* efficacy of new corticosteroids in humans.[5] Although proposed more than 20 years ago, this test is performed almost as originally proposed by McKenzie and Stoughton,[6] and even today it is considered a very powerful and reliable method to estimate relative potency of new corticosteroids.[5] This is partially caused by the high patient compliance, but is also due to the possibility of highly sensitive statistical comparisons between several corticosteroids.[5] The obvious advantages of cutaneous assays motivated several attempts to use the skin for toxicity studies or pharmacological testing. In particular, calculating relative efficacy of non-steroidal anti-inflammatory drugs (NSAIDs) in cutaneous systems has been attempted.[2,7] Originally the anti-inflammatory effects on ultraviolet (UV) light-induced erythema and inflammation were utilized, but recently less traumatic measures were suggested.[7,8] However, since cutaneous models are crude assays for different *in vivo* situations, it is very important to be familiar with both the experimental system and the tested compounds in order to analyze the obtained data critically.

Since the introduction of noninvasive procedures to measure cutaneous blood flow and erythema, some efforts to improve cutaneous assays have been disappointing, especially those using laser Doppler flowmetry.[1,9] However, studies in the author's laboratory, using an improved reflectance spectroscopic technique, showed enhanced ranking compared to visual evaluation and laser Doppler flowmetry.[2,5] This chapter will describe how reflectance spectroscopy has been used to analyze and improve both corticosteroid vasoconstriction assays and cutaneous *in vivo* NSAID testing.

II. MATERIALS AND METHODS

A. VOLUNTEERS AND EXPERIMENTAL CONDITIONS

Only after obtaining informed consent were subjects included. Experiments were performed under standardized indoor climate conditions (temperature = $22 \pm 1.0°$ C, 50 to 60% relative humidity) after a 20-min acclimatization period.

B. SKIN REFLECTANCE SPECTROPHOTOMETER

The modified reflectance skin spectrophotometer (RS) has been described in detail elsewhere.[10] Briefly, polychromatic light from a xenon short-arc lamp (Osram® XBO 150) was guided to the skin through a flexible light guide (Hirschmann,® Germany). Skin reflectance was collected in an integrating sphere and guided through a similar fiber to a monochromator (Jobin Yvon®, H 20, France), which split the light into 5-nm bands in the spectral range from 355 to 700 nm.[10] Skin reflectance was detected by a photomultiplier (Hamamatsu®, Japan) and analog-to-digital converted for further processing and computer (IBM-AT) storage. Each chromophore was detected by analyzing skin reflectance within specific ranges of the measured spectrum (355 to 700 nm).[11] First melanin content was calculated in the range from 360 to 390 nm, which corresponds to its main *in vitro* absorbing area in the recorded spectrum. Oxygenized and deoxygenized hemoglobins (oxyhemoglobin and deoxyhemoglobin, respectively) are subsequently analyzed using all data points from 515 to 610 nm and incorporating the previously determined melanin amount. By linear regression analysis the optical amount ($c_x \times d_x$) of the chromophores was calculated, where c represents the concentration and d the optical distance (thickness) of the chromophore layer in the skin.[11] Chromophore content is given in arbitrary units (AU), calculating relative changes as the percentage of chromophore content in control skin.[11]

C. LASER DOPPLER FLOWMETRY

Laser Doppler flowmetry (Perimed®, Sweden) is recognized as a noninvasive technique to measure blood flow changes in the microcirculation of the skin.[12] A helium-neon laser emits light at 632.8 nm onto the skin surface using fiber optics. The intensity of light frequency changes, caused by moving erythrocytes in dermal vessels, is detected through other fibers and used as a measure of cutaneous blood flow in arbitrary units.[12] Computerized data collection was performed using specially developed software (Brainware®, Skive, Denmark) allowing variable sampling. After readings were stabilized, measurements were performed 20 times per second in a 30-s sampling period; the average flow value was then calculated.

III. RESULTS

A. CORTICOSTEROID VASOCONSTRICTION ASSAY
1. Blanching—Potency Study

Corticosteroids of varying potency in alcohol solution were applied topically under occlusion to 16 panelists, and cutaneous blanching was investigated using visual scoring, laser Doppler flowmetry (LDF), and reflectance spectroscopy (RS).[5] In uninvolved back skin the amount of hemoglobin was determined; deoxyhemoglobin $= 0.35 \pm 0.04$ AU (61% of total hemoglobin) and oxyhemoglobin $= 0.23 \pm 0.11$ AU (39% of total hemoglobin), as measured by RS, and using LDF the blood flow was 1.86 ± 0.41 AU. The application of alcohol decreased total hemoglobin by 10%; deoxyhemoglobin $= 0.33 \pm 0.07$ AU and oxyhemoglobin $= 0.20 \pm 0.08$ AU. However, there was a corresponding $8\% \pm 27\%$ increase in blood flow and a visual blanching of 0.12 ± 0.29. Blanching data of the different potent corticosteroids are given in detail in Table 1 (in percentage of uninvolved control back skin) and are shown in Figures 1 and 2.[5]

2. Blanching—Kinetic Study

Using a specially developed betamethasone-17-valerate patch (BIO-PSA®) to ensure time-constant corticosteroid release the dynamic vasoconstriction reaction was studied in five volunteers. Visually, blanching increased with time exposure and the most pronounced blanching reaction was seen 12 h post application, with increased blanching at increasing exposure times. After 24 h no visual whitening was present for exposure times ≤12 h. At

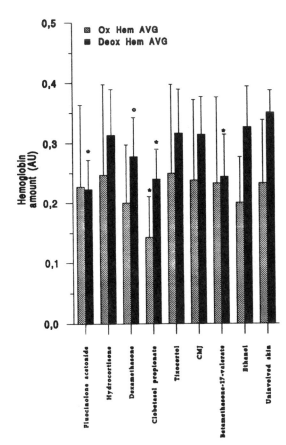

Figure 1 The figure shows topical corticosteroid vasoconstriction measured by reflectance spectroscopy displaying changes in arterial (oxyhemoglobin [Ox Hem]) and venous (deoxyhemoglobin [Deox Hem]) hemoglobin (y-axis; AU = arbitrary units) as a function of corticosteroids (x-axis). Clobetasol propionate was the most potent vasoconstrictor, and it decreased deoxyhemoglobin (30%) and oxyhemoglobin (33%). Fluocinolone acetonide, betamethasone-17-valerate, and dexamethasone only decreased deoxyhemoglobin significantly. Tixocortol, CMJ, and hydrocortisone acetate showed no significant blanching, although deoxyhemoglobin was decreased compared to the alcohol control ($^\circ$ $p < 0.05$; * $p < 0.01$.)

Table 1 Corticosteroid vasoconstriction—a potency study*

Solution	Average							
	Fluocinolone Acetonide	Hydro-cortisone	Dexa-methasone	Clobetasol propionate	Tixocortol	CMJ	Betamethasone-17-Valerate	Ethanol
Deoxyhemoglobin	*0.64 (0.15)	0.90 (0.21)	°0.80 (0.20)	*0.70 (0.18)	0.92 (0.23)	0.90 (0.18)	*0.71 (0.22)	0.94 (0.23)
Oxyhemoglobin	1.01 (0.46)	1.11 (0.57)	0.92 (0.45)	*0.67 (0.31)	1.11 (0.58)	1.06 (0.44)	1.06 (0.50)	0.94 (0.36)
Laser Doppler flowmetry	0.99 (0.26)	1.00 (0.24)	1.04 (0.29)	*0.82 (0.12)	0.99 (0.22)	0.95 (0.18)	0.97 (0.23)	1.08 (0.27)
Visual	*1.27 (0.93)	0.15 (0.30)	*0.88 (0.74)	*1.92 (1.21)	0.42 (0.62)	0.27 (0.42)	*1.04 (0.72)	0.12 (0.29)

Clobetasol propionate was the most potent vasoconstrictor inducing significant visual blanching and decreased Deoxyhemoglobin (30%), Oxyhemoglobin (33%) and BF (18%) ($p < 0.01$). Fluocinolone acetonide, betamethasone-17-valerate and dexamethasone also caused visual blanching and decreased only Deoxyhemoglobin (*$p < 0.01$; °$p < 0.05$).

Note: Numbers in parentheses are standard deviations. Clobetasol propionate was the most potent vasoconstrictor, inducing significant visual blanching and decreased deoxyhemoglobin (30%), oxyhemoglobin (33%), and blood flow (18%; *$p < 0.01$). Fluocinolone acetonide, betamethasone-17-valerate, and dexamethasone also caused visual blanching and decreased deoxyhemoglobin (*$p < 0.01$; °$p < 0.05$). CMJ = 21-thiol-9a-fluoro-11b, 17a-dihydroxy-16a-methyl-3,20-dione-21-acetylamino cysteine (Laboratory of Pharmaceutical Chemistry, University L. Pasteur, Strasbourg, France).

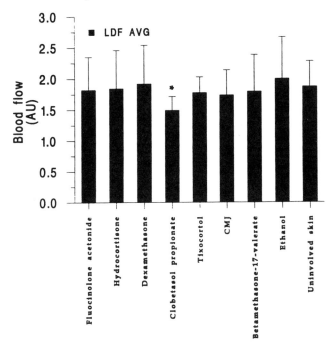

Figure 2 Topical corticosteroid vasoconstriction measured by laser Doppler flowmetry (LDF), showing changes in blood flow (y-axis; AU = arbitrary units) as a function of corticosteroids (x-axis). Clobetasol propionate, the most potent vasoconstrictor, decreased blood flow (18%). No other corticosteroid caused any significant decrease in blood flow. (*$p < 0.05$.)

increasing post application time intervals no individuals showed signs of skin blanching, but from 48 h some individuals developed a telangiectatic redness; this phenomenon had been present in all subjects after 72 h.[13] The visual findings were verified by reflectance spectroscopic measurements, and average changes for all subjects in venous (deoxyhemoglobin) or arterial (oxyhemoglobin) hemoglobin content ± standard deviation (SD) are given in detail (Table 2, Figures 3 and 4).[13]

B. OTHER CUTANEOUS CORTICOSTEROID AND NONSTEROIDAL ANTI-INFLAMMATORY DRUG RANKING TESTS

1. Ultraviolet B-Induced Inflammation

Normal skin responds acutely to UV light exposure with complex inflammatory mechanisms. In the present experiment UVB irradiation ranging from subclinical erythema doses to twice the minimal erythema dose (2 × MED) (24 to 96 mJ/cm^2) was delivered to the skin of eight volunteers.[14] Pre-irradiated sites were then immediately exposed to a 24-h occlusive patch containing one of four anti-inflammatory agents or the vehicle control. The resultant change in erythema (vascular reaction) was measured objectively using LDF and RS. The four anti-inflammatory compounds reduced the UVB-induced vascular reactions in different and dose-dependent ways ($p < 0.001$). Betamethasone-17-valerate and diphenhydramine were most effective at the 24 mJ/cm^2 dose site, and indomethacin and acetylsalicylic acid were more effective at sites ≥48 mJ/cm^2. The anti-inflammatory agents reduced UVB erythema in specific dose-dependent ways due to modulation of different inflammatory mediators present in the UVB inflammation. This investigation emphasizes the importance of incorporating

Table 2 Corticosteroid vasoconstriction—a time dynamic study

Time of Observation	Corticosteroid Exposure Time								
	1 h	2 h	3 h	4 h	6 h	8 h	12 h	24 h	28 h
Deoxyhemoglobin									
1 h	1.27 (0.20)								
2 h	1.11 (0.24)	0.85 (0.25)							
3 h	1.20 (0.13)	1.00 (0.11)	1.22 (0.19)						
4 h	0.86 (0.25)	0.95 (0.16)	1.25 (0.35)	0.92 (0.26)					
6 h	1.07 (0.32)	0.94 (0.18)	0.96 (0.16)	0.97 (0.12)	0.99 (0.33)				
8 h	0.80 (0.17)	0.79 (0.18)	0.80 (0.15)	0.80 (0.19)	0.89 (0.27)	0.90 (0.33)			
12 h	0.76 (0.11)	0.92 (0.21)	0.87 (0.10)	0.72 (0.10)	0.84 (0.37)	0.91 (0.20)	0.80 (0.20)		
24 h	1.02 (0.21)	1.06 (0.11)	0.86 (0.15)	0.68 (0.19)	0.86 (0.24)	0.75 (0.28)	0.87 (0.24)	0.64 (0.17)	
28 h	0.89 (0.17)	0.88 (0.30)	0.74 (0.20)	0.77 (0.30)	1.04 (0.24)	0.84 (0.29)	0.88 (0.44)	0.68 (0.28)	0.79 (0.14)
32 h	1.07 (0.41)	0.93 (0.21)	0.73 (0.15)	0.77 (0.16)	0.63 (0.40)	0.86 (0.26)	0.82 (0.35)	0.87 (0.33)	0.75 (0.15)
48 h	0.91 (0.07)	0.78 (0.09)	1.00 (0.19)	0.72 (0.13)	0.96 (0.29)	0.89 (0.29)	0.65 (0.23)	0.75 (0.21)	0.82 (0.14)
52 h	0.99 (0.24)	0.93 (0.06)	0.99 (0.17)	0.71 (0.30)	0.97 (0.17)	1.00 (0.18)	0.80 (0.24)	0.75 (0.31)	0.73 (0.07)
72 h	1.06 (0.12)	0.97 (0.22)	0.94 (0.29)	0.75 (0.22)	0.89 (0.20)	0.99 (0.50)	0.74 (0.16)	0.73 (0.36)	0.95 (0.23)
Oxyhemoglobin									
1 h	0.91 (0.24)								
2 h	1.17 (0.37)	1.11 (0.22)							
3 h	0.83 (0.24)	0.94 (0.11)	0.75 (0.19)						
4 h	1.13 (0.22)	0.73 (0.30)	0.77 (0.10)	0.88 (0.28)					
6 h	0.84 (0.32)	0.69 (0.28)	0.68 (0.13)	0.63 (0.32)	0.64 (0.25)				
8 h	0.88 (0.28)	0.63 (0.32)	0.64 (0.22)	0.53 (0.14)	0.58 (0.22)	0.84 (0.26)			
12 h	1.00 (0.46)	0.52 (0.11)	0.48 (0.26)	0.50 (0.31)	0.56 (0.21)	0.50 (0.27)	0.49 (0.24)		
24 h	0.87 (0.30)	0.86 (0.17)	1.21 (0.73)	1.22 (0.34)	1.20 (0.31)	1.18 (0.22)	1.72 (0.86)	1.01 (0.19)	
28 h	1.08 (0.29)	1.08 (0.56)	1.25 (0.31)	1.13 (0.59)	0.80 (0.32)	1.22 (0.39)	1.33 (0.36)	1.14 (0.29)	0.80 (0.30)
32 h	1.16 (0.35)	1.03 (0.41)	1.27 (0.27)	1.07 (0.19)	1.61 (0.48)	1.22 (0.33)	1.43 (0.21)	1.17 (0.24)	1.12 (0.17)
48 h	1.32 (0.88)	1.43 (0.62)	1.05 (0.83)	1.84 (1.23)	1.40 (1.18)	1.11 (0.47)	1.99 (1.70)	1.27 (0.53)	1.00 (0.31)
52 h	1.34 (0.64)	0.75 (0.34)	0.94 (0.30)	1.68 (0.95)	2.19 (1.46)	1.60 (0.87)	1.16 (0.55)	1.43 (0.39)	1.17 (0.61)
72 h	0.97 (0.21)	1.25 (1.08)	1.75 (0.71)	1.70 (0.91)	2.02 (0.63)	1.36 (0.25)	1.40 (0.60)	1.83 (1.04)	1.27 (0.41)

*Both cutaneous vaso-constriction and post constriction hyperemia were caused by changes mainly in Oxy Hem but also Deoxy Hem. Maximal skin blanching was found 12 h post application at the 12 h site. All exposure loci except 6 h and 12 h sites still showed some vaso-constriction 24 h post application, but from time intervals ≥ 32 h a paradox increase in Oxy Hem was found at all exposure sites ≥ 3 h.

Note: Numbers in parentheses are standard deviations. Both cutaneous vasoconstriction and postconstriction hyperemia were caused by changes in oxyhemoglobin, but also in deoxyhemoglobin. Maximal skin blanching was found 12 h postapplication at the 12 h site. All exposure loci except the 6 and 12 h sites still showed some vasoconstriction 24 h postapplication, but from time intervals ≥32 h a paradoxical increase in oxyhemoglobin was found at all exposure sites ≥3 h.

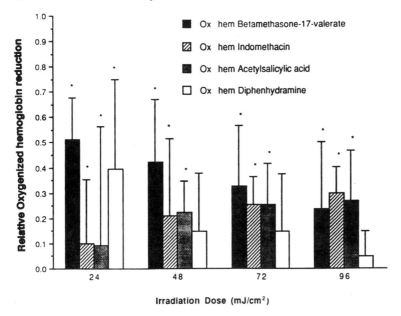

Figure 3 For all subjects (average + SD) the UVB dose-dependent anti-inflammatory effect for each compound is given as relative reduction [1 − (irradiated compound-treated site/irradiated vehicle-treated site)]. The decreased oxyhemoglobin (Ox Hem; (y-axis) is calculated as a function of increased UVB dose (x-axis). Betamethasone-17-valerate (r = 0.98) and diphenhydramine (r = 0.90) reduced UVB erythema in an inverse linear fashion showing the highest effects at lower UVB doses. Indomethacin (r = 0.80) and acetylsalicylic acid (r = 0.85) reduced erythema in a linear manner, with the highest anti-inflammatory effects at doses ≥ 72 mJ/cm^2 (*$p < 0.05$).

UVB dose ranges in *in vivo* experiments of UVB inflammation to evaluate the inflammatory response and offers insight into the complexity of pharmacological intervention.[14]

2. Postocclusive Hyperemia

Postocclusive reactive hyperemia describes the temporary increase in blood flow following transitory vascular obstruction and recently has been proposed as an *in vivo* method for ranking topical corticosteroids.[1] The *in vivo* vascular reactions before and during postocclusive hyperemia using LDF and RS were measured.[2] Using a randomized 24-h occlusive exposure in ten healthy volunteers the effects of corticosteroid (betamethasone-17-valerate), NSAID (indomethacin), antihistamine (diphenhydramine), or vehicle were studied before and during postocclusive hyperemia. The 24-h vehicle exposure decreased total hemoglobin, with a low increase in oxyhemoglobin ($p < 0.001$) and a higher decrease in deoxyhemoglobin ($p < 0.005$): oxyhemoglobin = 0.23 ± 0.18 AU and deoxyhemoglobin = 0.28 ± 0.12 AU. The blood flow increased 7.1% to 28 ± 8 AU ($p > 0.05$). Postocclusive hyperemia increased total hemoglobin maximally at the first observation time: oxyhemoglobin = 0.63 ± 0.13 AU and deoxyhemoglobin = 0.31 ± 0.11 AU ($p < 0.001$). Oxyhemoglobin increased maximally (178%) at 30 s (oxyhemoglobin = 0.64 ± 0.16 AU), and deoxyhemoglobin decreased throughout the observation period. The blood flow in vehicle-exposed sites increased 193%, also with a maximal increase at the 30 s reading. At all postocclusive observation intervals the increase in oxyhemoglobin was significantly reduced by betamethasone-17-valerate-, indomethacin-, and diphenhydramine-exposed sites. The greatest decrease in postocclusive oxyhemoglobin was measured for betamethasone-17-valerate. The increase in blood flow

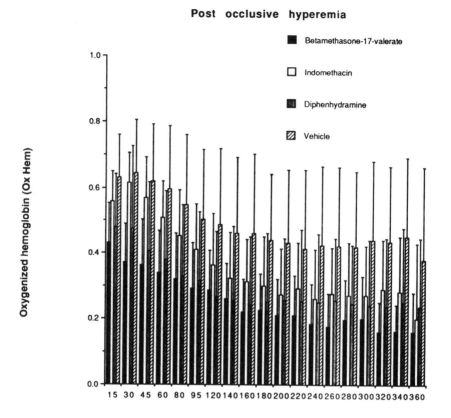

Figure 4   Oxygenized hemoglobin (Ox Hem) is shown as a function of postocclusive time. Postocclusive total hemoglobin in vehicle-exposed skin was maximal at 15 s and increased to 0.63 ± 0.13 arbitrary units (AU), but the increase in oxyhemoglobin was maximal after 30 s (0.64 ± 0.16 AU). The effect of each compound was compared to the vehicle control. In all postocclusive observation intervals betamethasone-17-valerate reduced oxyhemoglobin significantly, and indomethacin reduced oxyhemoglobin less than the corticosteroid. The reduction in oxyhemoglobin at diphenhydramine-exposed sites was significant only in the first and second minute, but was on average higher than the oxyhemoglobin reduction found for indomethacin.

was also significantly reduced in all postocclusive observation periods for betamethasone-17-valerate. Indomethacin also decreased blood flow, but the decrease was significant only in the 2- and 3-min observations. No significant blood flow changes were found after diphenhydramine exposure.[2]

IV.   DISCUSSION

Partially motivated by the growing concern about animal testing in general, human cutaneous models are used increasingly. Therefore, the development of both reliable *in vitro* cell culture tests and ethically acceptable human models is constantly desired. Although improved bioengineering techniques have enhanced our knowledge of skin biophysics, further investigations are needed to rectify cutaneous pharmacological assays.[1-7] The relative potency assigned to a compound in cutaneous systems is influenced by several properties, including the ability to penetrate the skin barrier after release from the vehicle, binding to macromolecules or receptors, intrinsic activity, and rate of clearance from the application site. Since cutaneous models are crude assays for different *in vivo* situations, it is very important to be familiar with both the experimental system and the tested compounds in order to analyze the obtained data critically. However, at the present state-of-the-art very useful data can be obtained in human models partially because of the possibility of doing multicompound comparisons using the subject as a statistical control.[1-4] In order to improve different cutaneous assays, refined bioengineering techniques have been included to analyze both new and well-established *in vivo* compound testing systems.[11,12]

A. CORTICOSTEROID VASOCONSTRICTION ASSAY

Using the improved reflectance spectroscopic technique it becomes clear why previous laser Doppler attempts to improve the corticosteroid vasoconstriction assay have failed.[2,5] Since the corticosteroid-induced blanching was predominantly due to venoconstriction and only the most potent steroid caused a significant decrease in oxyhemoglobin and blood flow, a moderate vasoconstriction, although easily seen by the naked eye, may not be detected by the flow-dependent laser Doppler.[5] Although LDF and RS measure different biophysical changes, several measurements on erythema showed good correlation,[15] leading to a general acceptance that the two techniques overall measure identical biological phenomena. However, the mentioned bioengineering methods measure different physico-optical phenomena. LDF detects the Doppler shift, caused by *moving* erythrocytes, of a monochromatic laser beam (632.8 nm) directed onto the skin surface and thereby measures cutaneous *blood flow* in the dermis.[12] RS measures the fraction of incident polychromatic light (350 to 700 nm) not absorbed by skin chromophores in the epidermis or upper dermis. Using computerized analysis the *amount* of skin chromophores (melanin, carotene, deoxy- and oxyhemoglobin) can be calculated and the vascular status expressed as arterial and venous components.[11] Decreased amounts of cutaneous venous blood may be easily seen visually and measured by RS,[11] but cannot be detected by the flux-dependent laser Doppler flowmeter.[12]

In the dose-kinetic study, maximal skin blanching was induced 12 h post application, but plasma corticosteroid concentrations peaked 32 h post application when a telangiectatic vasodilation started to appear.[13] These observations may imply a binary and probably time-related action mechanism of corticosteroids at their dermal vessel receptors where 6 to 12 h of corticosteroid exposure caused vasoconstriction, but as the exposure time increased ($\geq$24 h) a paradox vasodilation was induced despite rising tissue concentrations.[13]

Although the McKenzie-Stoughton vasoconstriction assay and various modifications are widely used bioassay systems to compare bioavailability of new topical corticosteroids or formulations, the exact blanching action mechanism is unknown. However, the relative potency assigned to a corticosteroid by this assay correlates very well with the potency calculated in much more time-consuming clinical trials, making the assay a very powerful tool to perform useful multicompound comparisons.[1,3,5]

B. OTHER CUTANEOUS CORTICOSTEROID AND NONSTEROIDAL ANTI-INFLAMMATORY DRUG RANKING TESTS
1. Ultraviolet B-Induced Inflammation

A human model combined with accurate noninvasive bioengineering techniques was used to study the complexity of pharmacological ranking models during subclinical and low-dose UVB inflammation.[14] Several inflammatory mediators have been found from 4 h post-UVB irradiation,[16] although *in vivo* vascular changes have been detected within 1 h.[15] Elevated histamine levels have been detected in human suction blister fluid from 4 to 8 h after UVB irradiation.[16] Prostaglandins (PGE_2 and $PGF_{2\alpha}$) were detected after 6 h, and 12-hydroxyei-cosatetraenoic acid (12-HETE) reached peak levels between 18 and 24 h after irradiation.[16,17] The erythema, however, persisted beyond the return of the prostaglandin levels to normal.[15,16] Leukotriene B_4 levels were unchanged.[16] Furthermore, interleukins (IL-1 and IL-6) have been found to be elevated starting from 12 h and lasting beyond 72 h.[18,19] These experiments differ from this study by employing higher UVB doses (3 to 4 × MED), which causes a severe skin reaction including erythema and edema.[15] In the present study, different anti-inflammatory kinetics of betamethasone-17-valerate, indomethacin, acetylsalicylic acid, and diphenhydramine were demonstrated showing specific dose-kinetic anti-inflammatory patterns for the different compounds in the UVB dose range, from subclinical erythemas to moderate erythemas (2 × MED).[14] The study shows the importance of analyzing anti-inflammatory activities at different UVB doses when this model is utilized for *in vivo* pharmacological testing, since ranking of the four compounds was dose dependent.[14]

2. Postocclusive Hyperemia

Recent studies favor the myogenic theory[20] but other experiments indicate the presence of vasoactive mediators and metabolites during postocclusive hyperemia.[1,21] Although the vascular mechanism of postocclusive reactive hyperemia remains unclear, our data reveal advantages of the method to rank both corticosteroids and NSAIDs. This study showed the strongest decrease in postocclusive oxyhemoglobin for the corticosteroid, but topical NSAID or antihistamine also reduced postocclusive oxyhemoglobin or blood flow. This might imply that prolonged tissue hypoxia increased the amount of several inflammatory mediators.[1,21] Postocclusive hyperemia was recently proposed as a possible corticosteroid ranking technique, but the present data imply that the model also may be used for *in vivo* NSAID ranking.[2]

Other types of erythematous skin have been suggested for both NSAID testing and corticosteroid vasoconstriction studies. Recently, methyl nicotine-induced vasodilation was used to rank NSAIDs,[22] and heat-induced vasodilation was shown to be reversed by corticosteroids.[23] The increased use of cutaneous models in product testing provides motivation for further experiments to describe the exact nature of the model in order to better understand the obtained results. The combined use of reflectance spectroscopy and laser Doppler flowmetry has increased our knowledge of the vascular changes during several cutaneous tests and has improved product ranking especially in the corticosteroid vasoconstriction assay, but also in erythematous skin models.[2,5,14]

ACKNOWLEDGMENT

The author wishes to acknowledge K. Abrams, P. Broichmann, K. Kubota, K. Milioni, and H. I. Maibach, all of the Department of Dermatology, University of California, San Francisco.

REFERENCES

1. Bisgaard, H., Kristensen, J. K., and Søndergaard, J., A new technique for ranking vascular corticosteroid effects in humans using laser Doppler velocimetry, *J. Invest. Dermatol.*, 86, 275, 1986.
2. Andersen, P. H., Broichmann, P. W., and Maibach, H., Corticosteroid, NSAID and antihistamine modulate in vivo vascular reactions before and during post occlusive hyperemia, *Br. J. Dermatol.*, 128, 137, 1993.
3. Feather, J. W., Ryatt, K. S., and Dawson, J. B., Reflectance spectrophotometric quantification of skin colour changes induced by topical corticosteroid preparations, *Br. J. Dermatol.*, 106, 436, 1982.

4. Queille-Roussel, C., Poncet, M., and Schaefer, H., Quantification of skin colour changes induced by topical corticosteroid preparations using Minolta Chroma Meter, *Br. J. Dermatol.*, 124, 264, 1991.

5. Andersen, P. H., Millioni, K., and Maibach, H., The cutaneous corticosteroid vasoconstriction assay: a reflectance spectroscopic and laser Doppler flowmetric study, *Br. J. Dermatol.*, 128, 660, 1993.

6. McKenzie, A. W. and Stoughton, R. B., Method for comparing percutaneous absorption of steroids, *Arch. Dermatol.*, 86, 606, 1962.

7. Poelman, M., Piot, B., Guyon, M., Deroni, J., and Leveque, J. L., Assessment of topical non-steroidal anti-inflammatory drugs, *J. Pharm. Pharmacol.*, 41, 720, 1989.

8. Ahluwalia, A. and Flower, R. J., Topical betamethasone-17-valerate inhibits heat-induced vasodilation in man, *Br. J. Dermatol.* 128, 45, 1993.

9. Berardesca, E. and Maibach H. I., Cutaneous reactive hyperemia: racial differences induced by corticoid application, *Br. J. Dermatol.*, 120, 787, 1989.

10. Bjerring, P. and Andersen, P. H., Skin reflectance spectrophotometry, *Photodermatology*, 4, 167, 1987.

11. Andersen, P. H. and Bjerring, P., Non-invasive computerized analysis of skin chromophores in vivo by reflectance spectroscopy, *Photodermatol. Photoimmunol. Photomed.*, 7, 249, 1990.

12. Nilsson, G. E., Tenland, T., and Oberg, P. Å., Evaluation of a laser Doppler flowmeter for measurement of tissue blood flow, *IEEE Trans. Biomed. Eng.*, 27, 597, 1980.

13. Andersen, P., Kubota, K., Lo, E. S., Hutting, G., and Maibach, H., A time correlation study between reflectance spectroscopic cutaneous vasoconstriction and plasma corticosteroid concentration, *Br. J. Dermatol.*, 1994.

14. Andersen, P. H., Abrams, K., and Maibach, H., Ultraviolet B dose-dependent inflammation in man: a reflectance spectroscopic and laser Doppler flowmetric study using topical pharmacologic antagonists on irradiated skin, *Photodermatol. Photoimmunol. Photomed.*, 9, 17, 1992.

15. Andersen, P. H., Abrams, K., Bjerring, P., and Maibach, H., A time correlation study of UVB induced erythema measured by reflectance spectroscopy and laser Doppler flowmetry, *Photodermatol. Photoimmunol. Photomed.*, 8, 123, 1991.

16. Gilchrest, B. A., Soter, N. A., Stoff, J. S., and Mihm, M. C., The human sunburn reaction: histologic and biochemical studies, *J. Am. Acad. Dermatol.*, 5, 411, 1981.

17. Black, A. K., Fincham, N., Greaves, M. W., and Hensby, C. N., Time course changes in levels of arachidonic acid and prostaglandins D_2 E_2 $F_{2\alpha}$ in human skin following ultraviolet irradiation, *Br. J. Clin. Pharm.*, 10, 453, 1980.

18. Granstein, R. D. and Sauder, D. N. Whole-body exposure to ultraviolet radiation results in increased serum interleukin-1 activity in humans, *Lymphokine Res.*, 6, 187, 1987.

19. Kirnbauer, R., Köck, A., Krutmann, J., Schwarz, Urbanski, A., and Luger, T. A., Different effects of UVA and UVB irradiation on epidermal cell IL-6 expression and release, *J. Invest. Dermatol.*, 92, 459, 1989.

20. Ryan, T. J., Reactive hyperaemia and response to changes in tissue, in *The Physiology and Pathophysiology of the Skin*, Vol. 2, Jarrett, A., Ed., Academic Press, London, 1973, 714.

21. Wilkin, J. K., Cutaneous reactive hyperemia: viscoelasticity determines response, *J. Invest. Dermatol.*, 89, 197, 1987.

INDEX

A

Absorption, 35–36, 74, 75, 260–261
Accuracy, 58
Acetylsalicylic acid, 74, 76, 77
Acroscleroderma, 86
AC signal processing, 42–44
AD, see Atopic dermatitis
Allergens, 78, see also specific types
Allergic dermatitis, 23, 69
Amethocaine, 76
Amputation, 159
Anatomy, 3–20
    laser Doppler flowmetry and, 9–20
    organization and, 3–9
    topographic mapping and, 12–14
Anesthetics, 76, see also specific types
Angiotensin converting enzyme (ACE), 80
Angiotensin converting enzyme (ACE)
        inhibitors, 80, 165, see also specific
        types
Animal studies of contact dermatitis, 69
Antibodies, 88
Anti-inflammatory drugs, 68–69, see also
        specific types
    in vivo assays of, 281–290
    irritancy and, 209
    nonsteroidal, 76–77, see also
        specific types
    in vivo assays of, 281–290
    in scleroderma treatment, 90
Applications
    of laser Doppler flowmetry, 134–136
    of light reflection rheography,
        178–180
    of photoplethysmography, 50
Argon lasers, 187
Arterial capillaries, 3, 5
Arterial insufficiency, 156–163
Arterioles, 3–5, 7
Arteriosclerosis, 23
Artifacts
    in laser Doppler flowmetry, 40, 41, 45,
        46, 48
    in photoplethysmography, 50
Atherosclerosis, 134, 140–141
Atopic dermatitis, 80
Autoantibodies, 88
Axon reflex vasodilator response, 136

B

Backscattered light, 33, 40–41
Barrier function, 262–263
Basement membrane, 5
Benzoic acid, 251
Betamethasone 17-valerate, 217, 220, 221, 287
Bradykinin, 80
Burns, 23, 24

C

Calcitonin-gene-related peptide (CGRP), 79, 80
Calibration
    in laser Doppler flowmetry, 47–48, 50, 51
    in photoplethysmography, 51
Capillaries, 3, 5, 48, 89, see also specific types
Capillaromicroscopy, 25, 161, 162
Capillaroscopy, 89, 90, 98–100, 106, 155
Capillary loop, 6–7
Capillary pressure pulse amplitude (CPPA), 87
Capsaicin, 80, 136
Captopril, 165
CGRP, see Calcitonin-gene-related peptide
"Chinese restaurant syndrome", 77
Chloride, 210
Chromametry, 235, 247–251, 259, 260
    advantages of, 278
    description of, 26
    hardware for, 247–248
    irritation test evaluation by, 269–278
    measuring principles of, 248–250
    principles of, 248–250
    reproducibility in, 251, 271–272
    sensitivity in, 251
    in vascular diseases, 155
Chromatography, 263, see also specific types
Cigarette smoking, 77, 134
Clinical evaluation, 66
Clobetasol-17-propionate, 78
CMBC, see Concentration of moving blood
        cells
Cognitive test, 135
Cold exposure, 89, 108
Collagen, 88, 90
Colorimetry, 67, 184, 259, 260
Concentration of moving blood cells (CMBC),
        15–20
Connective tissue, 88